Holography Marketplace

7th Edition

*The reference text
and sourcebook for
holography worldwide.*

The original version of this book contained holograms from the vendors and many of them are no longer in business. Therefore this version of the book contains everything that was in the original version but it has no holograms.

Edited by Alan Rhody and Franz Ross

Copyright © 1998 Ross Books

Library Catalogue Information
Holography MarketPlace - Seventh Edition
Bibliography:
1 . Includes Index.
2. Holography
3. Directories - Holography Industry
4. Photography
5. Physics
ISBN 978-0-89496-100-7
Printed in Canada

About The Cover of HMP 7

The hologram featured on the cover of this book was a collaborative effort among four companies: Dimension 3 (image origination and digital modeling), Chromagem (mastering the hologram), DuPont (who provided the photopolymer material which the hologram was reproduced on) and Krystal Holographics (who mass-produced the hologram). A special thanks to Doug Miller of Krystal Holographics for helping to coordinate the project. To find out how the hologram was produced in more detail, see Appendix A (page 221).

The book's cover art was designed and produced by Linda Law of Linda Law Holographics (who provided the cover art for last year's book, too.) Her company specializes in the design and origination of digitized artwork for holography. She is an experienced artist and holographer.

This hologram and the cover provide excellent examples of how new masterpieces can be created using a combination of the latest digital and optical imaging technologies. More importantly, it demonstrates methods that are commercially available today! These are not technologies that are locked in the research lab. They are affordable and feasible to use now. We encourage you to read the book to familiarize yourself with current production methods and then contact these companies to discuss your ideas with them.

Foreword

Welcome to the 7th edition of the *Holography Market-Place*. This edition includes new information that will interest professionals who work within the holography industry, as well as those in other fields that need to know more about holograms and their applications. *HMP 7* also provides basic knowledge that will appeal to the curious businessperson, student or casual reader. We hope you enjoy reading this edition and find it a very useful resource.

New Articles: Potential users of display holograms will benefit by reading John Perry's article in the *Introduction to Holography* chapter, which clears up some common misconceptions about the medium. For hologram designers and production managers, we have added a discussion of dot matrix technology to the *Artwork & Mastering* chapter, as this is becoming a very popular way of producing imagery for embossed holograms. "Recombining" is another hot topic which we've addressed in this chapter. Both novices and experts should appreciate an article by Hans Bjelkhagen in Chapter Three, *Color Holography*, as it summarizes developments in full color reflection holography.

For many holographers, the most important issue of the year is the changes taking place in the field of silver-halide emulsions. Agfa-Gevaert, formerly the industry's leading supplier of holographic plates and films has ceased production. Chapter Four, *Silver-Halide Recording Materials*, introduces the leading new suppliers to the industry and provides detailed product specifications. Experienced holographers T.H. Jeong, Jeffrey Murray and. Chuck Paxton review these materials and offer useful processing tips. Since some of the newer emulsions require relatively slow exposure times, we have included an article in Chapter Six, *Optics and Equipment*, about Jeff Odhner's "fringe-locker system" which can alleviate some potential problems faced by holographers using these new emulsions.

Photopolymer, DCG, photoresist and photo-thenTIoplastic films are covered in Chapter Five, *Other Recording Materials*. Chapter Eight, *Lasers - Current Trends*, describes the newest lasers available to the holography industry. Of special mention are articles by David Ratcliffe and Ron Olson regarding their respective pulsed laser systems.

International Business Directory: Our database begins with an alphabetical listing of holography-related businesses (it includes updated company contact information and a brief description of the goods and services the company offers.) Use the accompanying lists and the cross-index tables to quickly find the businesses and people you are interested in. Please note that many phone numbers, emails and website addresses have changed since our last edition.

Holograms: Once again, we are proud to include a wonderful and useful "sample kit" of holograms that represents the state of the industry today. No other book offers this feature. When you contact our advertisers, please let them know you appreciate seeing actual samples of their work.

Thanks to everyone that assisted with this edition.

Franz Ross & Alan Rhody
Berkeley, CA - Winter 1998

Table of Contents

Index to Advertisers

Note- **Bold** typeface indicates sample hologram(s) included in ad.

For information regarding advertising rates and availability contact:

Advertising Det. Ross Books ph. 510-841-2474 fax 510-841-2695 or email staff@rossbooks.com

RARE & COLLECTABLE EDITIONS

The following titles are now available in limited quantities. Each volume contains pertinent reference material, historically significant information and collectable holograms!

Now only **$19.95** each or **$50.00** for entire set of **2 - 5** !

HOLOGRAPHY MARKETPLACE 2nd EDITION (1990)

Includes the hologram*: Statue of Liberty (American Bank Note) embossed (2"x3")* Plus: Introduction to Holography; Emulsions and Recording Materials; Holographic Optical; Elements; Color Holography; Business of Holography; Holographic Distribution Process; Embossed Holograms; Computer Generated Holograms; Holographic Non Destructive Testing; Database of Businesses; Database of Individuals; Bibliography; Glossary

HOLOGRAPHY MARKETPLACE 3rd EDITION (1991)

Includes these holograms: *Brain Skull (Polaroid) two channel image photopolymer (3"x3"); Earth/Space/Grid (American Bank Note) embossed (3"x3"); Floating Alphabet (American Bank Note) embossed (3"x4"); Magic Wizard (American Bank Note) embossed (1"x1"); Woman/Fruit/Flowers (Light Impressions) embossed true color still life (4"x4"); Prehistoric Man (Bridgestone Graphic Technologies) embossed (3"dia.); Space Shuttle in Orbit (Archeozoic/Polaroid) photopolymer (1"x1")* Plus: Introduction to Holography; Varieties of Holograms; Recording Materials; Lasers; Holographic Optical Elements ; Non Destructive Testing; Computer Generated Holograms; Holography in Education; Embossed Holograms; Business of Holography; Businesses by Category; Database of Individuals; Bibliography; Glossary

HOLOGRAPHY MARKETPLACE 4th EDITION (1993)

Includes these holograms*: Transamerica Pyramid (Polaroid) photopolymer (2.5"x3"); Earth/Grid (AD2000 Inc./ABN) embossed (3"x3"); Disney Characters (Holograms De Mexico) color embossed (2"x3"); Ghostbusters (American Bank Note) 2 channel embossed (4"x5"); Butterfly (Holography Presses On) embossed diffraction (.5"); Egyptian King (Holopress) embossed (2"x2"); 4 Image Montage (The Diffraction Company) embossed (4"x4"); Inaugural Invitations (CFC Applied Holographics) embossed (1.5"x1.5"); Bouquet (CFC Applied Holographics) full color embossed (2.5"x3.5"); Folding Package (Transfer Print Foils/Light Impressions) embossed (3"x5"); Earth/Lab (Global Images/Chromagem) embossed (5"x5")* Plus: Sales and Distribution; Direct Mail Marketing; Model Making; Holography Basics; Advanced Principles; Holographic Optical Elements; Heads Up Display; Computer Generated Holograms; Holographic Non Destructive Testing; Embossed Holograms; Lasers; Recording Materials; Main Business Listings; Bibliography; Glossary

HOLOGRAPHY MARKETPLACE 5th EDITION (1995)

This edition features *TV- a limited edition photopolymer hologram which was produced using a newly developed process that incorporates digital, optical, and holographic technologies to create a truly amazing image!* Plus: *Harry 4"x5", Chinese Lion Dancers 3.5"x5" (The Lasersmith); Mount Rushmore 6"x9", Rock Solid (TPF); Space Shuttle, Flag, Eagle, Fireworks 5.5"x11" (Crown Roll Leaf); Map, (Hologramas de Mexico); Initials (HPO); Wizard 2"x3" polymer (Lazer Wizardry); Butterfly 3"x3", Valid (CFC/Applied); Tiger (Krystal Holographics); Matrix (Dimensional Arts)* Plus: Sales and Distribution; Holographic Stereograms; Embossed Holograms; Lasers; Recording Materials; Business Listings, Appendix and more.

ORDERING INFORMATION

Continental USA	Single $25	Entire Set 2-5 $65	All prices **include** Shipping and Handling
Alaska, Hawaii, Canada, Mexico	Single $30	Entire Set 2-5 $75	VISA, MC, AMEX, check or money
All Other Overseas Countries	Single $40	Entire Set 2-5 $85	order (US$) Payable to: ROSS BOOKS

P.O. Box 4340 Berkeley, CA 94704
Toll Free in USA 800-367-0930
ph. 510-841-2474, fax 510-841-2695
email sales@rossbooks.com

1

Introduction to Holography

This chapter assumes that you have no prior knowledge of holography and its uses. It begins with a discussion of the unique attributes of this fascinating visual medium, provides a simple explanation of what holograms are, and describes some commercial applications of the technology. Potential users of display holograms will benefit from reading the article on page 14, which debunks some common myths about holograms and provides information about how they are best used. The chapter concludes with an explanation of how basic types are produced.

Holograms Are Everywhere

You wake up in the morning and go out to breakfast. The restaurant is giving away hologram trading cards with each meal. You grab one for your son because you know he collects them. You finish your breakfast and head off to your job. Not paying close attention to the road at this early hour, you exceed the posted speed limit. The friendly Highway Patrol Officer who pulls you over takes your driver's license and examines it closely in the sunlight. He is looking for a hologram laminate that indicates your license is authentic.

At work, you open a new package of computer software. The box is sealed with a hologram label. On the way home from work you stop at the music store to buy a CD for your daughter. A 3-D holographic portrait of her favorite group is on the CD's cover and, even more amazing, the band members appear to dance a bit as you tilt the box back and forth! You pick out a greeting card to go with the gift. It has colorful holographic foil decorating the front. Then you pay for your purchases with a credit card that has a hologram on it. Once home, you pour yourself a martini. The bottle has a holographic tax stamp across the cap.

This scenario is an example of the frequency with which we see holograms in our everyday life. Unlike a decade ago, holographic images and holographic materials have entered the mainstream and are commonly used for a variety of commercial applications. This development is partly due to the standardization of manufacturing methods and the inevitable maturation of the entire industry. More important is the fact that the clients who commission holographic originations are achieving the results they desire.

Holography has proven that it is a viable technology that can be successfully integrated into a wide range of commercial endeavors, including advertising campaigns, marketing promotions, security programs and retail sales. The medium has established a track record of deterring counterfeiters, attracting shoppers, and adding value to products. As the technology evolves further and potential users become more informed about the holography marketplace, a host of other applications will certainly arise.

Major Attributes of Holograms

Although often compared with photography, holography is really a completely different medium. Holography is based on optical principles that are different than photography's and holograms have different physical attributes than photographs. They are comparable only because both are ways of recording an image onto a piece of photosensitive material (film), and because, at times, similar equipment and materials are used in making them.

The most obvious difference between a photograph and a hologram is image dimension and image depth. For instance, when we look at a photograph and move it from side to side we are unable to see "around" the scene or perceive any depth. We cannot see over or under the image. We only see a flat (two-dimensional) picture displayed on the surface of the film. The picture is actually only a collection of light and dark shapes that we recognize as a particular subject. Our memory might remind us that the subject really has dimension and depth, but this visual information is not recorded in the photograph.

Parallax, Dimension and Depth

A hologram is also flat, but the picture "on it" is not. When we look at a hologram and move it from side to side we can see many different views of the scene. We can also look behind foreground elements to see things in the background. This property is called **parallax** and it is closely tied to the process of visual perception. The Random House Dictionary defines parallax as "the apparent displacement of an observed object due to the difference between two points of view." Our two eyes see slightly different things. Our brain automatically uses these multiple views to create image dimension and image depth.

The relevant point here is that if we see at least two views of the same object, we can perceive a three-dimensional image. A hologram records and displays many views of the same object - a photograph is limited to only one. (In fact, the word "**hologram**", coined by physicist Dennis Gabor in 1948, is commonly defined as "whole picture" based on its Greek roots. The terms "**holography**" and "**holographic**" are typically used to discuss anything related to holograms, though the word "holograph" actually has another unrelated meaning.)

To better illustrate the difference between the two media, imagine looking out through a small window at a particular view. If the window represented a photograph, you would be frozen in front of it in one position. Consequently, you would only see one perspective of that scene and would have no way of perceiving depth. However, if the window were a hologram, you could move around in front of the window and see different parts of the scene from many different viewing angles. The scene would look three dimensional and display depth. To further elaborate on this analogy, if this window could retain the scene, and you could carry it around and show it to someone else later - they would be seeing a hologram of that scene.

Stereograms

It is possible to create an image with depth using two similar photographs taken from slightly different angles if these pictures are presented to our eyes in just the right way. The two pictures are called "stereo pairs." Perhaps you are familiar with stereogram posters, "Viewmaster TM" binocular viewers or virtual reality helmets. These artificial methods work, but they require people either to look at things unnaturally or to employ special viewing equipment. For many commercial applications it is more practical to use holography when dimensional imaging is required.

Images That "Project" Out of The Hologram

Another difference between holograms and photographs is that holographers can position their images to "project off" or "float over" the surface of the film. One popular holographic image is of a water faucet that projects a foot or so out of the picture frame. A viewer can reach right out and put his hand right through this apparently solid image. Other holographic images can be positioned to appear some distance "behind" the picture frame. Still others straddle the surface of the film (called the **image plane**). No photograph can do that!

Other Properties and Considerations

Holograms do have two properties which should be ad- dressed, as they affect commercial applications. One is viewing angle. Unlike a normal picture hanging on the wall or in a magazine, holographic images can only be seen within certain viewing parameters. If a viewer moves too far off-center, the holographic image will disappear.

Another problem is lighting. All holograms need to be properly illuminated in order to be seen. The lighting conditions that exist in many display environments are not the best for viewing holograms. Potential users must anticipate how their holograms are going to be displayed before starting production. There are ways to solve these problems using good design practices, as well as technological approaches which are currently being developed by researchers worldwide. Both these issues are covered in greater detail throughout this book.

How Holograms Are Recorded

The degree to which you can look around an object and the distance the image forms in front of or behind the film depends on how the hologram was made. This brings up another major difference between the two media - the procedures used to record imagery. We are all familiar with photography; we need a camera, film and an adequate amount of light. The light reflects off our subject, passes through the camera lens and exposes the film. Bright subjects expose the film a lot, darker subjects expose the film less. The picture we see is composed of varying tones.

Holography is different. It records an image in an entirely distinct way. Film is still employed; however, conventional cameras and ordinary lighting are not. Instead, to make a hologram the film needs to be exposed by a beam of **coherent light** - that is, light which is composed of lightwaves that have identical frequencies and which are vibrating together "in phase." Light from the sun and from lightbulbs will not work. These sources emit light of varying frequencies with randomly varying phase.

Lasers do emit such light, and are therefore utilized. Since most of these lasers are not portable, and many delicate optical components are used in the process to further ma- nipulate the laser beam, most holographic recordings are made in a darkened holographic studio, where conditions can be precisely controlled.

The Holographic Process

The process used to produce commercial holograms consists of three main parts -

1) recording the image,

2) regenerating the image so that we can see it, and

3) replicating it, if it looks good.

This process is somewhat analogous to the process used in making an audio recording. (Obviously, holography deals with visual images rather than sound, but the basic production steps can be compared.) We begin by recording the performer during a session in a recording studio using spe-

(continued on page 10)

Some Major Commercial Applications of Holography

Security/Authentication: Holograms are difficult to produce and/or duplicate by the average criminal and therefore have become integrated into many government and commercial security programs. They have been attached to official documents , currency, tax stamps, event tickets, ID cards, credit cards and product labels. To further deter professional counterfeiters, hologram manufacturers have added sequential numbering, tamper-evident materials and other covert features to their security holograms.

Packaging: Attracting attention is the name of the game and holograms and holographic materials help differentiate one product on the shelf from another. The unique dimensional images and/or colors inherent to the medium catch a shopper's eye. Billions of square inches of holograms have been generated for this purpose. Many packaging companies (especially those associated with the labeling and hot stamping industries) have successfully integrated holographic materials into their existing production lines. It is common to see holographic designs on foil wrapping, plastic films, cardboard containers and other paper products. Holograms are often combined with more traditional printing methods to achieve the best effect.

Advertising / Promotion: In a world swamped by merchandise displays and marketing gimmicks, holograms still attract attention long enough to convey a message - the goal of any advertiser. Holograms have been used in a variety of promotional campaigns including direct-mail, P.O.P. and even billboards. They have appeared on flyers, posters, magazine inserts, ad premiums, tee-shirts and executive gifts. Ad-industry trade journals have repeatedly reported successful and measurable results from companies using holograms in their advertising programs.

Value-Added Decoration: Textiles, paper products and even candy have been embellished with holograms. Many companies use holograms to spruce up their stationery, annual reports and product catalogs. Others decorate their products with attractive holographic materials. Still others produce actual holograms to increase the value of items, such as collectable trading cards and figurines.

One of the fastest growing decorative applications is in the fashion industry. For instance, one company, Spectratek, has developed threads of holographic materials that can be woven into fabrics using household sewing machines. Another company, Imagen, has developed techniques to hotstamp holographic accents on bolts of cloth in large quantities at the factory. Holography Presses On is involved in marketing holographic iron-on transfers which can be washed and dried repeatedly. Sommers Plastics has integrated holographic materials into its products by laminating them in a special manner. One researcher, Dr. Munzer Makansi, has even developed a method by which diffraction gratings and holographic images can be embossed directly on synthetic materials, such as nylon.

Signs: From indoor point-of-purchase displays to roadside billboards, holograms of all sizes and types are being used to deliver information to the viewer. Fully dimensional pictures, animated holographic images, and holographic portraiture are being employed to complement, and even replace, printed signs, photographs and transparencies.

Illustration: By providing dimensional information on a flat surface, holograms are able to illustrate books, catalogs and magazines in new ways. Although the educational uses have been mainly limited to children's books, the potential exists for instructional applications in medical, scientific and industrial publications.

Giftware: Over the past decade, a sizable retail market has developed for hologram pictures (wall decor),jewelry, watches and,related optical novelties. Once confined to museum shops and specialty stores, holograms have gone mainstream. Although the sales boom may have slowed as the initial novelty has worn off, unique hologram products are constantly being introduced, as the giftware industry requires new merchandise each year.

Trade Show, Museum and Lobby Displays: Over the years, numerous holograms have been produced for corporate clients and museum exhibitions. They are usually required to be large, measuring several square feet or more. When properly displayed, the effects are indeed remarkable. Images can project up to several feet in front of, and behind, the hologram. Fully detailed, multi-color images can be produced. Animated effects are possible as well.

Since holograms capture microscopic details, museums have recorded archival images to exhibit in place of the actual objects, which are safely stored away. Objects too impractical or too heavy to move (like industrial equipment) can be recorded on a hologram, rolled up in a tube and displayed in full detail at the next trade show.

Art: Some of the most distinctive and spellbinding art in the past thirty years has been produced by holographers. Still in its infancy compared to photography, and seemingly surpassed by the proliferation of digital technologies, holography still can do things other mediums can not. As the technology becomes even more accessible, more artists may start utilizing the unique attributes of holography to express their ideas.

Holographic Optical Elements (HOEs): One of the attributes of holograms is that they can direct light in a desired path. Therefore, industrial designers are using HOEs to replace bulky and breakable optical devices, such as glass lenses and mirrors. A flat HOE can be produced on lightweight plastic that duplicates the properties of the glass elements. These holograms are much less expensive to manufacture and service. They are now being used to enhance the viewability of LCD screens, watch faces, and other displays. HOE solar collectors are being researched.

cialized equipment. Next, we "play back" the original recording, using related methods. If we like what we hear, we can duplicate the original recording (stored on a "master tape") onto cassettes, CDs, etc. for sale to the public.

Let's consider the first step in the holographic process-recording the image. (This step is discussed in greater detail elsewhere in this book, but for beginners the following explanation should suffice.) As mentioned, holographers must use the coherent light from a laser to Illmuninate the subject, or object being recorded. The recording is captured on a photosensitive material, which is typically a sheet of special, high resolution black and white film.

In most holographic recording setups, the single beam of light leaving the laser is immediately divided into two smaller beams of equal length. One of these smaller beams travels directly to the film. The other beam reflects off the object and back onto the same film. As the two beams of light converge on the film they interact and combine to form a complex pattern. This pattern is called an **interference pattern**. An interference pattern created in this way and recorded by any means is called a **hologram**.

The interference pattern is recorded onto the film during an exposure that lasts from fractions of a second to min-, utes, depending on the film's sensitivity and the amount of laser light reaching it. The film is then developed and processed in order to permanently store the interference pattern: i.e., a hologram is produced. But if we looked at it, we would only see an indecipherable microscopic pattern of closely spaced overlapping lines. No recognizable image is seen, just a recording of the interference pattern.

To see the image which was recorded, we must perform a second procedure - image "playback." By properly illuminating the film (in this case, with the laser light that originally exposed it), a viewable image will appear. The image that results is commonly referred to as a hologram, even though the phrase "holographic image" is more correct.

The image that is produced is a replica of the unique set of light waves which bounced off our subject and exposed the film during the recording process. How does this work? The interference pattern stored on the film interacts with the incoming laser light to reconstruct an image of our subject. Whatever the film "saw" during the recording process, a viewer looking at the finished hologram would see. If we made a hologram of a physical object, we would see a three-dimensional image of that object. Since it is impractical to use lasers to play back all holograms, methods were developed that allowed holograms to be seen and enjoyed under ordinary lighting conditions.

The third step of the process is replication. If the holographic recording process was successful, the original hologram can then be mass replicated in a variety of ways for commercial applications (usually on plastic, paper or glass).

This simplifies the process quite a bit, but the basic facts to learn are that: a laser is needed to illuminate our subject and thereby expose the film; an interference pattern is re corded onto the film (not a visible picture); once this film is developed and illuminated correctly, a three-dimensional image of the subject is generated.

Two Types of Lasers Can Be Used

We will discuss lasers in depth later in this publication, but it is important that we touch on the subject in this introductory chapter, too. The type of laser you use affects what subjects you can record.

There are two kinds of lasers used by holographers; the Continuous Wave (CW) laser and the Pulsed Laser. The CW laser emits a steady wave of laser light, whereas the pulsed laser emits laser light in bursts. The CW laser is by far the most common laser used in holography. The power of a CW laser is typically measured in watts (W). In holography labs, most of these lasers fall in the 5 to 50 mW (milliwatt) range.

Remember that what we are recording on the plate are two laser beams converging (or interfering) with each other at the plate. If the object moves even a microscopic amount (on the order of a fraction of a wavelength) from one moment to the next, we will record two different interference patterns and the holographic image will look blurry or will not even appear.

An exposure with a CW laser can take from less than a second to several minutes. Because there cannot be any motion at all during the exposure, we need to eliminate any vibration coming from the ground. To do this we make or buy a vibration isolation table on which to put our laser, optics, and objects. Since it is absolutely critical that we have no motion at all, the subjects that we holograph with CW lasers have to be "dead" or immobile objects.

Pulsed lasers, quite the opposite of CW, emit extremely quick bursts of very powerful laser light. The output is measured in joules. Consequently, the exposure time is much shorter than a CW laser. Exposures can be made in nano-seconds (one nanosecond is one billionth of a second). You do not need a vibration isolation table for the pulsed laser. What can you shoot? Anything you want. You can shoot people, splashing water, animals. Why such freedom? Because your subject cannot move significantly in a nanosecond.

What are the drawbacks of pulsed lasers? Why doesn't everyone use one? The answer is expense. They typically cost tens of thousands of dollars or more, and require a lot of extra overhead and care. Lasers don't last forever and when a pulsed laser burns out it is costly to fix. Holographers are anxiously awaiting a low-cost, easily maintained pulsed laser. Progress is slowly being made in this area.

Recording Materials

Although we have discussed how holograms are made, we have not discussed the photosensitive materials on which they are recorded. In photography, the most common item used to capture images is a silver-halide emulsion coated onto a film base. In holography, there are a number of materials used to record your image. The most common recording media are:

1) silver halide

2) photoresist

3) photopolymer

4) dichromated gelatin

Note that we use the phrase "recording materials" instead of emulsions. This is because not all the items used to capture holographic images are emulsions. In an embossed hologram, for example, the holographic image is literally stamped into clear plastic or foil using a mechanical , rather than an optical, process. Holograms have even been re- corded on chocolate candies and lollipops. A discussion of the different recording materials requires a chapter of its own, which you will find later in this book.

Artwork Origination

Another important topic that we should cover in this introductory chapter is what kinds of subject matter you can make a hologram of. We will cover this in detail in the next chapter.

As with any creative art form there are many choices available, and much depends on what you want to accomplish. To simplify matters, we will list some of the most common things that are used for holographic subjects.

1) *3-D objects* (sculptures, miniature models, or actual objects) .

2) *2-D objects* (flat graphics, illustrations, photos, etc.).

3) *Stereographic composites* (specially shot motion pictures, video, or computer graphic files that are arranged to produce 3-D imagery).

Although some of the above topics are covered in more detail later in this book, we will give an overview of each now.

3-D objects: These are the most common subjects recorded in holograms. What your model is, of course, depends on the type of laser you are going to use for your exposure. With a pulsed laser, as we have already mentioned, you can shoot live subjects and just about anything you wish. Most holograms, however, are made with a CW laser and immobile objects are required as subjects. Holographers will frequently commission sculptors to create highly detailed miniatures ofthings which can't fit in the holography studio.

2-D objects: You will see flat artwork being used with great abundance in embossed holograms. Camera-ready art for 2-D holograms is created much the same way you would create art for a conventional printer, except that the graphics can be positioned in layers to create an image with depth.

2-D/3-D models: You also can have a combination of photos, line art and 3-D objects in your final hologram, although the depth of the 3-D object is often limited due to practical considerations .

Stereographic composites: The holographic stereogram is one of the most exciting compositions in holography today. It allows artists to incorporate a wide range of visual effects in their images - especially animation and dimension. Today's computer technology is making the production process more accessible and the finished products more

Lighting Holograms Properly

To fully grasp what holograms look like, and to understand how much their look depends upon proper illumination and viewing angle, one should examine the holograms in this book under various light sources. (All the holograms in this book are designed to be lit with a light source positioned above and on the reader's side of the hologram. Hold the book in front of you and tilt it back and forth until the holographic images appear brightest and most distinct.)

Use a Single, Bright "Point" Source of Light

An overhead, single beam of bright light (such as **direct sunlight**) works best. However, sunlight is not usually the most practical light to use. There is a whole range of other light sources with which to view a hologram; some are better than others. Good light sources cast sharp shadows. Ideally, holograms require a light source that mimics the laser light that originally made the hologram - the beam should contain the original exposure wavelength, should have enough intensity to replay the hologram, and should emanate from a single "point" source (such as **a very small light bulb**) . Fluorescent tube lights that are found in most offices do not have all these attributes.

Thus, whenever you go into a shop that specializes in holograms, one usually finds that the shop has subdued overhead lighting with a single spotlight focused on each hologram, or group of holograms. This serves the dual purpose of creating a pleasant lighting environment, as well as providing a proper illumination for holograms. People who display holograms in their homes find that an inexpensive way to illuminate them is by using a clear (unfrosted) light bulb with a single, small filament inside. Bulbs with vertical filaments often work better than bulbs with horizontal filaments. **Halogen spotlights** work very well, too. These bulbs are available at any shop with a large selection of lamps and lighting supplies. Put them in a ceiling lamp or lighting arm.

Avoid Multiple Lights and Diffuse Lighting

When a hologram is illuminated with light coming from different places (at different angles), each light source makes a separate image. This mixture of multiple images makes the hologram look blurry. Try to block off every source but the one positioned directly overhead. Then adjust the illumination angle. Diffuse light sources (such as fluorescent lights, frosted lightbulbs, light from behind lampshades and clouds) also create blurry, indistinct images. Avoid using them.

Whether or not a hologram can be seen well in ordinary room light also depends on how that hologram was designed. If the designer knows the hologram will end up in a room with lots of lights or under fluorescent lights, he will use flat graphics as artwork or record a 3-D subject with very little depth. The multiple light sources still will create multiple images, but the flat, shallow images will overlap more and the image still will be viewable. These shallow images are also more recognizable under diffuse lighting.

refined. We should point out that several techniques are used to make holographic stereograms. Sorting out the jargon can become confusing. Some of the names you will hear that refer to holographic stereograms are:

Holographic Stereograms - This is probably the most common, safe, and inclusive name. It is used to name any hologram that belongs to the group of holograms that are designed to achieve their effect by utilizing a human's capacity for stereo vision.

Integral Holograms - In general, this is a term that refers to a finished image that is constructed from many discrete units .

Multiplex Hologram - This phrase describes a hologram produced using a system developed by Lloyd Cross and refined by the Multiplex Company that utilizes stereographic and integral techniques. This is probably the first commercial holographic stereogram process developed and it involves filming a subject on a rotating stage.

A major advance in display technology was the introduc- tion of the LCD (Liquid Crystal Display). It did not take long for holographers making stereograms to see the benefits of using the LCD as a source for the image being recorded" LCD origination substitutes graphics displayed on a Liquid Crystal Display screen for cinematic footage. This allows digitized images (with all their advantages) to be easily incorporated into a hologram. A wide variety of cinematic, video and still images can be scanned in a computer, manipulated, and displayed electronically using LCD technology. Computer assisted design and origination of artwork will be discussed further in a later chapter.

Display Holography - Common Misconceptions, Myths and Reality

by John Perry (Holographics North)

Editor's Note: This article was originally written to address the misconceptions surrounding large-format stereogram holograms, which the author's company produces. However, many of his points also apply to other types of display holograms., so we have included his article in this introductory chapter.

Most people recognize a large hologram when they see one, and love it, in spite of a heap of bad information that has been presented to them. I recently witnessed this as I was installing two of our large holograms, 32"x 42" (80 x 105cm), at a local shopping mall. As I worked, there was a steady audience of people stopping, pointing excitedly, calling to their friends down the corridor, and murmuring various remarks including the one common word, "hologram."

The mythology surrounding the medium, nevertheless, is extreme. Some of it was spawned in Hollywood and still more can be traced to honest confusion over 3-D imaging technologies. Most insidious is the advertising of non-holographic products under the name of holography, hoping to tap some underground market reservoir.

Misconceptions are dangerous, especially in the large format arena, where many potential users have no direct experience. In 14 years of talking with clients, I have compiled a hit parade of the seven most popular holographic myths in an effort to finally put them at rest.

Myth #1- Holograms project images into empty space, to be seen from all sides.

"Do you also produce the projection type holograms?" is a question that became suddenly more common around late summer, 1996. It probably came from a movie. We have all seen bluish-green "Leah 's" in the movies, hovering in space, presumably being "projected" from some ridiculous apparatus in the background, and looking for all the world like 3- D apparitions. Though they're called holograms in the screenplay, they're not 3-D, they're not produced by the baffling equipment shown in the movie, and they're not holograms. They're 2-D, and they're spliced in by any number of Hollywood film tricks.

Real holograms are, in many ways, a lot better. The holographic image can, in fact, be focused, or projected, well out in front of the film, up to about 20 feet, or 6 meters, even though the angle of view may be narrow. And there is no fancy equipment. It is only the "seen from all directions" part that is missing.

A hologram is usually a piece of film which focuses light into an image and then straight through to the viewer's eyes. Since light travels in straight lines, the image can only be seen when it is framed by the outline of the film. That is, the image must be either directly in front of the film or behind it. It can only be seen from all sides if the film wraps all the way around. These "cylindrical" holograms were in fact once popular, but are now only made occasionally by one or two suppliers. They are feasible only in small sizes, the image must be contained within the cylinder of film and suffers from distortions that cannot be eliminated.

For the more common, and higher quality, flat film holograms, deep imagery, both in front of and behind the film, can be sharp and bright, but only seen from certain directions. As the viewer walks past, these images will slide across the film due to perspective, and eventually disappear off the edge.

We have found in our deep image pieces that the nature of the subject and the composition of the deep elements within the viewing field are crucial to the success of the illusion. Large objects way out in front will enlarge by perspective to spill out of both sides of the frame at once. Straight lines can easily distort, or "flag", if the illumination of the piece is not right. And, of course, the larger the hologram, the larger the angular field of view will be for deep images. It is usually unwise to make an image of one meter depth in a hologram smaller than about Y2 meter in width.

Myth # 2 - A holographic installation requires a lot of equipment.

There are two objects in a holographic installation - the hologram and the lamp. A large hologram, say 40 x 50 inches (l x 1.25m), will usually consist of a piece of film laminated to a 1/8 inch thick acrylic sheet. all weighing about 8 lbs. Our preferred lamp is a 75 watt halogen MR-16 type spot. This is a fixture about the size of two fists , weighing several pounds and costing about $100. They are common equip- ment in museums and trade shows. A small acrylic mirror may also be recommended in some situations to reflect the light to the hologram.

The height of the film is important. as well as the angle of light striking it. A lighting diagram showing the positions of the hologram and the light are packed with every piece we ship. But for animated and still image holograms alike, there are no machines, lenses or moving parts of any kind. They ship very easily and set up quickly.

Myth #3 - A laser is needed for display.

This misconception derives from the first holograms, 25 years ago, which were of the laser transmission type. This means they were viewable only in laser light, transmitting through the film, or in other words, rear illuminated. Trying to view these holograms in ordinary white light yields a stunning, but unrecognizable, blaze of spectral color. These images are strikingly realistic. It can be difficult to believe that the subject is not sitting there, behind the film. But there is little flexibility for multiple colors, image overlays, and animation. And most of all, it is extremely inconvenient, expensive and dangerous to display large images of any kind with lasers. We never recommend them for public display.

Virtually every display hologram today is white light viewable, meaning no laser. There are basically two types- white light transmission (or "rainbow"), and white light reflection. Small halogen track lights are excellent for holograms, but direct sunlight and slide projector beams are also good.

Confusion also arises because lasers are, indeed, required in the production of a hologram. The microscopic pattern that is exposed onto the film requires the precision of coherent, or laser, light. Clients are often concerned that their subjects may be exposed to dangerous visible radiation, and this is a valid concern in some cases. Usually live subject imagery is produced with the stereogram process, which originates with a series of photographic images. No laser light is involved at this step. In fact the shooting can be accomplished in daylight and on site.

Occasionally a pulsed laser will be used to image a live subject directly into a hologram. This produces an expo- sure time short enough to freeze any subject movement to within the required fraction of a wavelength of light. This procedure is indeed quite dangerous, if the proper precautions are not taken to instruct and protect the subject. All other use of laser light is in the confines of the holography production lab, under the supervision of trained staff.

Myth #4 - A dark environment is preferred

When showing and viewing a hologram. The 1970's and 80's bristled with holography galleries and museums which were dark and often mysterious. The mysterious part was unfortunate, but the dark surrounding's were important because the film, chemistry and knowhow of that era were not yet producing bright images.

Today's images are dazzling. Any hologram made within the last five years should be the brightest thing in the room. If it's not, it's either poorly made or poorly lit. The test I like to run on our large format pieces, before we ship them, is to look at them in the subdued light of our showroom with the fluorescent lights off - about like a typical living room. Then I turn on the surging fluorescent and see if the holograms actually look better. Usually the slight haze of the film surface vanishes, and the image is still clear and bright.

Our showroom walls are, by the way, white. Another mistake in this age is to paint the environment black or dark grey. Light grey is lovely, but there is no need to look at holography in a "coffin."

Myth #5 - Those 3-D photographs with ribbed plastic surfaces are holograms.

These are called lenticular photographs, and are made by a few manufacturers, They consist of several photos taken from different perspectives, diced up into strips, and viewed through a lens array, the ribbed plastic surface, so that the left and right eye are always getting a stereo pair.

Basically, these photographs have between four and ten perspectives, whereas a hologram contains an infinite number, in a continuum. The pros and cons of each technology are many. They are not at all the same thing. They look different, are made differently, and are good for different purposes.

Myth #6 - Holograms Of Any Size Can Be Mass Produced.

Unfortunately, this is not so at present, although the next two years should see major changes in this area. Several labs are now working on the technology to mass produce holograms in sizes up to 32" x 42" (60 x 120cm).

Mass produced holograms, as discussed elsewhere in this volume, come in two types - embossed foil and photo polymer film. The images from each are extremely clear and bright, and the replication costs are competitive with other imaging media. The problem traditionally has been size, the maximum for each process being about 10" x 14" (25 x 35cm). Larger work must be produced by hand, driving the cost to levels prohibitive for large volume. We do not currently accept large format mass production orders, but I anticipate taking and filling these orders, starting sometime in 1998.

Myth #7 - Large holograms can be displayed in shallow light boxes.

Airports are filled with wonderful (but flat) photographic images in large transparency formats, displayed in fluorescent lit, diffusing light boxes. The boxes are generally six inches deep. Many an advertising executive has dreamt of covering those light boxes with holograms, and brightening the dreary tedium of air travel... and so, we get calls.

These are difficult opportunities to give up. But these light boxes simply will not work for large holograms today. The hologram must be illuminated from a specific angle, and the lamp must be a minimum distance away, about five feet for a large hologram, to avoid serious distortions. We have designed lighting schemes so that only 18 inches of horizontal backspace are required for a 32" x 42" (80 x 105cm) piece. But rarely are light boxes that thin. Rear lit holo- grams can also be mirror-mounted to allow front lighting. But most light box environments, like airports, do not allow external lights to be installed on the walls or ceilings. At present, we attempt to find 'alternatives to the light boxes, and this is often possible.

Conclusion

Occasionally I am disheartened when I see someone walk by a hologram without even noticing. I consider how far we have to go before our product is as versatile and polished as a color transparency. Then I am blessed by a crowd of enthusiasts, frozen in their tracks as they slosh through a mall, marveling at a couple of holograms getting installed. And when I consider how incredibly far we've come in, say, the last seven years, I am increasingly optimistic about what the future holds for display holography.

In the Holography Studio

The following information is provided in order to introduce the reader to what is involved in making ordinary holograms in a holography studio. It is by no means a complete description of what occurs; however, it should prove useful to beginners and potential users of the technology. More detailed instructions are available in the "Holography Handbook - Making Holograms the Easy Way", also published by Ross Books.

The Two Major Categories of Holograms

The terms "transmission" and "reflection" are one of the primary ways used to categorize all holograms. Very simply, the terms refer to how the hologram is illuminated during the viewing process.

Transmission holograms require that the illuminating beam of light pass through the hologram in order for an image to be seen. Therefore, these holograms must be backlit. Sometimes positioning a light behind the hologram is difficult.

Conversely, **reflection holograms** require that the illuminating beam emanates from a source on the viewer's side of the hologram. The light reflects off of the hologram back to the viewer's eyes. Reflection holograms are often favored by the general public because they can be easily hung on a wall and illuminated with a ceiling light or a lighting arm.

Most of the embossed holograms in this book seem to be reflection holograms; however, they are actually transmission holograms with a mirror attached to the back. The mirror sends the light back through the recording material and creates an image. Very thin coatings of metal are used in place of glass mirrors. This is a clever and practical way to solve the rear lighting problem while retaining the advantages of using transmission holograms.

Within the two major divisions of holograms (reflection and transmission), there are many variations. Like any other specialized field, holography has its own lingo, and in some cases the same hologram can be described using more than one name.

In the Holography Studio

Suppose one enters a holography studio where a simple hologram is about to be made. The first thing you will notice is a special vibration-free table in the room called an **isolation table**. During a typical exposure (which can last seconds to minutes) a movement smaller than a wavelength of light can ruin days of work. Therefore, the table is designed to isolate the holography set-up from the smallest vibrations in the surrounding environment. It is usually quite massive and built to float on a cushion of air.

On the table is a laser, some mirrors and a piece of photosensitive material positioned in a **plate holder**. This photo-sensitive material is typically a silver-halide emulsion coated onto a glass plate, and thus will be referred to as the **recording plate**. Since the recording plate is sensitive to light, the studio must be darkened when it is out of its box. Sometimes certain safelights can be used so that the holographer can see, but extreme care must be taken not to pre-expose the recording plate.

Everything on the table is arranged in a carefully measured manner. The object to be holographically recorded is positioned in front of the plate holder in the middle of the table.

As mentioned, to record a hologram we use a laser that emits a single beam of light at one wavelength. We cannot use the sun or just any light as our source because the light from common light bulbs or daylight contains many constantly-changing wavelengths: it is not coherent. If we make the exposure using incoherent light, the changing wavelengths would create a multitude of interference patterns and the resulting holographic image would be completely blurred and useless.

Making a Transmission Hologram

To make a transmission hologram, first we turn on the laser and aim the beam at Mirror 1. Due to the fact that Mirror 1 is only partially reflective, part of the beam is reflected toward Mirror 3, and the other part passes through Mirror 1 to Mirror 2. Because the beam is split, Mirror 1 is referred to as a "**beamsplitter**." (See figure 1.1).

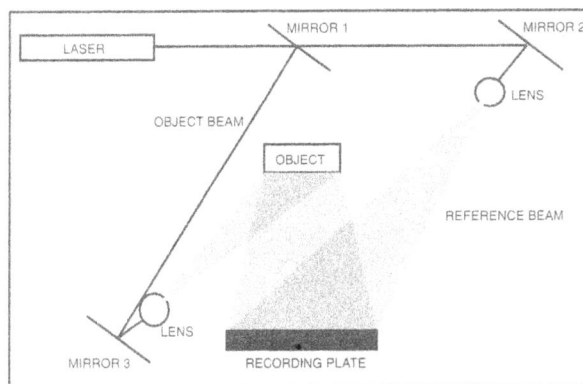

Figure 1.1. Transmission Hologram

The beam that passes through Mirror 1 to Mirror 2 is called the **reference beam**. The reference beam's path always ends at the recording plate without ever illuminating the object. For instance, in this set-up after the reference beam strikes Mirror 2, it is reflected through a lens toward the recording plate. The lens' function is to spread the beam so that it will cover the entire plate (in some cases, the lens is placed in front of Mirror 2; in either case its function is the same - to spread the beam).

At the same time, the other beam, which we call the **object beam**, reflects off Mirror 3 and also passes through a lens. This lens spreads the beam out so that it illuminates the entire object. The laser light reflects off the object (hence the name object beam) and strikes the photographic recording plate. *The two beams must travel exactly the same distance so that when they recombine at the recording plate they will be in sync with each other, and an **interference pattern** will be formed.*

After exposure, the photosensitive plate is developed using the appropriate chemistry and standard darkroom procedures (i.e. developer, bleach, fixer, washing, drying, etc.) The resulting developed plate is the **hologram**.

Holding the developed plate up to light, we see that the plate is semitransparent. On closer inspection we see that the darkness of the plate is caused by developed emulsion. The plate seems to have countless swirls of threadlike developed emulsion which are called fringes. The **fringes** look like the swirls that make up your fingerprints or the boundaries on topographic maps. There appears to be no order to the swirls. In fact, the fringe pattern is a recording of the interaction between the reference beam and the object beam.

Viewing the Image Using a Laser

To see the image, we put the recording plate back in the plate holder on the table in exactly the same place it was for the exposure. Then we remove the object and Mirrors 1 and 3 from the table. Now, when the laser is turned on again only the reference beam illuminates the plate. When you look through the plate, an image is seen of the original object, in its original place and at its original depth. (See figure 1.2.) This reconstructed image is indistinguishable from what you would see if the object was not removed! The first time holography students see this happen they are quite amazed!

Why We See an Image: A Simple Explanation

A detailed explanation of why this happens would occupy many pages. A simple explanation migbt go like this: the two beams strike the photosensitive recording material at the same time. Since they both originated from the same laser beam, and traveled equal distances, they are precisely in sync with each other. When two such waves of light recombine, their interaction produces an interference pattern.

This pattern is recorded on the photosensitive material during the exposure step. When we develop and process the recording plate, the interference pattern is stored on the plate as the fringes that we see. Because we are recording the interaction of lightwaves (which are quite small), the fringe patterns are microscopic (on the order of 1,000 or more fringes per millimeter).

After development, if we aim only the reference beam at the plate (at exactly the same angle that originally exposed the plate) the interference pattern which was recorded naturally causes the light waves passing by to change direction. This phenomenon is called diffraction. It occurs whenever light passes through small apertures, like the space between fringes . This diffracted light has exactly the same form as the lightwaves which were originally reflected from the object.

In other words, a properly illuminated hologram regenerates or recreates the way light was reflected from the object to the recording plate. You can see an image of the object without that object even being there. Under laser light, it looks exactly like the original thing being recorded.

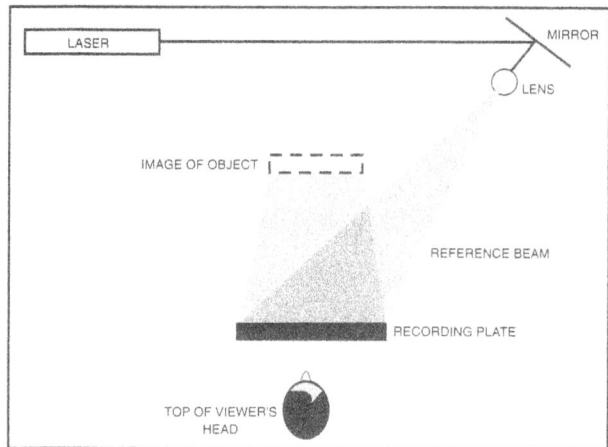

Figure 1.2. Viewing a transmission hologram

Making a Reflection Hologram

If we start with the basic setup previously depicted, but transfer the reference beam around with mirrors so it illuminates the recording plate from the side opposite the object beam, we create a reflection hologram. Remember that reflection holograms are meant to be illuminated from the "front", so changing the position of the reference beam will achieve this. (See figure 1.3.)

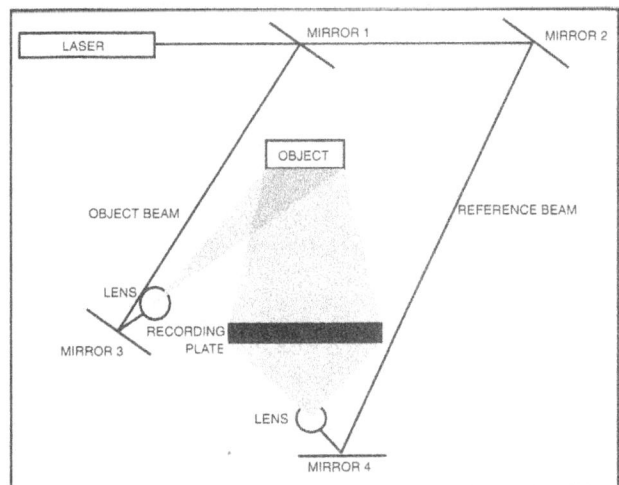

Figure 1.3. Reflection Hologram

*For more information,
visit our Internet Website:*
www.holoinfo.com

Positioning the Holographic Image

It is important that we cover the topic of the **H1** and **master hologram** in this introduction because it is a fundamental procedure in the making of almost every commercial hologram. H1 stands for "hologram one", which simply means it is the first hologram you make on the path to your desired final hologram. Sometimes the H1 is the master hologram from which you make multiple copies. Frequently, though, there is more than one hologram that needs to be made before you get the finished hologram from which you will make copies. If this is the case, the next hologram in the sequence is called the **H2**, and then **H3**, and so forth.

A question that immediately comes to mind is, "Why would anyone want to make an H2 ?" Well, historically one of the big problems that holographers had was placing the subject exactly where they wanted it. Suppose, for example, you want the object in the final hologram to appear half in front and half behind the recording plate. How would you do it? You obviously can't do it on your first shot because the object would have to be going right through your photographic plate.

This problem was solved by the following procedure:

• Make an H1 transmission hologram.

Since the H1 hologram creates an image of the object, why not use the *image* (generated by our H1) as our subject, and make another hologram (an H2) of it?

• In other words, make a hologram of a hologram. This H2 hologram can be a transmission or reflection hologram, depending on your need. It sounds strange, because you are making a hologram of an image and not an object. But it works. (See figure 1.4.)

• Now, since you can make a hologram of the H1's image, take time to move the image around to wherever you want it positioned. In this case, adjust the H2 recording plate so that the image of the object is half in front and half behind the plate and then make your H2. The problem of getting half of the object in front of the plate, and half behind, is solved.

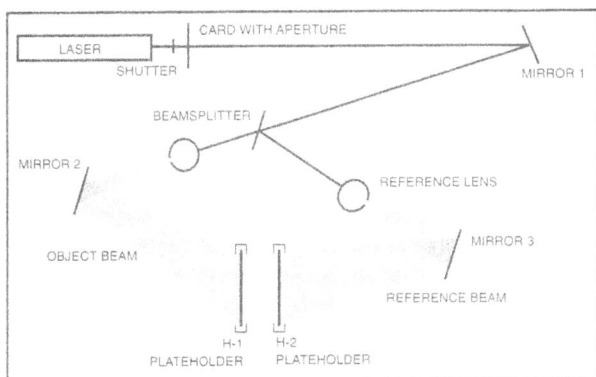

Figure 1.4 – Reflection H-2 being made from H-1

In short, there are at least three good reasons why an H2 should be made:

1. The H2 allows you to reposition the image of your subject. When you reposition your image from the H1, you may make your subject focus in front of the recording plate, behind the plate, or anywhere within the limits of your equip- ment (you are usually limited by the laser's ability and the quality of the optics). The creative potential here is enormous because you are able to move solid objects around as if they are ghosts. You can have two objects occupying the same space, etc. The process of moving the image around to make the H2 is called **image planing**.

2. It gives the holographer a chance to brighten up the image. Since you may move your image anywhere, you can focus the image right at the recording plate. This concentrates the light directly on the recording material and brightens up the image considerably. This is commonly done in silver-halide reflection holograms.

3. It saves time on remakes. If you develop the H2 and decide you don't like the position of your subject astride the recording plate, you don't have to find the Original subject and set it up again. This can be important if there are large costs in arranging the H1 shot.

Going through the pains of making H1 , H2, etc. to produce a master for commercial replication is usually required. It is technically possible to get results from the first shot - but most professional holographers shoot a series of holograms in order to end up with a suitable production tool.

Image Projection of Holograms

Although transmission holograms seem to be naturally designed to create a hologram with considerable projection, one can also make reflection holograms that have a great deal of projection. In fact, reflection holograms with considerable projection are a favorite among artistic holographers and the buying public. They are favored because they can be hung on the wall and illuminated just like a painting, whereas transmission holograms need to be lit from behind, often requiring a much larger viewing area.

Laser-viewable transmission holograms can demonstrate amazing depth and projection when the correct equipment is used to make and display them. It should be noted that the depth of the holographic image is not so much a function of the power of the laser as it is of the coherence length of laser light (you can read more about coherence length in Chapter 7 - Laser Fundamentals). Theoretically, the maximum image projection in front of the hologram plate can be as great as the projection in back of the plate (depth of the image). Unfortunately, it is difficult for our brains to make sense of greatly projected images. Because of this, and the fact that usually there are optical distortions created in the image planing process, projected distances in transmission holography usually are kept under four feet.

Laser transmission holograms have the widest parallax and display deep images best. There are laser transmission holograms, for example, of people and objects in a 4,000 cubic foot room, made by pulsed lasers. Not surprisingly, projected hologram images like this generate some of the highest shock and thrill responses from viewers.

Making a Rainbow Hologram

We mentioned earlier that although it is necessary to use a laser to make a transmission hologram, it is not always necessary to use a laser to see a transmission hologram. In fact, most transmission holograms can be seen in sunlight. This may seem confusing, because we have said that in order to see a holographic image you have to shine the laser beam that made the hologram, on it. This is true, but sunlight contains a multitude of wavelengths of light, including the one from the laser that we used to make our exposure. Also, the sun is such a great distance from earth that it appears to be a single beam of light shining on our plate. It would seem that we have only to position the plate at the proper angle, and we should see our image.

This is logical, but it also stands to reason that if sunlight passed through a transmission hologram we would also get images being formed by all of the other wavelengths that are somewhat close to the wavelength of the reference beam. These other frequencies of light would diffract at a some- what different angle than the original reference beam. The result would be a multitude of images forming right next to each other, creating a blur instead of a clear, crisp image.

That's exactly what does happen and it took a while for a solution to be developed. Around 1969, Dr. Steve Benton came up with a solution. The resulting hologram is sometimes referred to as a "Benton" hologram, or more frequently, a **rainbow** hologram.

Rainbow Holograms

Benton reasoned that since our problem is too much imagery at the point of reconstruction for our object, why not block off some of it? In other words, suppose we put up an opaque mask against the transmission hologram, with a long, narrow horizontal slit through which we view our transmission hologram. This would certainly clean out a lot of the annoying secondary images that are blurring the primary image's reconstruction. (See figure 1.5.)

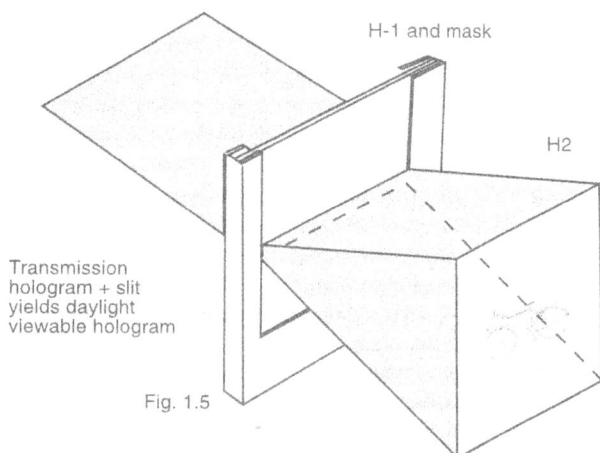

Transmission hologram + slit yields daylight viewable hologram

H-1 and mask

H2

Fig. 1.5

This "cleaning" comes at a price, however, because the mask causes loss of vertical parallax (the ability to be able to see over and under our object). We would, however, still have our horizontal parallax (ability to see side-to-side

around the object). Humans, with feet fixed on the ground and eyes on a horizontal plane, are actually more accustomed to horizontal parallax than vertical.

The procedure to produce this masked hologram is as follows:

1) First a normal transmission hologram is made.

2) Next, an H2 copy of the transmission hologram is made, but an opaque card with a horizontal slit is placed between H1 and H2.

To see the resulting holographic image (using the laser light we used to make it or light of an identical frequency) our eyes must be positioned at the "real image" of the slit. This viewing geometry will be apparent in the studio, but what happens when we want to see the hologram outside?

Imagine viewing this H2 hologram in two different colors (two different frequencies) of light. A hologram of the image made through the slit will be played back, but each of the two wavelengths of light will diffract through the hologram fringes at a slightly different angle. There will be two different images of the object, each a different color and each at a slightly different vertical position.

Next, think of the image in white light or sunlight. All of the wavelengths present will reconstruct their own image, all slightly displaced vertically with respect to one another. As you move up and down in front of the plate the color of the image will shift through all of the colors of the rainbow (hence the name "rainbow hologram").

As you move from side to side you will have horizontal parallax because nothing has been done to destroy it. By careful planning, the image can playback any desired color at the correct viewing angle, or even a combination of colors (a multi-color rainbow hologram).

In effect, the hologram is filtering the white light, while all that is sacrificed is vertical parallax, which, as we mentioned, our two horizontally-positioned eyes usually don't miss anyway. Also, these rainbow images are often extremely bright, because all of the frequencies in white light are being used to form the image.

So the rainbow hologram technique is a way of making a transmission hologram sunlight-viewable. Other names for this are "daylight-viewable" or "white-light viewable" hologram. They all mean the same - a hologram you can see without the need of a laser.

Making a Holographic Stereogram Most historians credit Lloyd Cross and his cohorts in San Francisco with the development of a process that resulted in the first reliable method for producing a holographic stereogram -- it resulted in a three-dimensional cinematic image that appeared to "float in space." Their method, developed in the early 1970's, allowed live subjects, life-size models and special visual effects to be incorporated into their holograms in a practical and affordable way, as expensive pulsed lasers were unobtainable.

In order to commercialize the endeavor, Cross and his colleagues manufactured a motorized display unit for their free-standing 360-degree version. They also developed a stationary wall-mounted unit that displayed 120 degrees of viewing angle as the person moved around it. The idea of creating a self-contained holographic display device was quite revolutionary and very admirable. The complete units, which incorporated an inexpensive light source (an unfrosted light bulb with a vertical filament) along with the hologram, sold for several hundred dollars. The Multiplex

Company has been producing units based on this process for over twenty years.

Here is a simplified description of the process:

1. Make a rotating stage.

2. Place an object or a scene with live actors on the stage.

3. Set up a stationary movie camera in front of the stage.

4. Film the subject as the stage rotates 360 degrees, making sure to shoot at least three frames for each degree of rotation. In addition to the stage moving, the subject is allowed to move slightly in a manner that will result in a smooth animated sequence. Rapid or uneven motion, however, will create undesired "blurring" effects.

5. Develop the movie footage in a normal manner.

6. HOP Transfer: We now want to make a hologram of each frame of the movie footage. These holograms will be sequentially exposed onto a sheet of film using a holography setup whose elements are collectively referred to as the "Holographic Optical Printer". The HOP setup illuminates each individual frame of movie footage with laser light. Another laser beam meets the beam that went through our movie frame at the emulsion by another path to create the hologram. Each frame is optically "condensed" into a narrow strip on the film using lenses and a mechanical slit aperture that restricts the image to one, narrow, vertical slit. The film is advanced and the process is repeated. A series of vertical slit holograms, running the length of the film, results. (See figure 1.6.)

7. After the process is complete, you will have a length of film with hundreds of thin vertical holograms on it. Once processed, you can take the film and wrap it into a cylinder shape. When the film is illuminated from inside the cylinder (behind the film) with an appropriate light source, the viewer will see an apparently solid image floating in space inside the cylinder! As the cylinder rotates, or the viewer walks around it, the image looks fully dimensional and appears to move!

These dramatic effects result from the fact that each of the viewer's eyes sees a slightly different image at the same time. Our brain then combines these images to give us a "stereogram" effect. One limitation to Cross's approach is that this technique creates images that display horizontal parallax only (i.e., you cannot see above and below the image) . This is very adequate in most situations because in life we generally inspect images by looking side-to-side and not over-and-under the image.

Subsequently, holographers produced variations of Cross's concept. Some made stereograms with different degrees of view, commonly 60 or 90 degrees. Others began shooting the sequence of frames by moving the camera along a track (instead of moving the stage). They went on to flattening out the cylinder, which allowed the holographic stereogram to be produced and handled more easily. Eventually, researchers embossed these holograms onto mirror-backed plastics or produced copies which allowed front lighting (which is more practical in most situations).

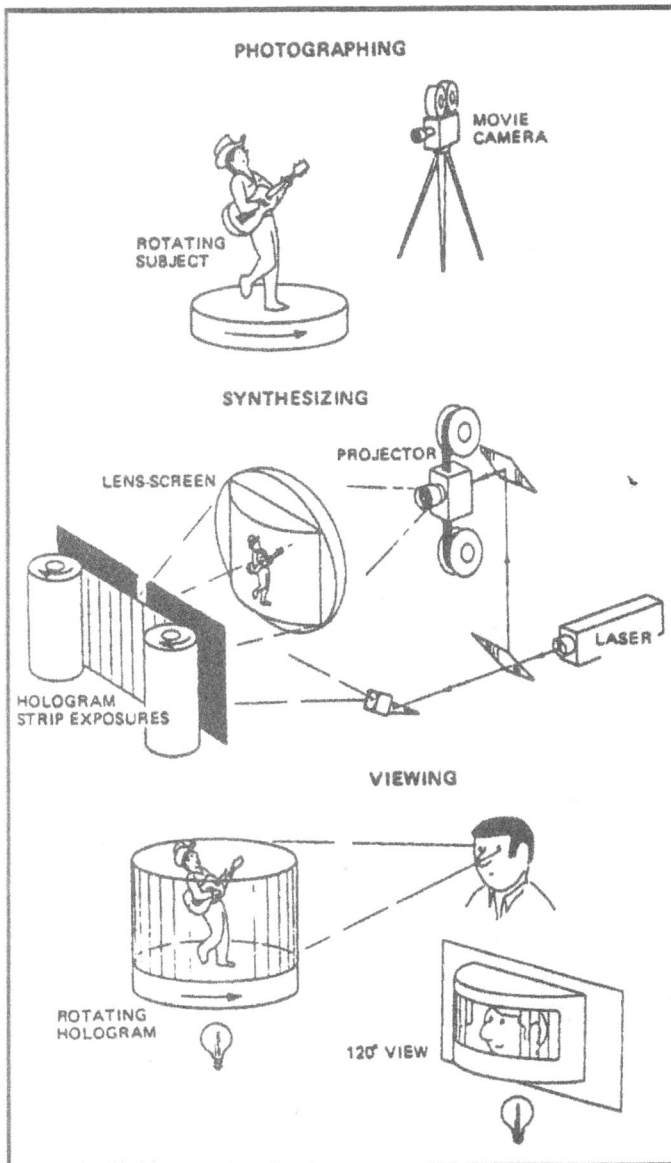

Figure 1.6. Making a holographic stereogram using film footage.

Linda Law Holographics

Computer Graphics for Holograms and Lenticular Images

2D & 3D Graphic Design for Print

2D/3D Graphics, 3D Animation

P.O. Box 434, Centerport, NY 11721
Tel: (516) 754-6121 Fax: (516) 754-9227
e-mail: llholo@i-2000.com

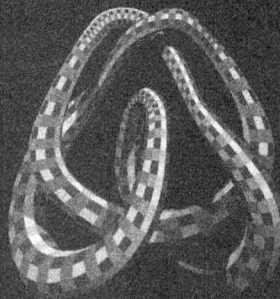

Graphics for Holography Marketplace cover created by Linda Law

2
Artwork Origination & Mastering

The first step in producing a holographic image is designing and preparing suitable artwork. This is a crucial step, as the type and quality of the subject matter recorded in the hologram will determine how the final image will look. This chapter discusses the different kinds of subject matter that can be supplied to the holographer and explains how each type is incorporated into the production process.

In addition, there is an overview of "dot-matrix" technology and an article about the "recombining" process. Both subjects are relevant to users of embossed holograms, as these production techniques are being used more and more frequently in the hologram origination process.

In the early years of holography, specially designed works of art or sculptures were commonly used as subject matter for holograms. These items are still used today. However, it is now possible for computer-graphic artists using readily accessible hardware and software programs to electronically generate "camera-ready" artwork that holographers can assemble into images that display dimension, depth, projection and motion. Holographer's are substituting this digitally originated artwork in place of the time-consuming drawings, hard-to-record physical objects and expensive cinematic shoots that they traditionally utilized. They are using the computer to increase flexibility and versatility in the design and production processes, as well as to cut production costs. This merger between electronic imaging systems and optical-based ones is resulting in new and profitable opportunities for all those involved, especially the artists and designers that are able to best utilize both media to achieve their client's goals.

In most instances, the "camera ready" artwork prepared by the design team is output as a series of computer graphics files which is sent to a hologram origination facility. These computer files correspond to various graphic elements of the holographic image being produced. In brief, the holographer uses these graphics files to generate a "master" hologram which is recorded on a high resolution photosensitive material using a laser and specialized optics. The master hologram can then be mass-replicated in a manner suitable for commercial applications.

In this chapter, we will discuss how artwork for your hologram can be produced and explain how it is integrated into the manufacturing process.

Common Methods of Making a Master

There are several different methods that can be used to make your master hologram in the production studio. A clear understanding of these methods will go a long way toward helping you understand the manufacturing process and helping you plan your project. In this chapter, we will list the four most common methods of making a master hologram and discuss each method.

1. 3-D Artwork - A physical object (or person) is used as the subject to be recorded in the studio.

2. 2-D/3-D Artwork - A computer image (such as a Photoshop or illustrator file) is created, output to film or paper and used in the studio shot. You can use several images to create the feeling of depth if you wish.

3. LCD Artwork and HOPs - The computer image in method 2 can be made and, instead of outputting to film, the image can be illuminated on a LCD (Liquid Crystal Display) screen. The LCD image becomes the object you make a hologram of. In the advanced form of this method you can create a series of images (a cinema) which is mechanically exposed in sequence onto your master hologram creating a mini cinema for the viewer. This method is very popular and the machinery used to expose these optical images is called a Holographic Optical Printer (HOP).

4. Dot Matrix Machines - Another recent method is to skip the LCD altogether and simply bum the image you want directly into the photosensitive emulsion. These machines are referred to as Dot Matrix holo-

gram machines and they are becoming more and more popular in embossed hologram production.

As you can imagine, there are pluses and minuses to each of the above methods, and the artwork used in some methods will not work in others. We will now discuss how each of the above methods work, and the type of artwork that should be supplied when using it.

Method 1: 3-D Model Making

Although live subjects can be recorded using specialized equipment, most holographers shoot inanimate models.

In thinking about models, it is important to note that the "depth of field" in your final hologram is closely tied to what type of hologram you intend to mass produce and how the hologram will be illuminated. Embossed and dichromate holograms generally reconstruct fairly shallow images (one inch or less), while photopolymer films can replay images of several inches in depth. Holograms produced on silver-halide glass plates can reconstruct images several feet deep under proper illumination. On the other hand, the unit cost for silver halide and photopolymer production runs is much higher than embossed runs. You obviously need to discuss carefully with your holographer the specifications of your sculpture before creating it.

Basically, a hologram can be thought of as a window through which you view a scene. One of the first steps in 3-D model making is to pretend you are standing directly in front of a window which is the size of your final hologram. This impresses on you the limits of the viewing space. Imagine yourself moving from side to side - which allows you to see around a given object and into areas of the room which would otherwise be hidden. Designers should pre- determine viewing angles with the holographer and communicate these measurements to the model maker in order to maximize the usable image area. Designers also need to take into account that a holographic image, has "volume" - it can have depth (an object can be behind the window) and/or projection (an object can appear in front of the window). Images should be designed to best utilize this front-to-back dimension. Many model makers make the most important visual components of their images focus right on the image plane (right on the surface of the hologram), which stays in focus under less-than-ideal lighting conditions.

Execution

After the design process comes execution, a step which utilizes a modelmaker's craftsmanship and technical proficiency. One craftsman who is quite experienced in this field is George Sivy of Richmond Development Group (formerly Gray Scale Studios). Sivy has been a holography model maker for eleven years. He worked for Polaroid during its first years in holography, making models utilized for custom and stock images (such as the popular "Brain/Skull" which appeared on the cover of the *Holography Marketplace's 3rd Edition*). In addition, he has collaborated on numerous commercial projects, ranging from embossed security holograms to photopolymer holograms designed for the giftware market.

In terms of materials, Sivy uses "what will solve a given problem." Although some materials and combinations that he uses are proprietary, he did mention the readily available Sculpey, a synthetic clay that can be baked. This ensures a stable model which is important since most holographic mastering uses continuous wave lasers that require absolutely no motion during exposure periods. Long exposure times might require even more stable materials, and Sivy may use Sculpey first, then make a mold which is used to generate an even more stable sculpture.

After the model is created, Sivy says that it is often painted to create contrasting areas of light and dark. Since most, but not all, holograms are intended to be reproduced as monochromatic images, the "coloring" process is quite different than in ordinary modelmaking. Less experienced model makers will often work under a safelight which duplicates the laser light which will illuminate the model during exposure. Good model makers will also utilize textures, shading and special effects to maximize the hologram's visual impact. Again, this requires communication among the design team, the model maker and the holographer.

There are a number of other professional model makers that work in the field of holography. If you that are interested, an interview with some of them appears in Holography MarketPlace 6th Edition. They discuss how they produce their work in more detail.

Method 2: 2-D/3-D Artwork

A computer image (such as a Photoshop or Illustrator file) can be created, output to film and used in the studio shot. You can use several images to give the illusion of depth if you wish.

The production steps used by one business that does a lot of 2-D/3-D work is listed on the next page to give you an idea of what is involved in the making of a master. The details of the steps involved vary from business to business.

The 2-D/3-D hologram method is used in making a master for photopolymer, dichromate, silver-halide or embossed holograms but we will restrict the examples in our discussion to the embossed hologram master because it illustrates the process best.

The creation of 2-D images obviously needs to be done with care. Imagine breaking a conventional print ad into three levels of related graphics - a foreground, a middle ground, and a background. Picture each element to be a separate flat graphic (a 2-D) or a photographic transparency. (For example, the foreground image might be a corporate logo, the middle image a picture of a product, and the background image a landscape.) Arrange the three elements front to back, yet separated from each other by a 1/4 inch or so of space (3-D). This array represents the multilevel imagery associated with a typical 2-D/3-D hologram.

Production Steps - Creating an Embossed Multi-level (2-D/3-D) Hologram

Image Design

1. Designer consults with holographer regarding job-specific design requirements.

2. Designer prepares client's artwork for holographic reproduction. Creates a multi-level image "on paper" consisting of black & white line art drawings and/or photographic images.

3. Drawings, photos and graphics that comprise the images are scanned into the computer using Adobe's Photoshop™. If the desired imagery must be copied from existing corporate artwork, Photoshop tools can be used to extract the desired graphics.

Digital Image Assembly

4. Digitized image is assembled and checked. Line art is cleaned up. (Black outlines should separate image components and all lines must be unbroken.)

5. Bit map images converted to postscript using Adobe Streamline™ (now included with Adobe Illustrator™)

6. Postscript images imported into Illustrator. Image is broken into multiple levels - one file created per image plane (i.e. an image with a foreground, a midground and a background requires three files).

7. Designer assigns appropriate colors to image components on each level.

8. Completed files sent to holography studio by diskette, SyQuest™, DAT, optical disk, e-mail, etc.

Creating a Production Tool

9. These files are reviewed and imported into new Illustrator "master" file standardized for that particular holography studio. Images are ganged if necessary; they are sized to fit production equipment; and pre-designed cut guides, registration marks, and TM symbols are added.

10. Adobe SeparatorTM is used to generate color separation "sub-files" for imagery levels that are multi-color, and composite "sub-files" for imagery levels of single color. These "sub-files" will be used to make masks that the holographer will use when exposing the recording material.

Photoshop in its RGB mode generates necessary separations if non-primary colors are being copied from existing artwork.

11. Output "sub-file" separations on paper to check color assignment.

12. Take Illustrator "sub-files" (on SyQuest drive) to image setter for output.

13. Image setter outputs film positive transparencies - one per"subfile." Colored areas are now solid black.

14. Holographer uses these film positives make glass negatives on Kodak HRP (color areas are now clear to allow laser light to pass through and expose plate at specified angle per selected color).

Holographic Recording

15. Holographer shoots one layer at a time, one color at a time. Each time masks are positioned to block offportions of the recording material that should not be exposed. Every time another color is required, the holographer must adjust the optical setup to change the reference beam's angle.

16. Holographer repeats the exposure process for each level of imagery until the master hologram is complete.

Holographic Replication

17. Finished hologram is checked for flaws and prepared for electroplating.

18. The hologram is metalized and stamping dies are created that reproduce the microscopic patterns on the original hologram.

19. These stamping dies are used to emboss the hologram on rolls of foil or plastic.

20. Holograms are die cut, finished, and sent off for application (hot stamping, packaging, etc.).

(Editor's note - Special thanks to Mike Grogan of Holographic Dimensions Inc. for information provided.)

Digital Design Tools

These "levels" of artwork can be easily generated using digital tools familiar to most computer graphics artists. Adobe's Illustrator, Corel's DrawTM and MacroMedia's FreehandTM can be used to create original drawings. Adobe's Photoshop can be used to import and touch up a client's existing artwork to make it suitable for holographic reproduction. Adobe's PostscriptTM or Microsoft's TruetypeTM font collections are often utilized to create logos. Kai's Power ToolsTM and BryceTM programs can be used to further enhance imagery. In short, a variety of software programs are capable of doing the job.

Once the artwork is finished, the digital files are sent to the hologram production studio either electronically or physically. There, an in-house designer will typically use Adobe's Separator, Photoshop or Quark's ExpressTM to break the image into the appropriate component layers, if the original artist has not already done so. These files will be sent to an image-setter (a machine that outputs film) which will generate film transparencies corresponding to the different levels of imagery. These transparencies will be copied onto rigid glass plates. These glass plates will then be stacked in a sequential array. This array will constitute the physical object which will be recorded.

Creating Dimension and Depth

Designing an image for a 2-D/3-D hologram is similar, but not identical, to designing an image for print. Since the artist is designing a multilevel image, subject matter obviously should be positioned to take advantage of these unique dimensional properties. For clarity, the most important elements of the scene are usually placed directly on the image plane. Foreground elements intended to float "above" the image plane should be easily recognizable, as they will blur out under less than ideal lighting. The same applies to background images with great depth. Drop shadows, textures and shadings are often incorporated to exaggerate dimensional effects.

Designers commonly 'arrange the different image levels in one of two ways:

·   So that in the finished hologram the foreground image appears to "float" slightly above the surface of the embossed material, the primary image is on the hologram's surface (called the "image plane"), and the background is behind the other two.

·   Or, the foreground image might be positioned directly on the image plane, with the middleground and background images underneath, which further exaggerates the apparent depth of the image.

Design Guidelines - Parallax

Designers must also consider take advantage of parallax - the ability to look around the sides of an image. It is important to note that due to standardized holographic production methods, most embossed holograms only display horizontal parallax, i.e. a side to side view, rather than an over and under one. This is usually adequate as it mimics the way we ordinarily look at the world.

To use parallax effectively, foreground images positioned to float off the hologram's surface need to be sized correctly so edges do not "cut-off' prematurely if a viewer moves off center. Since background imagery will be in sight when a viewer looks "behind" the foreground elements , the graphics for the background should extend completely from one side of the hologram to the other. For these reasons, and for finishing purposes, foreground and background artwork should be oversized in relation to the final size of the hologram. Only the imagery that is planned to appear on the hologram's surface should be actual size.

Although parallax considerably expands the viewing zone of the image, this attribute does have certain restrictions. If the viewer moves too far off center, the image will disappear. It is wise to consult with the holographer to determine the viewing parameters of a particular manufacturing process before starting design work.

Design Guidelines - Size

Although embossed holograms can be produced in a variety of sizes, cost considerations, manufacturing equipment, and marketing requirements generally favor making holograms of 6" x 6" or less. It can be quite a challenge to achieve the visual impact required by a client while working on such a small canvas. To save production time and "

expense, it is common to gang a number of smaller images on one "master" hologram. "4-ups", "9-ups" and "36-ups" are common arrangements . The entire set of images is replicated together, and then each separate image is die cut out during the finishing process.

Design Guidelines - Color

The designer should also always consult with the holographer beforehand to determine what colors are best utilized in a particular type of hologram. Embossed holograms create different colors by bending light to varying degrees. It is like creating a customized prism that will direct the light according to the designer's wishes. Therefore, the graphic artist needs only to assign colors to specific image areas in the artwork - the holographer then "colors it in" during the exposure process by adjusting his optical set up.

Another unique property of embossed holograms is that they do not display permanent colors - that is, as the viewer moves up and down in relation to the hologram, the colors will shift through the entire rainbow. In practice, it is common to take advantage of these unnatural color shifts to emulate movement and increase visual impact. Clients usually desire the brilliant, dynamic effects which result.

The designer only needs to specify the colors that are in- tended to play back when the finished hologram is viewed directly at eye level under proper illumination. Unlike standard four-color printing that utilizes combinations of cyan, magenta, yellow, and black (CMYK), the primary colors utilized in embossed holography are red, green, and blue (ROB). These three colors can be combined holographically to create yellow, white and a range of secondary colors (complex colors are a result of halftone screen mixing of ROB; different densities result in different colors). In holography, black results from unexposed areas of the film.

Some hologram mastering facilities are able to recreate "true color" using proprietary techniques based on pixilated renditions of the artwork. In some cases the studio will provide a palate of colors that designers can work from. Other studios claim that any color can be holographically duplicated. Again, we advise that designers consult with the production house to determine the most appropriate way to proceed. Ask to see samples that have been produced for other clients.

Experience dictates that lots of black and white colored image areas are visually unappealing, (matte white is commonly reserved for registration marks and trademark information). Drab colors should be avoided in favor of the bright colors inherent to this process - red, yellow, blue and green. Colors should be arranged to best contrast imagery and accent dimensional effects. Each area of the image should be assigned a color separate from its neighbor, and different from the areas that may layover or under it. Colors in the center of the color spectrum are usually recommended for major image components, as they are the brightest and will be the last to disappear as the viewer shifts position.

At the Holography Studio

Once the graphics work has been done, files should be organized, labeled and shipped to the holography studio for review. Most studios are capable of accepting a variety of file formats including diskette, DAT, SyQuest disks, and optical disks. At this point the designer's work is finished.

Once the holographer records the image the resulting master hologram can be proofed by the designer and/or client. If it is acceptable, it is sent to a production facility to be replicated. After replication, the holograms are finished in accordance with the client's wishes. Most embossed holograms are backed with an adhesive for hot stamping to paper stock or delivered on rolls for "peel and stick" applications. They work best when attached to a rigid material that has a flat and smooth surface, such as magazine covers, bank cards and cardboard packaging.

Method 3: A Series of Pictures

Instead of outputting the computer generated image used in Method 2 to film, the image can be illuminated on a LCD (Liquid Crystal Display) screen. The LCD image becomes the object you make a hologram of By using a series of pictures and special equipment, you can produce animated holographic images. The hardware which mechanically records a sequence of images onto the master hologram is called a Holographic Optical Printer (HOP). Such devices are gaining popularity among holographers due to the unique effects they are capable of producing.

In some of the first HOP devices, a frame of movie film was used for the object. (See figure 2.1.) Each frame of movie footage was sequentially projected through a slit placed in front of the hologram film. As the movie film was advanced one frame, the roll of holographic recording film was similarly advanced. In this way, an entire "mini-movie" was recorded onto the holographic film.

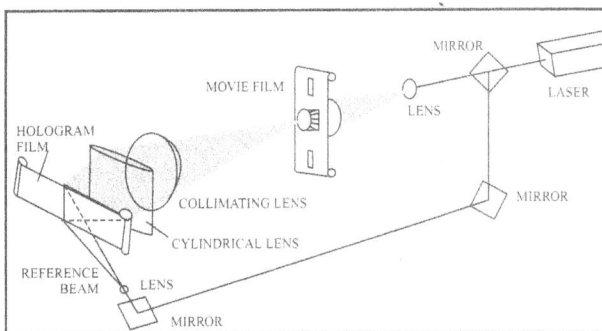

Figure 2.1. Diagram of an early HOP that used movie footage.

(Note - Today the movie projector is usually replaced by a LCD screen that is attached to a computer. This eliminates the cost of film and allows for easier artistic editing.)

The phrase Holographic Optical Printer obviously comes from the fact that we are taking a optical image, as opposed to a object such as a sculpture or 2-D artwork, and "printing" it on the film. If you are illuminating a sequence of images, and you filmed the sequence correctly, your fi-

nal result will be a stereoscopic cinema. This is called a holographic stereogram and it is how many "moving" holograms are produced.

The holographic stereogram does have differences from a traditional stereogram. As Steve Larson of Chromagem Inc. points out: "Similar in some respects to the old ViewmasterTM concept (a binocular-like 3-D slide viewer that displayed two near identical pictures taken from slightly different angles), a holographic stereogram differs in that the number of stereographic pairs can be in the hundreds, instead of the single pair that was provided with the Viewmaster. What that means to the holographer/artist, is that now, motion and time can be captured and displayed holographically."

Filming

There are different ways to shoot your film for holographic stereograms and professionals that do this usually have special stages or tools to do the filming. It is important to control the motion between frames because the viewer will be seeing two different images at a time and the brain will be rendering the two images into one to visualize depth. Obviously, you need to control the amount of motion between frames or the brain will not be able to render the two images into one 3-D image. One common way to have a large angle of view around your actors in the cinema and also control the motion between the frames is to have your actors on a stage which you can rotate in front of the camera (while the actors are acting). Another way is to have some variation of a railroad track in front of your actors, and move the camera along the track. In both cases, the intent is to create a horizontal "angle of view" in your final film, with a controlled amount of motion between frames.

Digital Modeling and Rendering

Most recently, affordable digital design, modeling, and layout programs have been used to create realistic looking images using no physical objects as subject matter. Hundreds of different perspectives of a "cyber-scene" can be rendered, combined and output to create fully-dimensional holograms.

Holographer Steve Smith states, "Early on, pricey work- stations such as the first series of Silicon Graphics machines were required to run expensive image modeling and rendering programs; these cost upwards of $20,000, generally out of the range of professional holographers. With the release of powerful yet lower priced graphics workstations, such as the Pentium™ and the PowerMac™, and with more capable yet affordable image modeling and rendering software, a new realm of imaging techniques are setting the stage become the way to create holographic stereograms."

In brief, the computer graphic artist first models a real or an imaginary object or scene "on screen." Lightwave 3- DTM, Byte by Byte's Sculpt 4DTMor Alias/Wavefront's Power AnimatorTM are good modeling programs to use. The computer is then used to render an appropriate sequence of graphics files which correspond to the various visual perspectives which the designer wants to appear in the finished hologram. Pentiums and PowerMacs can be used to render simple rotations. DeskStation Technologies'

Raptor™ rendering engine or SGI hardware have been used to render more complex imagery, i.e., hundreds of different perspectives of one image.

George Sivy, a model maker who specializes in holo- graphic design, elaborates on the aforementioned process. "I use an accelerated Amiga 3000 Tower computer running VideoToasterH1 to import and mix existing imagery. I use LightWave to model, light, surface and manipulate single objects or even entire scenes. In layout mode, I animate the object, and animate a virtual camera in accordance to the holographer's requirements. (I typically either rotate the object in relation to a stationary camera or pan the scene in a smooth horizontal path.) I use a Raptor rendering engine provided by DeskStation Technologies to generate 180 frames. This translates to six seconds of real time animation which may be incorporated into the final hologram. Next, the rendered frames are edited further using the VideoFlyer™ and copied onto S-VHS tape for proofing. Once approved, the files are downloaded onto an appropriate storage medium and shipped to the holographer."

Clients who want to generate their own imagery are advised to consult closely with the holographer before originating and delivering any artwork. Most holography studios are capable of accepting a variety of file formats which you can ship to them electronically or by diskettes, DAT, SyQuest disks, or optical disks. Since there are currently only a few computer graphics artists familiar with the procedures required to create animated stereograms, the holographer's in-house artists will probably need to review your files, clean them up and re-sequence them.

The Full Parallax HOP Grid Process

In the HOP we have discussed so far, the exposures are done through a vertical slit. There is another design being used by some holographers where, instead of having slits through which you expose your film, you have a grid like a chessboard and expose each square in- the grid. Chuck Hassen, of Holo Sciences, uses such a grid and he, along with Bob Hess, made the master hologram for the 5th Edition of the Holography MarketPlace using this HOP grid. Since each of the squares in the grid is a "separate" hologram, you can have movement of an object not just horizontally but vertically, too. In theory, you can have an object appear to tumble in space a full 360 degrees and be seen from any angle you wish.

The cover of the 5th Edition of the Holography MarketPlace was an animation of a set of nine televisions. Inside each TV screen was a video capture of an actual TV broadcast (Clinton and Dole in a debate, OJ Simpson driving, Michael Jackson performing, etc.). As you tilted the image top to bottom, the captured TV footage played in the TV sets, frame-by-frame. As you tilted the hologram to the top left, right, etc., you got a parallax view from that angle. Brian Kane of General Design did the video capture and rendering, which is still considered quite noteworthy. To give you some idea of how a full parallax grid hologram works, we've repeated some of the questions we asked Hassen and Hess in the 5th Edition.

Three Cells From the HOP Grid Used to Make The Hologram on the Cover of *Holography Marketplace 5th Edition*

The image viewed from straight ahead.

The image used to create the view from approximately 70 degrees up and 30 degrees right.

The view from approximately 70 degrees up and 30 degrees to the left.

An Interview With The Holographers

Bob Hess of Point Source Productions and Chuck Hassen of Holo Sciences collaborated on making the master for the hologram which appeared on the cover of HMP5. Their combined years of experience enabled Hess and Hassen to make the mastering devices and production tools which were used in this groundbreaking project.

Looking at the project from a commercial point of view, it is obvious that the holographers must communicate clearly with the design team (in this case, General Design) about what can and cannot be done. Here are some of the questions that the computer artists would need to ask, with answers from Hess and Hassen.

1) What is the practical limit (upper and lower) of rows and columns that we can use? What is the optimum number and why?

First, I should say that there is no single optimum configuration. The appropriate choice depends on the visual goals and the budget of the creator of the image. On the one hand, live action might be the only feature of interest, in which case we might recommend using as many as 30 or 40 frames in the vertical direction, without any horizontal parallax (one frame on the horizontal direction). This would be a format useful for converting existing film or video to 2-D animations. On the other hand, one might desire to present only horizontal parallax, but to do so with a large number of frames per degree of angle of view, to maximize the smoothness of the changes in horizontal view-point.

I should point out that in choosing our array parameters of 300 images, we put 20 frames of horizontal parallax on each horizontal strip, and 15 cells of animation on each vertical strip. This allowed us to produce a holographic viewing space with full parallax in the vertical and horizontal directions. (A typical horizontal-parallax-only integral hologram might use 180 frames to achieve the same angular spread in the horizontal direction. 15 times 180 is 2,700 frames, which would be the number recommended to produce our image according to conventional practices.)

We can certainly handle such a large number of frames . In fact, the hardware we built for this purpose divides the horizontal axis into more than 20,000 steps. One of our primary objectives in bringing this technology to the marketplace, however, is to cut the costs of production for our customers, while also improving the total image quality through the elimination of time smear, and addition of vertical parallax. Holo Sciences' Solid Light™ printing apparatus is capable of using any combination of aperture width which gives the step size on the horizontal axis) and height which provides the step size on the vertical axis) from I mm by 1 mm up to 20 mm by 20 mm, or even larger by request. The Solid Light Cybersculpture™ format we used inherently does not require as many frames of horizontal parallax as an animated HPO image to attain the same frame-to-frame smoothness, because there is virtually no difference in time between stereo pairs on any given horizontal row of cells. The high number of frames recommended in conventional stereograms is not actually necessary for a convincing illusion of three-dimensional form.

Along the vertical axis, the maximum number of rows depends on how much the subject is moving over the entire range of vertical angle of view. As one increases the number of rows per degree of view, one improves the smoothness of the vertical parallax displayed. On the downside, however, one may run into unacceptable levels of time smear if the subject is moving quickly and if the rows are so narrow that one perceives several rows simultaneously. In the general case of vertical animation, therefore, we recommend that the horizontal slits exceed the diameter of the pupil. In that way, a maximum of two rows will hit the visual field at any given time. In the case of a stationary subject, overlap between horizontal rows is less objectionable.

For the cover image, we employed the geometry shown in figure 2.2 (below). The integral hologram master is a planar grid of images, 20 cells wide by 15 vertical. We wanted to encode parallax on the horizontal and animation plus vertical parallax on the vertical axis. In the final hologram, the aperture plane is 14 inches from the image plane. Note that the view you present for "upper right" in this grid is the view from the upper left as you face the screen. Total angular displacement in the horizontal direction is 56.4 degrees (+ 28.2), and vertical is 46.4 degrees (+ 23.2).

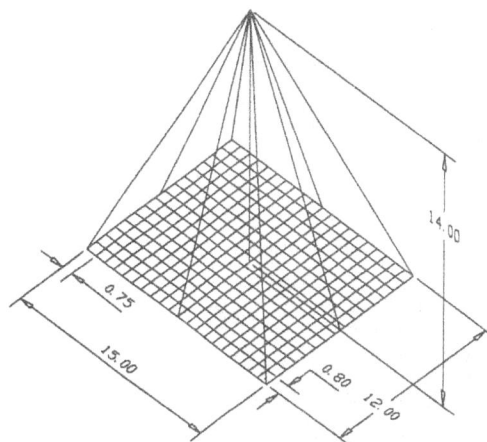

Figure 2.2. Diagram of a full parallax HOP grid geometry.

2) Since I will have to render each cell from a new perspective, how many degrees off-center is each new cell from the center square?

This is a good question. The answer is not a simple single number, but is instead a function of the geometry shown in figure 2.2 (above). In Brian Kane's rendering, we chose to ignore the small errors which result from approximating the planar displacements of 0.8" and 0.75" as constant angular displacements of 3.3 and 3.1 degrees, respectively. This is computationally the equivalent of placing the center of rotation at the middle of a sphere, and moving the camera a fixed number of degrees on the surface of the sphere for each horizontal parallax step [from geometry,

this is approximately tan^{-1}(0.75"/14.0") = 3.07 degrees]. Once a given row is complete, move the camera in the perpendicular direction by an angular displacement equivalent to one vertical step (tan^{-1}(O.8"114.0") = 3.27 degrees), and then repeat the horizontal scan. This is the equivalent of setting up our virtual object on a turntable, and mounting our camera at different elevations for each horizontal scan. In fact, the Solid Light™ integral process was designed to accept images created in exactly this way using a video camera. Point Source Productions even has constructed a 24" diameter stage which allows lights to rotate with the objects being filmed, so that shadows do not change with the angle of view.

Ideally, the concept is to match the changes in point of view so they correspond to the final locations of the hologram cells (called the "aperture plane") on the master plate. This is done according to the physical size of the master and its separation from the intended final image plane. For any user who would care to be more accurate, Holo Sciences will provide the angular coordinates for every cell in an arbitrary perspective grid to be rendered.

Many software packages allow one to specify the coordinates of a "virtual camera" with respect to the rendered scene. If this capability is available, one must remember to convert the scene scale to match the final hologram geometry. For example, if your image is a vehicle concept to be rendered at 1/12 scale, then one inch in real space equals one foot in the model space, and the camera would need to be located 14 feet away from the center of the image in model space for the perspective grid to end up accurate when it is recorded at a distance of 14" from the image plane of the final hologram in real space.

3) *What are the restrictions on lighting? Are dark shadows or bright highlights a problem? I know the dark image has more information for you to work with but it also tends to destroy the detail in shadow areas. A bright image is clear but you start to lose detail in the highlight areas. In light of this, is it better to have the image I deliver a little bright, or a little on the dark side?*

This sort of question has plagued people working with holographic images from the very beginning. The Solid Light™ format does release the user from many of the constraints of conventional (non-integral) holographic formats, because one may use standard lighting equipment and techniques on a real object, and the whole range of computational light modeling approaches on a virtual object in cyberspace. The degree of realism that one obtains in generating these images, and then merging them to form a hologram, depends on a number of factors:

Contrast: It is best to use a fairly narrow range of contrast within a given image. One may certainly employ clearly defined regions of high and low brightness, but the absolute difference in brightness between the highlights and the shadows should be kept fairly small.

Innovation Starts Here

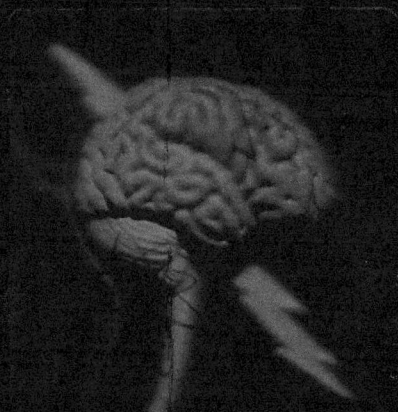

Finally a holographic technology with the power to match the creativity of the mind. KHI's photopolymer holograms are the brightest and most advanced holograms available. When used for promotion, security, or in product development, these high resolution holograms are superior in performance. Combining KHI's state-of-the-art imaging and replication technology with DuPont's Omnidex® film, KHI provides holograms ideal for today's creative challenge.

For more information, please contact Krystal Holographics International Inc.

**Krystal Holographics
Corporate**
555 West 57th Street
New York, N.Y. 10019
Tel: 212-261-0400
Fax: 212-974-2237

**Krystal Holographics
Manufacturing**
365 North 600 West
Logan, UT 84321
Tel: 435-753-5775
Fax: 435-753-5876

DYNAMIC
EXCITEMENT
FOR YOUR
BRILLIANT
IDEAS...
THAT YOU
CAN AFFORD

PROMOVISION™
HOLOGRAMS

CROWN
ROLL LEAF, INC.® &
HOLO·GRAFX™

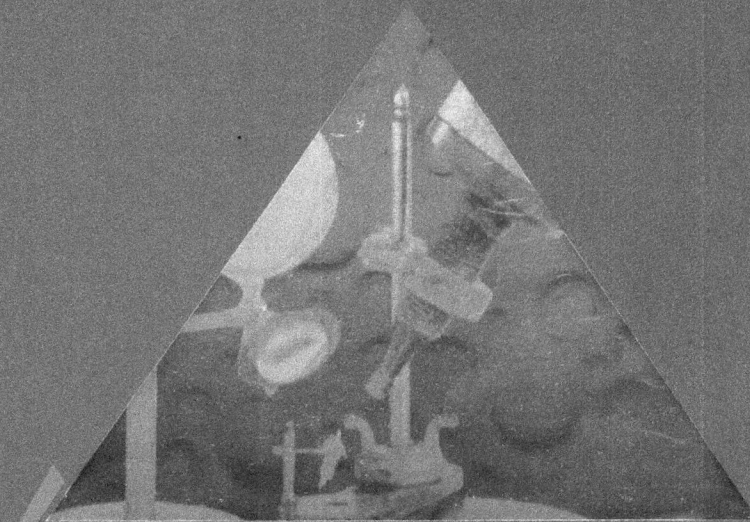

TURN OVER ▷

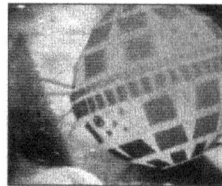

Holographic Masterpieces

As the nation's leading manufacturer of embossed holography, Transfer Print Foils, Inc. offers the widest range of holographic technique. Perhaps it is the reason that everyone, from people beginning their first holography project to industry experts, relies on TPF to see their project through to completion.

Three dimensional images, 2D/3D images and TransFraction™ (prismatic) patterns are all available from TPF. We also proudly offer "pixel" images and patterns as well as an in-house Design Center to help turn your concepts into reality.

Rock Solid Security

High-tech security products such as our exclusive HoloClear™ provide the highest level of Rock Solid Security available today! Counterfeiters will run when they see a document or ID card with the HoloClear™ overlay. Call us for samples of this fraud-stopper immediately!

With a security production facility, advanced research and development and "state-of-the-art" technology, we provide unique images that are used for security on drivers licenses, international ID cards, trading cards... the list goes on and on!

Transfer Print Foils, Inc.
9 Cotters Lane, P.O. Box 538, East Brunswick, NJ 08816, USA
(908) 238-1800, (800) 235-FOIL, Fax: (908) 238-7936

A HoloPak Technologies, Inc. Company

Sparkle™ I

Dot Matrix Mastering System

Applications:

- Packaging - multiple colors create colorful 3D packaging paper
- Promotional - 3D dynamic images create stunning effects
- Anti-counterfeiting - higher resolution provides clearer images
- Security - adjustable grating pitch increases security elements

Features:

- High speed and high resolution (15 dots/sec and 1300 dpi)
- Fully computer controlled
- Multiple colors can be user specified
- Powerful, friendly, easy-to-use GUI controlling program
- Handles any *.bmp file for easy downloading
- Realistic preview function
- Calculates optimal time to create master

Brief Specifications:

Light source	● He-Cd laser, wavelength 442 nm
System dimensions	● 900mm(L) × 900mm(W) × 1500mm(H)
Grating pitch	● software adjustable, six colors
Dot resolution	● 150 dpi to 1300 dpi
Exposure rate	● 15 dots per second
Exposure size	● standard 8" × 10" ~ maximum 300mm × 300mm
Yellow light space	● approx. 4m^2
System contents	● Pentium 166 or equivalent computer ● MS Windows™ 95/ MS Windows™ NT operating systems ● Kinetic vibration isolating optical table ● Optical bridge, XY stage and 5-axes controller ● Sparkle™ 97 GUI software (system control, pattern generation & preview) ● Adobe Photoshop 4.0, CorelDraw 7.0 & KPT 3.02 ● Electric power interrupt sensor and UPS

AHEAD OPTOELECTRONICS, INC.

Tel: (+886) 2- 2369-1520   Fax: (+886) 2- 2362-0485
B1, No. 130, Section 3, Keelung Road, Taipei, Taiwan
http://www.ahead.com.tw

AHEAD Optoelectronics, Inc.

AHEAD is a high-tech company specializing in holographic products and systems, optoelectronic and optomechanical systems as well as piezoelectric systems.

AHEAD's holographic specialities include diffractive optical elements, dot matrix mastering systems, embossed holograms and furrow overlay embossing. In addition, AHEAD also designs and produces high performance optical instruments including integrated sphere ellipsometers, optical wavefront metrology equipment, laser encoders and differential laser Doppler interferometers, etc.

AHEAD, is part of the Wah Lee Industrial Corp. Group of Companies which has over 800 employees worldwide with offices in 13 different countries.

AHEAD has ties to National Taiwan University, Industrial Technology Research Institute (ITRI), CSIST, as well as various other industrial worldwide high-tech firms in the semiconductor industry, DVD industry and hard disk industry.

AHEAD offers two dot matrix mastering systems (Twinkle at 150-1300 dpi and adjustable grating orientation and Sparkle at 150-1300 dpi with both grating pitch and orientation adjustable)

Twinkle

Dot Matrix Mastering System
(Basic System of Sparkle I)

Applications:

- Packaging: rainbow effect creates unique colorful packaging
- Promotional: adjustable grating orientation creates 2D/3D kaleidoscopic images
- Security: high resolution can be used as added security feature

Features:

- 150-1300 dpi adjustable grating orientation
- Fully computer controlled, easy alignment
- Rainbow effect and dynamic effect
- Powerful, friendly, easy-to-use GUI controlling program
- Handles any *.bmp file for easy down loading
- Calculates optimal time to create master

Brief Specifications: (other specifications same as Sparkle I)

Dot resolution	● 150-1300 dpi
Exposure rate	● over 15 dots per second
Orientation	● yes; adjustable
Grating pitch	● no; rainbow effect only

AHEAD OPTOELECTRONICS, INC.

Tel: (+886) 2-2369-1520 Fax: (+886) 2-2362-0485
B1, No. 130, Section 3, Keelung Road, Taipei, Taiwan
http://www.ahead.com.tw

**YOUR COMPLETE HOLOGRAPHIC
CONVERTING SOURCE
WITH OVER TWENTY YEARS EXPERIENCE**

~MASTERING~   ~EMBOSSING~

~SPECIALTY LAMINATING~   ~REGISTER DIE CUTTING~

~CUSTOM PACKAGING~   ~PREMIUM ITEMS~

~HOT STAMPING~   ~PRODUCT ASSEMBLY~

**WITCHCRAFT TAPE PRODUCTS, INC.
BOX   937   COLOMA,   MI.   49038
1-800-521-0731   FAX: 616-468-3391**

Bright object distance from image plane: A bright object tends to cause a region of over-modulation, or burn in, relative to the rest of the image. An example of such an image component in the cover image is the violin wiping its nose in the top row. Note that when the image of the TV screen pulls away from the image plane as you swing the hologram from side to side, a ghost image remains behind in the region where the light may be seen from most angles of view. The further behind or in front of the image plane one places a bright object, the more pronounced will be the separation of the bright object from its average position, when seen from extreme angles of view. That separation magnifies the perceived level of burn-in. There are two ways around this problem. You can place the bright object far behind the image plane, so that its average exposure is smeared out over a large area, or you can place it very near to the image plane, so that the parallax-induced separation of the image from the region of over-modulation is minimized.

Desired visual effect: Remember too, that you are not necessarily attempting to match reality photon for photon. Solid Light™ is a medium that demands experimentation and pushing the limits . One image's flaw may well become another's special effect. This is an area where consultation with someone who has experience producing holograms, like Bob Hess or Brian Kane, can really add value at the concept stage of a project. The cover image shows many of the effects that are possible with different types of 2-D subject motion.

Attention to shadows: Remember the direction from which your hologram is designed to be lit when played back. The TV hologram is lit from the top, so the shadows were made to appear as though cast by an overhead light source. Small deviations from the reference beam angle are acceptable, but large mismatches will confuse the viewer's eye.

Overall brightness: Although we are reluctant to provide any hard and fast rules, chances are that your hologram will be more visible if it is on the bright side, than on the dark side. If you feel comfortable with the issues listed above, try to remain close to your artistic vision and still stay within the guidelines. If you have any doubts, call a professional.

4) *Assuming I deliver an image such as the one on this cover to you according to your specs, how much do you charge for mastering? Will this price drop over time or with volume?*

The cost of creating a master from a set of full parallax animated images similar to the ones used for this cover would be approximately $7,500 for a single hologram contract. Simpler formats with fewer cell would cost less; more cells would cost a little more to produce. If a customer wished to produce a series of 4 or more images with similar grid geometries, all at once, that price could be cut by 30% today. This would also make sense from the point of view of creating a production tool "grid master" with more than one image at a time for replication on photopolymer. Setup costs are high in both steps, and one stands to gain very much from simultaneous production of series of images. We are still on the expensive side of the process

learning curve, and hope to be able eventually to offer masters from user-provided art for only a few hundred dollars each. We hope to grow in the long run by making this format accessible to any business, small or large , that has ever toyed with the notion of making a commercial hologram. (Although space here does not allow us to list them all, an interested customer may obtain more information about other Solid Light™ formats by contacting us.)

5) *Is there any special file format that my animator needs produce the files in?*

At this time our preferred file format is PC-based, PCX format. I am aware that many artists use other formats, and we do have the capability to perform batch conversions of most popular output formats for the PC and for the Macintosh™. We use a shareware program called Graphic Workshop™, which comes in a Windows™ compatible version, and in a DOS-only version. Another excellent graphics conversion package is called Hijack™, though I have not used it. We will be happy to do the conversions for a flat hourly rate. We can also sample directly from videotape, but please contact us before you begin to produce the frames on video. It will save everyone time and money.

Method 4: Dot Matrix Machines

This method does not use a physical object as subject matter and simply records a pattern of diffraction gratings directly into a photosensitive recording material. The collection of gratings produce an image the same way the dots in a photograph or pixels on a television screen do. The equipment that translates your artwork into these gratings is called a "Dot Matrix" hologram machine. They are designed to make the mastering process for embossed holograms easier.

The great benefit of the dot matrix method is that it is very cheap (no film at all) and totally digital. The drawback is that you will only be working in embossed holograms and the depth of field is very limited. To produce holograms, 2-D artwork is furnished to facilities that have one of these machines. A surface-relief master hologram results. Progress is being made on producing 3-D stereogram images . Because of the gaining popularity of the dot matrix machines, the following article describes the process in more detail.

Dot Matrix Hologram Machines:
Basic Theory and Some Existing Systems

A dot matrix hologram is, as the name implies, a hologram created from an array of smaller units (dots). Each dot is actually a separate hologram that produces a predetermined result, usually a simple diffraction grating. When illuminated with a white light source and viewed at the correct angle, the diffraction grating produces a tiny spot of color. The microscopic gratings can be arranged to produce a single composite picture, a sequence of related images, and/or a variety of "special effects".

Dot matrix technology allows for computer control of each individual pixel in a holographic image. Systems that generate dot matrix holograms currently have resolutions that range up to 2,000 dots per square inch (each dot measuring 16 microns), though resolutions of 25-600 d.p.i. are more common. This gives the holographer design options than are not available with traditional holographic originations. In addition, the process creates holograms that have some very useful properties, especially for commercial applications.

Major Attributes

Brightness - Control of each pixel in a holographic image allows the holographer to optimize the brightness of each and every image component in relation to a viewer. This results in images that are easily recognized under the less-than-ideal lighting conditions which are present in most commercial environments.

Viewing angle - Dot matrix holograms, unlike most other embossed holograms, have a very wide potential viewing angle. This is due to the fact that the diffraction gratings that make up the hologram can be positioned to redirect light in a variety of directions.

Kinetic effects - Dot matrix holograms can easily incorporate dynamic effects that make the image appear to move or change as the viewer shifts position or the hologram is tilted. For example, designs can be made to "pinwheel" (radiate in and out when the hologram is rotated) or "flash" on and off.

Subject matter - Any two-dimensional artwork that can be digitized can be used to make dot matrix holograms. Corporate logos are often reproduced. More often, repeating geometric patterns are generated. Many commercially available graphic design software programs (such as Adobe's Photoshop) can be used to generate the required digital files.

One-step Mastering - Aside from enhanced visual effects, dot matrix technology allows a wider variety of companies to originate holograms. Although minimal training

on the currently available systems is required and an extensive knowledge of embossed holography is certainly beneficial, these machines are intended to eliminate the need for a highly experienced holographer on staff. As of today, many dot matrix systems have been purchased by existing hologram manufacturing facilities, but these machines can be operated by less-specialized production departments. Operators should have experience with computer controlled mechanical devices and a knowledge of lasers (along with laser safety procedures).

Applications

Currently, most dot matrix hologram production systems are being used to create masters for holographic security labels or holographic packaging materials. The complex dot matrix designs that can be produced aid in document authentication. To further enhance anti-counterfeiting efforts, the proprietary computer software which translates the original artwork into a matrix of dots can encrypt security information in a way that is invisible to the casual observer. In addition, with some systems the hardware which creates the gratings can be set up to produce holograms with other signature effects (fingerprints) which can be traced back to the machine which made them.

Dot matrix holograms incorporated into embossed packaging materials help overcome many limitations of holograms created by other means. Dot matrix holograms are extremely eye-catching. They are bright and colorful even under fluorescent lighting. The animated designs attract attention. The repeating geometric patterns typically produced can be seen if the package is right-side up or upside-down. Many hologram manufacturers combine dot matrix holography with conventional optical mastering techniques in order to utilize the advantages of both systems - resulting in fully 3-D images with striking 2-D effects.

Producing Dot Matrix Holograms

Producing dot matrix holograms is a four step process: artwork origination, digital-to-dot transfer, mastering and replication. In brief, a digitized image is downloaded into a dot matrix transfer system which converts digital parameters into a corresponding set of mechanical positioning instructions. These instructions guide a laser-based machine that uses holographic techniques to produce thousands of separate diffraction gratings. These diffraction gratings are recorded onto a photoresist plate, creating a master hologram. The master hologram is plated, and embossing shims are produced. The shims are used to emboss the hologram into a chosen substrate by commonly used methods.

It is important to note that there are several other digital-to-optical transfer systems that create holographic images from an assemblage of smaller image components. However, they are not dot matrix systems. Many employ LCD screens and are based on holographic stereogram)1 theory. Although they have some properties that are similar to dot matrix systems, they produce very different results. (See Holography MarketPlace 5th Edition for a discussion of these "LCD/HOP" systems.)

Basic Theory

Dot matrix holography was developed in the mid-1980's by Frank S. Davis at Advanced Digital Holographics, Inc. In 1992, the patent was assigned to Dimensional Arts, Inc., a U.S. company that builds and sells a line of production machinery based on the Davis patent.

(Editor's Note - Other companies, including Cfc/Applied Holographies and Toppan Printing have patented methods that produce dot matrix holograms, but choose to keep their technology "in house". Ahead Optoelectronics claims to hold patents for dot matrix technology which is incorpo- rated in the machines that they sell. Spatial Imaging and Westmead Technology also sell dot matrix equipment. We encourage potential users of dot matrix technology to re-search this issue further, as patent protection may vary from country to country.)

A fundamental principle of the Davis method involves the use of a laser beam which is first split in two, and then recombined to create tiny interference patterns (holographic dots) on a recording material. This is the basic dot matrix patent. Changing the angle and orientation of the intersecting laser beams changes the interference pattern, which is the primary optical property this technology exploits.

Each interference pattern, once recorded on photoresist, becomes a surface relief diffraction grating. The diffraction grating operates like any split beam hologram - an illuminating beam coming in from the appropriate angle is redirected along a specific corresponding path (usually perpendicular to the surface of the hologram). It's like setting up an array of microscopic mirrors so that incoming light can be aimed back to the viewer in a precise way.

Creating Gratings

There are two important parameters for each grating:

1) The first parameter is the grating pitch, which is determined by the intersection angle of the two incoming coherent light beams. The larger the intersection angle, the smaller the grating pitch. More specifically, the grating pitch equals: the wavelength of the incoming laser beam / sine of the intersection angle.

2) The second parameter is the grating orientation, which is determined by the direction of the two incoming laser beams. (See figure 2 .3.)

Playback Control

Pitch and orientation of each grating pixel are the most fundamental parameters to consider when designing a diffraction grating. These two variables determine how the incident light beam can be diffracted precisely to where the observer is. The orientation of each grating pixel is a function of three variables:

1) the light source position,

2) the observing location, and

3) the position of each grating pixel.

The grating pitch of each grating pixel is influenced by another variable,

4) the color wavelength produced by each grating pixel in relation to other pixels in the same image.

Thus, a graphic designer needs first to decide the location of a fixed light source, the position of the observer, the color distribution and the desired viewing effect of every graphic pixel of the original graphic design. All of the data mentioned above must be entered into the design program. The software program automatically calculates the pitch and orientation of each dot.

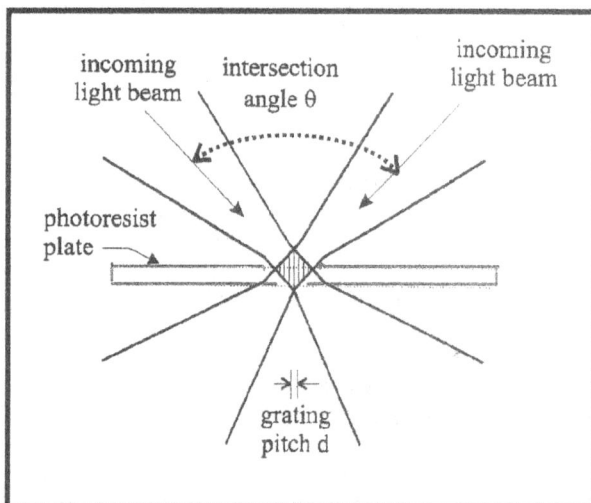

Figure 2.3. Creating a grating with two beams.

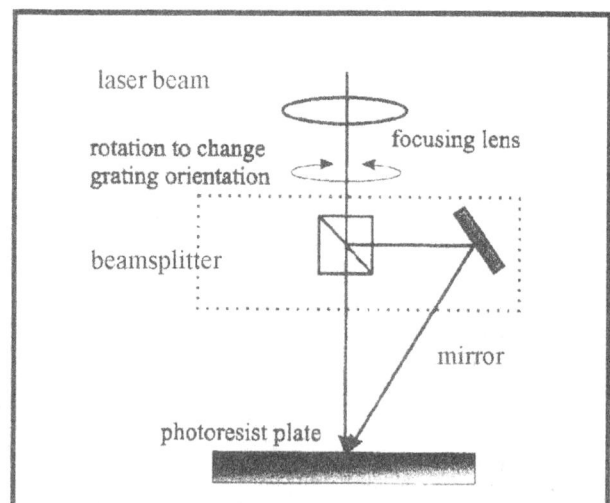

Figure 2.4. A traditional dot matrix writer.

Viewing Angle

The viewing angle associated with each microscopic grating in a dot matrix hologram is limited by the optical geometries used to create it. Angles of view are very small. However, when the observer is located within the designated viewing angle for that specific pixel, all the redirected light is concentrated within that small angle. This creates very bright dots.

Kinetic Effects

If many gratings are aimed at one viewing zone, very bright images result. If the gratings direct light away from the viewer, "dark" areas result. The ability to control a sequence of thousands of tiny "flashing lights" allows the designer to create a variety of special visual effects such as disappearing images, zooming-in, zooming-out or changing shape.

Color

Different colors can be produced by changing grating angles. Each dot matrix system has its own palate of colors available to the hologram designer. Like most embossed transmission holograms, colors are chosen in relation to a specific viewing zone. As the viewer moves out of this primary viewing zone (or the hologram is tilted) the intended color will shift to the next color in the spectrum. Prismatic effects result.

Some Existing Dot Matrix Systems

Some of the technical developments investigated during the last few years are reviewed below.

For the traditional two-beam interference dot matrix writer shown in Figure 2.4, the laser beam goes through the focusing lens, and is then split into two converged beams by the beamsplitter. The laser beam reflected by the mirror intersects with the other beam to form a small grating pixel on the photoresist plate. The beamsplitter and the" mirror are fixed on a rotational stage. The orientation of gratings and the size of each grating pixel recorded on the photoresist plate can be altered by rotating the stage and by moving the lens respectively.

In Figure 2.5, the beamsplitter, mirrors, and lens are all fixed inside a rotational stage. The laser beam source is first split into two beams by the beamsplitter. The two laser beams are reflected by mirrors and then converged by the lens. These two laser beams finally intersect and form a grating spot on the surface of the photoresist plate.

Figure 2.6 shows an optical configuration where the beamsplitting grating and lens are fixed inside a rotational stage. When the laser beam source goes through the beamsplitting grating, only a +1 order and -1 order of the diffraction beams will be selected to pass through the focusing lens. Finally, these two beams will converge and form an interference spot on the photoresist plate.

To achieve the goal of altering the grating pitch within each grating pixel, the intersecting angle between the two interfering laser beams must be changed. The main difficulty in designing and constructing a system that can change the grating pitch on the fly lies in the fact that a focusing spot will move in and out of the photoresist plate whenever the intersection angle of the two recording laser light beams are changed.

The Dimensional Arts System

Dimensional Arts, now part of the HoloCom group of companies, was the first company to successfully market dot matrix technology. The company sells the "Light Machine™", a basic dot matrix production system for tiled patterns and imaging, as well as an enhanced version specifically built for security applications. According to company president Ken Harris, the company sells the only patent protected dot matrix system commercially available in North America. In a notable development, Harris reports that his company recently has been awarded another, more extensive patent covering "holographic dot" technology. "Further patents pending will cover three dimensional holograms made using stereo pairs of holographic dots and additional color-control methods," he adds.

Harris notes that dot matrix technology has proved reliable and popular since his company placed its first machine over four years ago. He points to a list of major em-

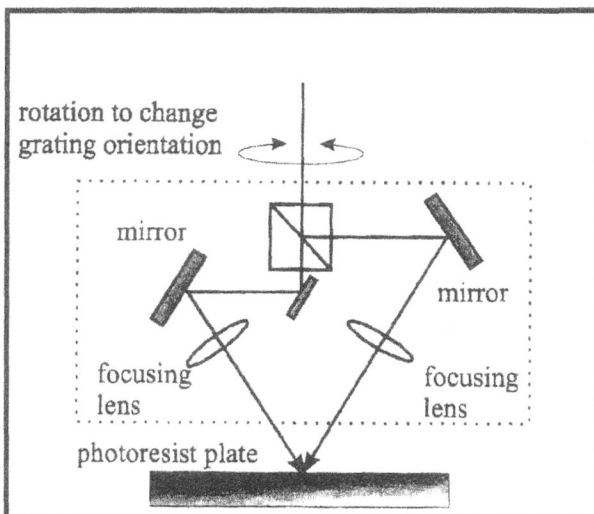

Figure 2.5. Another configuration of a dot matrix writer.

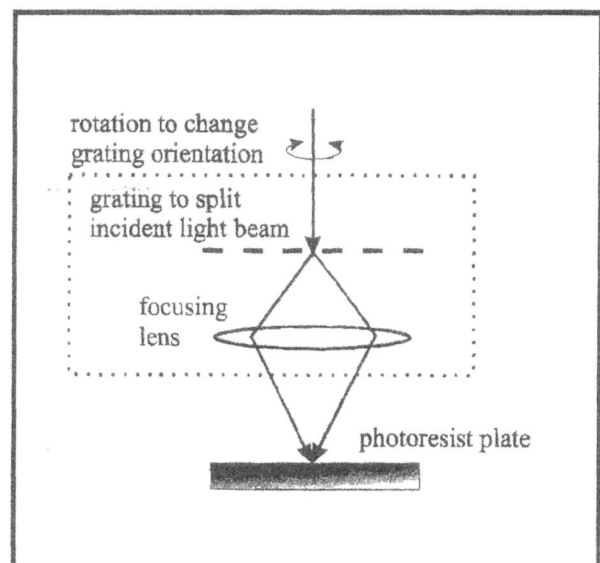

Figure 2.6. A grating is used instead of a beamsplitter.

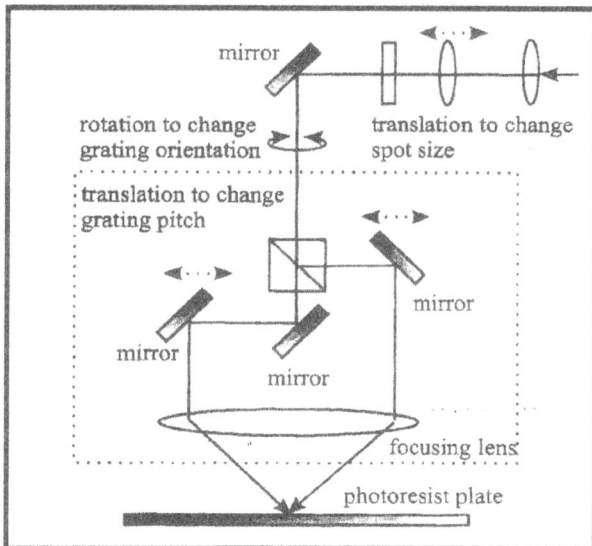

Figure 2.7. A dot matrix writer designed by AHEAD.

bossed hologram manufacturers that use his machines for everything from small security holograms to large poster sized decorative holograms (including American Bank Note Holographics, Transfer Print Foils, Hologramas de Mexico, and Crown Roll Leaf, among others).

The AHEAD System

According to company literature from AHEAD Opto-electronics, a Taiwanese company that builds and sells dot matrix machines, the lack of grating pitch variation capability in other systems prevents the graphic designer from specifying colors in order to achieve more vivid effects. In addition, the absence of color specifying capabilities actually prevents an accurate preview function in such systems. "Many hologram manufacturers and designers are thus forced to create a look-up table to compensate for the incorrect results," states a company report.

Company spokesperson Julie Lee, elaborates: "One of the optical configurations that not only keeps the advantages of the traditional design but also overcomes drawbacks associated with other designs, is our Sparkle™ machine. (See figure 2.7.) In AHEAD's design, the incident

laser beam is split by the beamsplitter to become two incident laser beams. The two coherent light beams are focused onto the photoresist plate by a specially designed focusing lens. The grating pitch of the grating pixel formed on the photoresist plate can be varied on the fly by translating a set of stages to change the distance of the two coherent light beams before impinging on the focusing lens.

As is shown in figure 2.8, the larger the distance between the two coherent light beams, the smaller the grating pitch. The focusing lens shown in figure 2.7 and 2.8 has been especially designed and fabricated in order to prevent the spherical aberration of the focusing lens to have the focusing dot location changed whenever the distance of the two incoming coherent laser beams is changed.

"The grating pixels generated by this new class of dot matrix writers are able to have spot sizes, grating orientation, and grating pitch completely specified by the designer. The images constructed by this type of dot matrix writer can exert many pre-specified colors at a specific viewing angle. In addition, all of the special effects that can be created from other traditional two-beam interference dot matrix writers when an observer continuously varies the viewing angle, can certainly be generated."

Conclusion

Dot matrix technology is continuously being advanced and improved. Limits are being pushed towards higher resolution, different shapes other than dots, higher speed, more security features, better color selection, smaller overall machine size, and lower cost. Ease of use and cost effective production methods should result in more dot matrix holograms being used in a variety of applications.

For further information about the original dot matrix machines and current dot matrix technologies contact: Dimensional Arts Inc. ph. 505- 527-9183, email arts@holo.com.

For information about other dot matrix systems contact: AHEAD Optoelectronics,Incph. (886) 22369-1520, email ahead@ahead.com.tw.

Other companies that produce dot matrix holograms and sell dot matrix machines can be found in the International Business Directory.

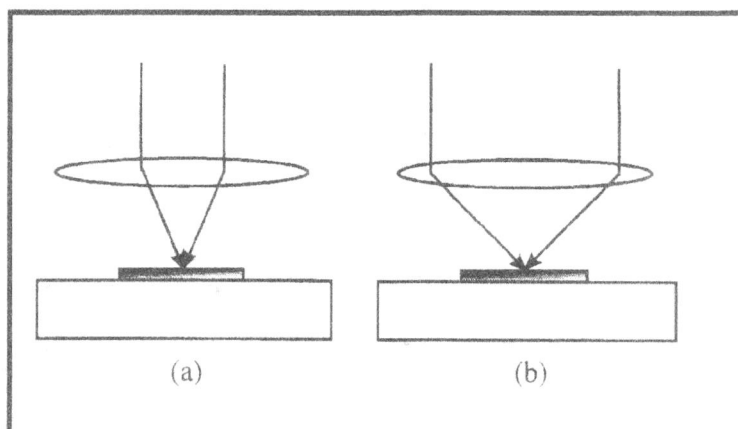

Figure 2.8. Changing the pitch of the grating.

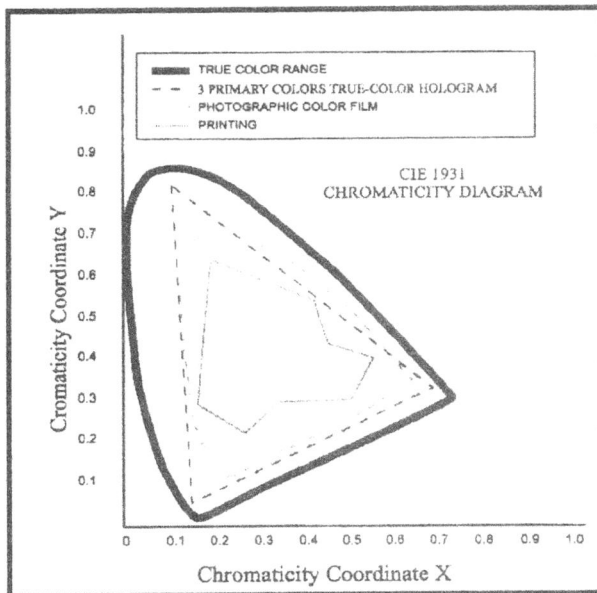

Figure 2.9. Color range comparison.

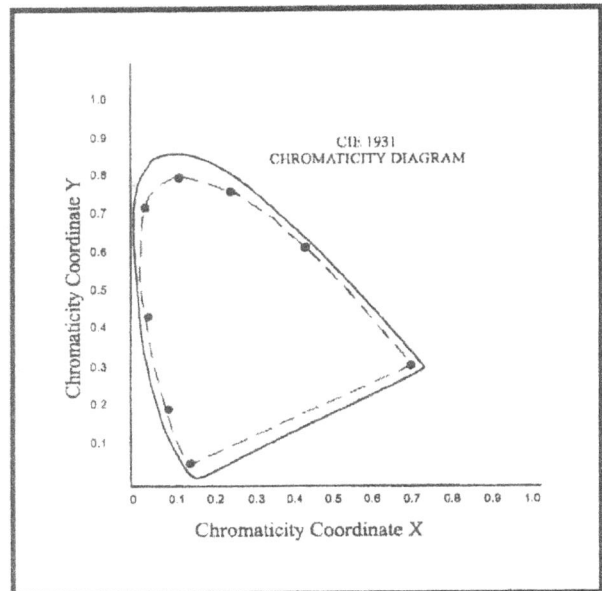

Figure 2.10. Color range of dot matrix holograms.

Dot Matrix Hologram Related Research and Technologies

by Julie T Lee, John Hsu (AHEAD Optoelectronics) and c.K. Lee (National Taiwan University)

Color Reconstruction

Color reproduction technology is one of the most important techniques used in printing, computer monitors, television monitors, color printers, color films, etc. The major color reproduction method used in printing is the CMYK (Cyan, Magenta, Yellow, Black) system, while the RGB (Red, Green, Blue) system is used for computer monitors and color film technology. More specifically, both color methods separate images by using three predetermined primary colors (i.e: CMYK or RGB) and then color-subtracts or color-adds to reproduce the desired color of the original images. Unfortunately, these two methods are not able to reconstruct all the colors seen by the human eye, especially metallic colors.

The CIE (Commission Internationale de L'Eclairge) chromaticity diagram, which was developed in 1931 to describe how object colors are perceived under different lighting conditions seen by a typical observer, breaks the barrier mentioned above. Taking the 1931 CIE chromaticity diagram (figure 2.9) as an example, the fundamental viewpoint behind the design of this CIE chromaticity standard lies in the fact that a different optical spectrum can be perceived as the same color under a specific lighting condition. This also means that the same spectrum guarantees the same color, but a different spectrum under different lighting conditions may still be viewed as the same color.

In the CIE chromaticity diagram, the two chromaticity coordinates x and y coupled with hue and chroma can be used to describe the locations of the various colors. The inner region of the half-elliptical shaped curve in figure 2.9 represents all the visible colors that can be perceived

by the human eye. The boundary of the half ellipse is a combination of fictitious colors that represent "pure" colors which don't occur naturally but which are equivalent to the primary colors RGB or CYMK of the color separation systems. More specifically, the primary colors of RGB or CMYK are in fact extreme points located within the true color range.

Another major concept unveiled from this chromaticity diagram is that we can assign certain extreme points to represent all the colors that can be generated by a combination of the extreme colors. Using this concept, we can identify the maximum color range obtainable from conventional printing technology by examining the location of the primary colors (CMYK) of the traditional four-color printing method. Similarly, the color range accessible by a television set can also be examined by measuring the chromaticity coordinates (i.e., chroma and hue) of the three primary colors (RGB) of television monitor pixels.

For example, if three primary colors are used to create a hologram, the process can display only colors located within the region encircled by these three extreme points. The color region that can be displayed by a photographic color film and by a conventional printing method are also shown in figure 2.9 for the reader to compare the color region obtainable for each color reconstruction process. The wider the color region, the more powerful the color reconstruction process.

Compared to the color separation models such as RGB, CMYK, and HSB (Hue, Saturation, Brightness), the CIE color model has the advantage of quantification and linear superposition. More specifically, the CIE color model can completely cover all visible colors and can precisely describe color coordinates as hue and chroma. In addition, according to this CIE color model, we can use a linear superposition to convert any color into different ratios of the primary colors.

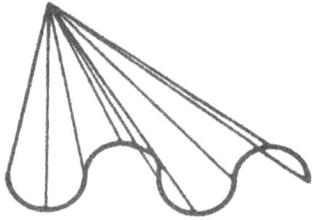

WAVEFRONT TECHNOLOGY

RECOMBINING
- Images
- Patterns

SPECIALTY EMBOSSING
- Diffusers
- Diffractive Optical Elements
- Holographic Optical Elements
- Computer Generated Holograms

LARGE FORMAT EMBOSSED HOLOGRAMS
- Up to 28"x 50"
- 2D/3D

15149 Garfield Avenue, Paramount, Ca. 90723
Phone (562) 634-0434 Fax (562) 634-0434   *wavfrnt@IDT.NET*

A traditional true-color hologram reconstructs an object color by adding the three primary colors RGB. In other words, a color addition process is used to recreate colors in the hologram. This process is completely different from the color subtraction process adopted in a printing process. According to the experimental data performed at AHEAD Optoelectronics, Inc., the color reproduction range that can be displayed by combining several grating pixels with different pitch will cover almost the entire domain of the CIE color model (See figure 2.10.) As the CIE color model displays all the colors which can be seen by the human eye, this finding indicates that dot matrix holograms using more than three primary colors as shown in figure 2.10 can have the widest color display region of any color reconstruction scheme.

E-beam Systems
In addition to the two interference beams methodology shown in figure 2.3, the other common approach in dot matrix systems is to apply e-beam lithography to create each pixel. As e-beam lithography is commonly used in making photoresist masks for today's semiconductor processing, the equipment is widely available within the semiconductor processing industry. Dot matrix holograms generated bye-beam lithography are commonly referred to as e-beam holograms. Since e-beam lithography can specify line width, line orientation, and line location of every pixel, a dot matrix hologram created by using e-beam lithography can generally achieve an even higher resolution and brightness than other dot matrix systems.

However, the data files needed to generate an e-beam hologram are typically in the range of gigabytes, which is 109 data bytes. In addition, using an e-beam machine is extremely expensive. A regular e-beam machine can cost more than $10 million. Generally only hologram manufacturing companies that belong to a very large consortium or organization which either owns or has easy access to an e-beam machine can afford to produce e-beam holograms.

AHEAD Optoelectronics, Inc. has developed an e-beam holography program called Radiant™ which serves as an add-on option to their high-security dot matrix machine Sparkle™. The program allows users to take any MS Windows™ graphic file from packages such as Corel Draw™ or PhotoShop™ and converts them into the so-called GDS-II data file format that an e-beam machine can read. With Radiant™, designers can design their e-beam hologram by using a graphic program that is familiar to them and then convert the design into the data files required to create the photolithography masks using e-beams.

In such a process, a standard semiconductor processing equipment called mask aligner (which has an UV light source and a set of optics to transfer the photolithography mask into a pattern on a photoresist plate) is used to create the e-beam hologram master. Once an e-beam hologram master is created on the photoresist plate, the traditional hologram manufacturing process such as electro-forming, mechanical recombining, embossing, etc., can be used to mass-produce the holograms.

Recombining Images for Embossed Holography

Recombining refers to an important step in the embossed hologram production process that often takes place between HI origination and replication. In brief, it is a way of combining pre-existing holographic images into a new "master hologram" in the lab, without incurring the expense and time it would take to achieve identical results in the holography studio.

Recombining techniques are closely guarded by the companies that develop them, but they all eliminate multiple exposure optical compositing and the many problems associated with it. Instead, a special type of embossing method is used to duplicate existing surface relief patterns which are then recombined into a working production tool. The tool is made into an embossing shim. The basic technique dates back 50 years to early experiments with diffraction gratings for military-related purposes.

Recombining is not always necessary, but it is frequently required by companies embossing high-volumes of small holographic images and/or manufacturing larger-sized embossed holographic materials. The process saves the factory (and, hopefully, its client) time and money.

Applications

There are numerous instances where it is very useful to create a composite hologram for embossed replication. Here are several:

One image repeated - As the embossed hologram industry has grown, bigger and faster embossing machines are being employed by hologram manufacturing plants. Six inch wide "narrow-web" embossing used to be the norm. Now, many facilities are capable of embossing onto much wider webs of materials. However, most customers require finished images that measure several inches square, or less. To best utilize their equipment and personnel, many factories employ the services of a recombiner.

For instance, if a client has an image that measures one inch square, and the factory has a machine that can emboss six square inches at one time, it is much more efficient to position (or "gang") thirty-six copies of the image into the allotted space (arranged six across and six down) and emboss them all at once, rather than singly. Since it is very difficult to expose thirty-six identical images on one master hologram and retain quality, the factory sends the client's single image to the recombiner, who then creates the appropriate production tool without shooting another hologram. After embossing, the individual holograms can be cut out of the larger sheet and separated. This drastically reduces production time and the associated expense.

Joel Petersen and Robert Carey of Wavefront Technology, a U.S. company that specializes in holographic recombining, elaborate: "Embossing companies have different types of machines that vary in size, so the finished size

of the recombine will vary. Recombine sizes from 3" square to 32" x 54" are common. We have worked with single images as small as .125" square, up to images 40" square."

In addition, the company has developed proprietary techniques that allow it to create apparently " seamless" recombines. This continuous-composite technique is especially useful for holographic packaging applications (such as decorative films and foils) that employ repeating prismatic patterns, graphics or logos. In the past, an edge or outline often appeared between each image component, which was rather unattractive.

Combining different images - Sometimes a factory might need to gang several different images on the same master. These images might be different sizes and/or different types (i.e., 2-D, 3-D, color, achromatic, etc.). For instance, a client might want to reproduce a hologram from an earlier production run along with a more recent one. Or the factory might want to run two jobs at the same time.

Here again, the different image components can be assembled in the recombining lab easier than if a holographer were commissioned to reshoot a whole new master. The factory, working under tight deadlines, cannot afford the time it would take to do so even if were cost-effective - which it is not. In addition, reshooting is not just impractical; in many cases it is impossible as the original artwork used no longer exists.

Enhancing pre-existing images - To save a client more money and give him a wider variety of holograms to utilize, pre-existing stock images also can be customized with 2-D surface graphics in the recombining lab. A number of different effects can be added. They include: several frequencies of grating or rainbow; soft or crisp colors; dot matrix patterns; computer generated textures; and the use of several shades of matte or white textures.

These 2-D effects are "dropped-in" on top of the existing hologram, creating effects which are difficult to achieve by traditional methods. It is possible for recombiners to add simple text (as small as l.pt.), complex patterns (to thwart counterfeiting), or logos in this fashion. The customized composite image then can be sent to the embosser without incurring additional mastering charges.

The Wavefront Technology Process

Petersen and Carey recommend that the "hologram components which need to be recombined be supplied on a photoresist glass plate or nickel shim. When time is critical, photoresist is the preferred option. Most types of photoresist can be used, but materials should be identified and labeled to insure that no unforeseen problems arise while the process is underway. The normal thickness for the nickel shim is 6 to 10 mils, but other thicknesses can be used. Silver nickel masters should never be used because this

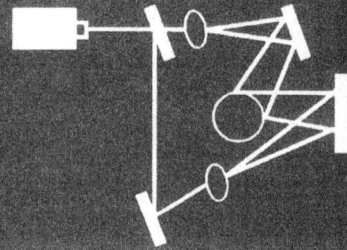

may render the original master useless for future electroforming ."

Petersen and Carey point out that with their process, "recombining time can be further reduced by eliminating the supplier's need to electroform the photoresist plate - saving up to three days of production time. Instead, the image is stepped directly into plastic. A 'racing stripe' for align- ment and registration marks for die cutting of the holo- grams can be added during this process."

In addition, the method they use provides time for an intermediate product check. "Before the embossing process begins, a hologram film proof of the recombine can be made. This allows us to inspect the image for defects and check for final layout problems. This reduces unanticipated costs and increases customer satisfaction. The recombine is then packed to ensure damage does not occur prior to electroforming. The total time from receiving the image until shipment of the recombine can be as little as eight hours. Complex and multiple image recombines can take a few days," they explain.

Questions to Consider

Since hologram production can be maximized by wise use of the available space on the embossing shim, the recombiner must arrange the image components in the most efficient manner. To determine how the given parts should best be positioned, the recombiner asks the client the following questions:

• What is the final image size?

• What is the exact shape of the finished piece?

• What are the spacing requirements between images?

• Is the final product a pressure sensitive sticker, a hot stamped product, or another finished substrate?

Once these questioned are answered, an engineering layout is made to show the precise location of each image and any registration marks. This drawing should be approved by all parties involved in the production process before the job is started. Designing the hologram to fit a standardized layout will save additional expense. Careful arrangement of the components on the shim will eliminate the need for the embosser or finishing company to purchase custom tooling later in the process.

Conclusion

Recombining reduces production costs while increasing the manufacturing efficiency. It also allows for better utilization of expensive equipment both in the origination studio and at the reproduction plant. The process allows the hologram designer and production team greater opportunities to create high quality embossed hologram in less time than ever before.

Major portions of this article were contributed by Joel Petersen and Robert Carey. For further information contact: Wavefront Technology, Paramount, CA ph. 562- 634-0434

3 Color Holography

One of the goals of holographers worldwide has been to develop cost-effective techniques that result in affordable full color, fully-dimensional images. The ability to produce such realistic-looking hologram images outside research laboratories will undoubtedly aid all aspects of the medium.

Commercial holographers would certainly benefit - clients have been asking for (and expecting) true color images for years. Many jobs have been lost when the words 'monochromatic', 'multi-color' or 'pseudo-color' have been uttered by holographers to potential customers. Technical achievements that we know are extraordinary and viable are less exciting to those expecting to see images like those created by Hollywood's special effects teams.

Holographic artists would also benefit if they could easily add realistic colors to their three dimensional images -- the market for holographic fine art might even be resurrected. Imagine the frustration of painters and photographers if they were confined to using only shades of red or green or gold in their work...and if creating combinations of these few colors proved difficult and expensive, if not impossible. It's no wonder the majority of holographic artists do not prosper.

If the technology to produce realistic images were affordable and accessible, many new applications for holography would certainly result. The entire field would regain some of the attention it once held among visual artists and the general public (and which it seemingly lost over the past decade to the fast moving digital imaging industry).

Steady Progress Has Been Made

Over the past two decades, great strides in true color imaging have been made by holographers working in their various fields. Embossed holographers have gradually increased their palate of available colors to the point where, theoretically, almost any hue can be generated using existing technology. More importantly, embossed holographers have been able to accurately reproduce the colors typically found on printed materials and computer monitors. In practice, even the most difficult colors to depict (e.g., flesh tones) can be, and have been, reproduced.

Holographers working with dichromated gelatin (DCG) have also produced notable achievements in color reproduction. Unfortunately, further research in this area has been slowed by economic restraints -- there seems to be no point in pursuing such labor intensive manufacturing techniques, since they result in relatively small, glass-encased holograms that are impractical to mass produce for most commercial applications.

The silver-halide reflection camp has consistently struggled to make more colorful display holograms. Years ago, a handful of determined artists developed tricks that produce a reasonably realistic-looking apple, pear and banana, using only HeNe lasers and red sensitive emulsions. As methods became perfected, many worthwhile images were produced on glass and film - but customers willing to order large volumes of these holograms were rare. Even with today's technology these multi-color holograms remain difficult and expensive to make, especially in the sizes requested by corporate clients.

True color reflection holograms were sporadically displayed at technical conferences but never mass produced. However, new developments in panchromatic emulsions should invigorate holographers interested in working with silver-halide. Affordable materials are becoming available. This is a very exciting time for holographers with access to the right lasers.

Another very promising field of research in full color holographic imaging revolves around photopolymer recording materials. Although successful results were achieved a decade ago in the lab, corporate managers decided it wasn't yet cost effective to bring the existing technology to market. New developments in panchromatic photopolymers and more efficient mass replication equipment may change the situation. Many holographers are now working closely with photopolymer suppliers to refine production methods and make them more feasible to utilize.

To give our readers a progress report about the state of color reflection holography, we've consulted with an acknowledged leader in the field, Hans Bjelkhagen. His article follows,

Progress in Color Reflection Holography

by Hans I. Bjelkhagen

The possibility of producing true color volume holograms in large quantities is just around the comer. So far, most holograms are of the pseudo-color or multi-color hologram types. There exist methods of creating colors that give an impression of a true "Color in the" finished image; e.g., multiple recorded stereograms or rainbow holograms, although the recordings could have well been made from objects with completely different colors, or from multiple sets of color-separated photographs or movie recordings.

By using the rainbow technique, it actually has been possible to mass-produce embossed holograms with either "natural" or artificial colors. It is also very common among artists to make multiple-exposed color reflection holograms using a single-wavelength laser, where the emulsion thickness has been changed between the recordings of special objects.

Covered in this review is the technique for the recording of objects, which will result in color holograms. These holograms are recorded on holographic panchromatic materials. Ultra-high resolution, single-layer silver-halide emulsions and new photopolymer materials are used for this purpose.   .

Setup for Recording Holograms

Color reflection holography presents no problems as regards the geometry of the recording setup, but the final result is highly dependent on the recording material used and the processing techniques applied. ' The single-beam Denisyuk recording scheme has produced the best results so far.

Three laser wavelengths (red, green and blue) are needed for the recording. Suitable primary wavelengths can be selected from different CW lasers currently in use in holographic recordings: argon ion, krypton ion, diode-pumped frequency-doubled Nd:YAG, helium neon, and helium cadmium lasers. In particular, the selection of the wavelength in the blue part of the spectrum is critical. There are several possible wavelengths: 458, 476, and 488 nm obtained from the argon-ion laser. The green wavelength, 532 nm, from a CW frequency-doubled diode-pumped Nd:YAG laser, is most suitable for the green primary wavelength. The 531 nm wavelength from an argon-krypton ion laser is an alternative. In regard to the red wavelength, the 647 nm of the krypton-ion laser offers high output power, important for large-format color holograms. The 633 nm wavelength from the helium~neon gas laser can be used for small holograms.

The color hologram recording setup is illustrated in figure 3.1. Three laser wavelengths are employed for the recording: 476 nm, provided by an argon ion laser; 531 nm, provided by a mixed argon-krypton ion laser; and 633 nm, provided by a helium-neon laser. Two dichroic filters are used in combining three color-laser beams. The "white" laser beam goes through a spatial filter, and a color reflection hologram of the Denisyuk type is recorded. The hologram recording setup is arranged on an optical table. The lasers are installed on an independent vibration-isolation

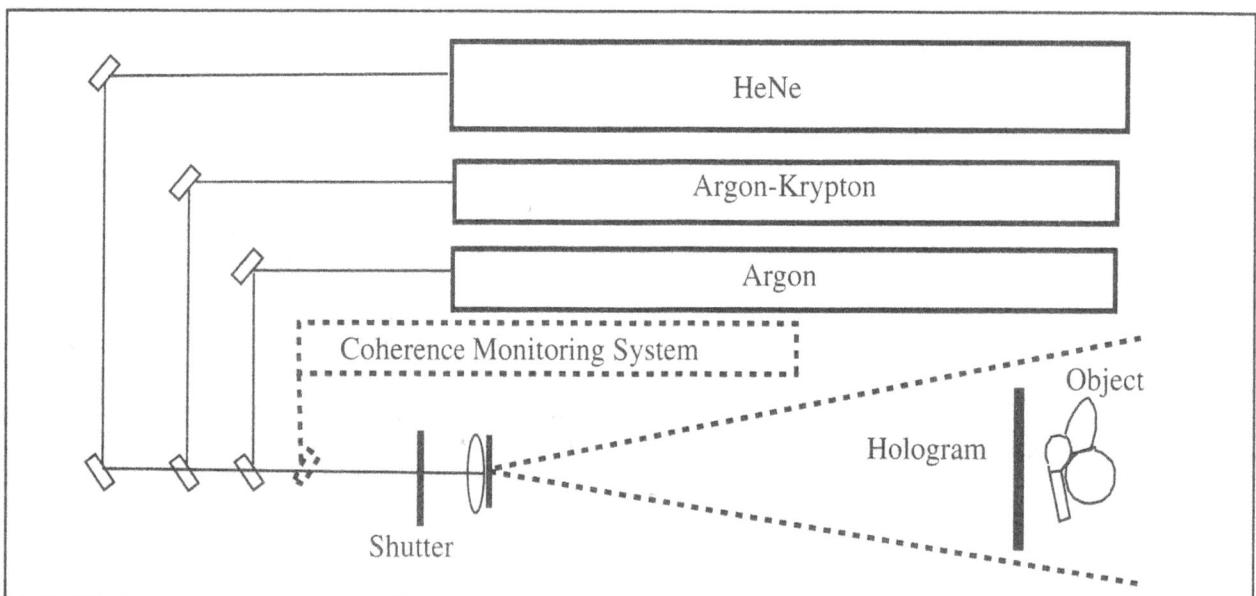

Figure 3.1. The set-up for recording color holograms.

system isolated from the table surface with the Denisyuk hologram recording setup. Using the dichroic filters in combining laser beams demonstrated multiple advantages compared to the previous beam-combining mechanism with movable mirrors. Without changing mirror positions between exposures, the dichroic filter approach shortens and simplifies the exposure procedure tremendously. The light intensity and RGB ratio on the recording plane remain undisturbed after the setting. Therefore, the check-up and calibration between hologram recordings are not absolutely necessary. The improvement is revealed more significantly when producing large quantity color holograms.

Besides the simplification and convenience, the most important advantage of using the dichroic filter beam combination is to perform simultaneous exposure recording. In contrast to the sequential exposure method used before, the simultaneous exposure approach makes it possible to independently control the light RGB ratio and overall exposure energy in the emulsion. The light RGB ratio can be varied by changing the power outputs of the lasers, while the overall exposure energy is controlled solely by the exposure time.

Specially designed test objects (the 1931 CIE chromaticity diagram, and the Macbeth ColorChecker@) were used for the color balance adjustments, color rendition tests and exposure tests . To record holograms with deep scenes , long coherence is absolutely required from all wavelengths. A special system has been designed in order to observe the temporal coherence of the laser wavelengths.

Recording Color Holograms on Silver-Halide Materials

In order to record color holograms special silver-halide materials are needed.[1]   Holographic color plates (PFG-03c) are produced by the Slavich photographic company outside Moscow[2] The sizes used range from 4" x 5" format up to 12"x16". Slavich can coat 60 cm x 80 cm size glass plates, which currently represents the largest format. Since there are great variations from batch to batch of this material, it is rather difficult to make a detailed characterization of the emulsion itself. However, the silver-halide grain size is the most important parameter of this material and is the main reason for the obtained quality of the holographic images. Some characteristics of the Slavich material are presented in table 3.1 (below):

Table 3.1: Characteristics of SLAVICH color emulsion	
Silver halide material	PFG-03c
Emulsion thickness	7 µm
Grain size	12 - 20 nm
Resolution	~10000 lp/mm
Blue sensitivity	~1.0 - 1.5 × 10⁻³ J/cm²
Green sensitivity	~1.2 - 1.6 × 10⁻³ J/cm²
Red sensitivity	~0.8 - 1.2 × 10⁻³ J/cm²
Color sensitivity peak at: 633 nm, 530 nm	

By composing special processing chemistry and processing baths, it has been possible to obtain high-quality color holograms, first reported by Bjelkhagen and Vuki-evi[3] Bjelkhagen et al. published a paper on the development of color holography and, in particular, the new volume reflection holograms in a single-layer silver-halide emulsion, including a bibliography on color holography.[4]

The RGB sensitivity values of the recording plate are determined experimentally. Using the simultaneous exposure approach, the overall exposure of about 3 mJ/cm² is needed.

Recording Color Holograms on Photopolymer Materials

The color holography photopolymer materials from E.I. du Pont de Nemours & CO.[5-8] are very interesting and easy to use for color holography. In particular, this material is the main candidate for mass production of color holograms. Although less sensitive than the ultrahigh-resolution silver-halide emulsion, it has its special advantages of easy handling and d,y processing (only UV-curing and baking).

The DuPont color photopolymer material has a coated film layer thickness of about 20 im. The photopolymer film is generally coated in a 12.5" width on a 14" wide Mylar® polyester base which is .002" thick. The film is protected with a .00092" thick Mylar® polyester cover sheet. Such experimental materials have been manufactured by the company, e.g., HRF-700X071-20. However, these materials are not yet introduced on the market. The latest experimental version, HRF-800XOOl-15, has a film thickness of 15 im and is improved over the earlier film.

The recording of a DuPont color hologram is rather simple. The film has to be laminated on a piece of clean glass or attached to a glass plate using an index-matching liquid. (Some practice with laminating techniques is necessary in order to not get any air bubbles or dust particles trapped in between the film and the glass plate.) To obtain the right color balance, the RGB ratio depends on the particular material, but typically it is about 4: 1:2. It is difficult to obtain high red-sensitivity of the photopolymer film. Simultaneous exposure is the best recording technique for photopolymer materials.

Holograms can be recorded manually, but in order to produce large quantities of holograms, a special machine is required. For hologram replication the scanning technique can provide the highest production rate. In this case, three scanning laser lines are needed, which can be adjusted in such a way that all three simultaneously can scan the film. The color photopolymer material needs an overall exposure of about 10 mJ/cm².

After the exposure is finished, the film has to be exposed to strong white or UV light. DuPont recommends about 100 mJ/cm2 exposure at 350-380 nm. After that, the hologram is put in an oven at a temperature of 1200 C for two hours in order to increase the brightness of the image.

Processing of Color Holograms

The dry processing of color holograms recorded on photopolymer materials has already been described. The process is simple and very suitable for machine processing using, e.g., a baking scroll oven.

The processing of the silver-halide emulsion is critical in order to obtain good results. The Slavich emulsion is rather soft, and it is important to harden the emulsion before the development and bleaching takes place. Emulsion shrinkage and other emulsion distortions caused by the active solutions used for the processing must be avoided. In particular, when recording master color holograms intended to be used for photopolymer replication, shrinkage control is extremely important. The general processing of Slavich color holograms has been reported previously. The processing steps are summarized below:

1. Tanning in a Formaldehyde solution   6 min.
2. Short rinse   5 sec.
3. Develop in the CWC2 developer   3 min.
4. Wash   5 min.
5. Bleach in the PBU-amidol bleach   ~5 min.
6. Wash   10 min.
7. Soak in acetic acid bath   1 min.
8. Short rinse   1 min.
9. Wash in distilled water   1 min.
    (with wetting agent added)
10. Air dry the holograms.

Some modifications have been introduced in order to produce holograms without any shrinkage. Such color holograms can be suitable as masters for mass production of color holograms on photopolymer material. The photopolymer materials from DuPont have a shrinkage of about 5% after processing, making these materials less suitable for color hologram mastering.

Evaluation of Recorded Color Holograms

Color holograms recorded on silver-halide materials of the test targets have been evaluated using a PR-650 Photo Research SpectraScan SpectraCalorimeterTM. The illuminating spotlight used to reconstruct the color holograms was a 12-Volt 50-Watt Phillips halogen type 6438 GBJ lamp. This type of spotlight is normally used for the display of color holograms. The particular lamp used to reconstruct a color hologram is much more critical than lamps for monochrome hologram display, since the color balance during the recording of a color hologram is determined by the type of spotlight that is going to be used for the display of the finished hologram.

Figure 3.2 shows the normalized spectrum, which means that the diffraction efficiency for each color component is obtained by assuming a flat spectrum of the illuminating source. The noise level, mainly in the blue part of the spectrum, is visible and low. The measured hologram is processed in such a way that no shrinkage has occurred. The three peaks are exactly at the recording wavelengths for this hologram; i.e., 647 nm, 532 nm, and 476 nm.

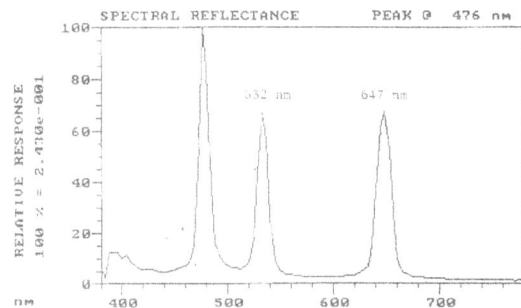

Figure 3.2. Normalized spectrum from a white area of a color test target hologram.

In table 3.2 some results of the Macbeth ColorChecker® investigation are presented. The 1931 C.LE. x and y coordinates are measured at both the actual target and the holographic image of the target. The measured fields are indicated in the table by color and the corresponding Macbeth field number.

Color photographs of the color holograms are presented in the color pages section of this book. (See figures 3.3 - 3.6 for black and white versions of the same pictures.) The photographs of the reconstructed color holograms were recorded using the above-mentioned halogen spotlight, positioned at the correct distance from the hologram and illu-

Object	White #19	Blue #13	Green #14	Red #15	Yellow #16	Magenta #17	Cyan #18
CIE xy	x/y	x/y	x/y	x/y	x/y	x/y	x/y
Target	.435/.405	.295/.260	.389/.514	.615/.335	.517/.450	.524/.322	.285/.380
Image	.354/.419	.335/.362	.337/.449	.476/.357	.416/.437	.448/.338	.295/.366

Table 3.2. Chromaticity coordinates from color hologram recording tests using the Macbeth ColorChecker

minating the hologram at the correct angle (according to the recording geometry).

Shown in Figures 3.3a and 3.3b is the left and right views of a 4" x 5" color hologram of a French house. Another 4" x 5" color hologram is featured in Fig. 3.4: a fish brooch. A color hologram of chocolate (mounted in an opening of a chocolate box) is shown in Figure 3.5 . The size of the hologram is about 8" x 10". A 12" x 16" color hologram of a stuffed parrot is reproduced in Figure 3. 6.

Discussion

The problem of emulsion shrinkage and the resulting wavelength shift, as well as the color desaturation problems, make holographic color reproduction difficult. The white-light reconstruction of a color hologram shows a decreased signal-to-noise ratio and an increased bandwidth, compared to the wavelengths used at the recordings. Desaturation is caused primarily by noise, but also partly by the increased bandwidth. Although good color rendition can be obtained, problems connected with color desaturation still remain to be solved.

The development process has been further improved in order to avoid emulsion shrinkage and non-uniform development. Other limitations concerning the recording of colors in a hologram include the fact that some of the colors we see are the result of fluorescence which cannot be recorded in a hologram. There are some differences in the recorded colors of the Macbeth color test chart. However, color rendition is a very subjective matter. Different renditions may be preferred for different applications, and different people may have varied color preferences.

In regard to the silver-halide materials, the research on processing color reflection holograms is still in progress. We are working on techniques to increase diffraction effi-

ciency by SHSG processing. SHSG means silver-halide sensitized gelatin and represents a method for converting the silver-halide hologram into a dichromated gelatin-type hologram.

Figures 3.3a&b. French house, left and right views. (B&W versions. See color pages.)

Figure 3.4. *Fish brooch hologram. (B&W version. See color pages.)*

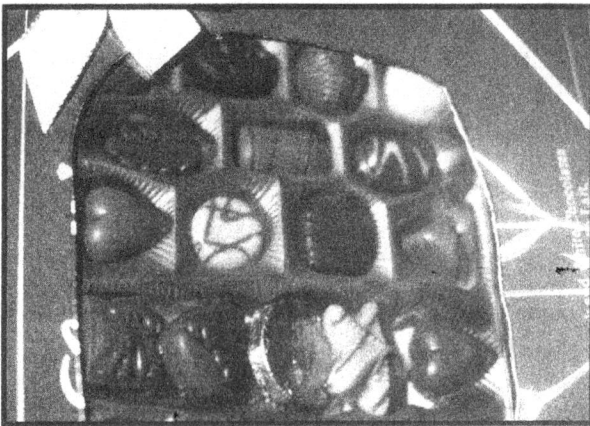

Figure 3.5. *Chocolate box hologram. (B&W version. See color pages)*

Figure 3.6. *Parrot on perch. (B&W version. See color pages)*

DuPont's color photopolymer film is still in the development phase. However, the monochrome materials (OmniDex®) are commercially available and have been used successfully for many years. Photopolymer film can become a very suitable recording material for mass replication by contact-copying color holograms and color HOEs recorded on silver-halide masters.

The virtual color image behind a Denisyuk holographic plate represents the most advanced image of an object that can be obtained today. The large field of view adds to the realistic illusion of viewing an image which will really not differ from viewing the actual object. The wavefront reconstruction process accurately recreates the three-wavelength light 'scattered from the object during the recording of the color hologram. Such an imaging technique will have many obvious applications: in particular, for displaying unique and expensive artifacts. There are also many potential commercial applications of this new feature of holography. Holography may well become next century's new and highly recognized imaging technology which has its unique applications, and will become a complement to modem photography.

References:

1. HI. Bjelkhagen: Silver Halide Recording Materials for Holography and Their Processing, Springer Series in Optical Sciences, Vol. 66. Springer-Verlag, Heidelberg, New York,(1993).

2. SLAVJCH Joint Stock Co., Micron Branch Co., 2 pl. Mendeleeva, 152140 Pereslavl-Zalessky, Russia.

3. Hi. Bjelkhagen, D. Vuki-evi: Lippmann color holography in a single-layer silver-halide emulsion, in Fifth [nt'l Symposium on Display Ho- logaphy, ed. by TH Jeong. Proc. SPJE 2333, 34-48 (1994).

4. HI. Bjelkhagen, TH Jeong, D. Vuki_evi_: Color reflection holo- grams recorded in a panchromatic ultrahigh-resolution single-layer sil- ver halide emulsion. 1. Imaging Sci. Techno!. 40, 134-146 (1996).

5. WJ Gambogi, WK. Smothers, K. W Steijn, SH Stevenson, A.M. Weber: Color holography using DuPont holographic recording jilm, in Holographic Materials, ed. by J. Trout. Proc. SPIE 2405, 62-73 (1995).

6. TJ. Trout, WJ. Gambogi, SH Stevenson: Photopolymer materials for color holography, in Applications of Optical Holography, ed. by T Honda. Proc. SPJE 2577, 94-105 (1995).

7. K. W Steijn: Multicolor holographic recording in DuPont holographic recording jilm: determination o f exposure conditions for color balance, in Holographic Materials II, ed. by J. Trout. Proc. SPIE 2688, 123-134 (1996).

8. S H Stevenson: DuPont multicolor holographic recording jilm, in Practical Holography XI and Holographic Materials Ill, ed. by SA. Benton, TJ. Trout. Proc. SPIE 3011, 231-241 (1997).

Is this the first published report of holography in the popular press?

From an article entitled "**Why Lasers Don't ---Yet**"

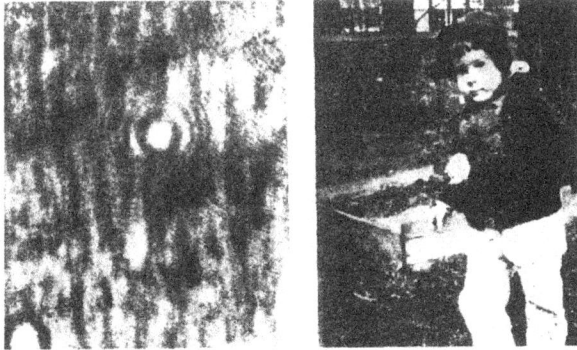

"Lensless Photography: Helium-neon laser light beamed at an original snapshot of little girl resulted in blurred negative, left. Laser beam trained on the negative in a lensless projector produced the clear image at right."

"...One of the newest applications of laser techniques, photography without the use of a lens, was recently announced by two engineers at the University of Michigan, Emmett N. Leith and Juris Upatnieks. While the concept sounds simple (all that is needed apparently is a mirror, photographic film and a holder, plus a gas laser), it's not likely to become part of the amateur's darkroom for years to come – if ever. At present, it can be used only in the laboratory.

"The new technique has two steps. First, the laser light is trained through a transparent object, such as a photographic negative, a color slide or a microscope slide; a portion of the laser beam is directed around the object with mirrors. All of this laser light is trained on a lensless cameralike device which contains ordinary film. The resulting exposed film (negative) is an unrecognizable blur.

"Second, the negative is then placed in a lensless projector and a laser beam trained on it. A screen or a piece of light sensitive paper placed at a precise distance from the projector picks up the laser beam and reproduces a sharp, clear, magnified picture of the object.

"The lensless photography technique may eventually result in sharp, well defined X-rays or better enlargements of ordinary photos; it will be used in such instruments as microscopes, both light and electron, for studies of individual molecules."

Popular Mechanics magazine,

March 1964, p.105

Museum of Holography Chicago

Founded in 1978, the museum houses the world's most extensive collection of artistic and commercial holograms for exhibition, education and research. Holography classes are also offered. Visit or call today!

**1134 West Washington Blvd.
Chicago, IL 60607
ph 312-266-1007**

4

Recording Materials: Silver-Halide

Last year, the primary manufacturer of silver-halide holographic recording materials (Agfa) ceased production. Since that time, several other companies around the world have expanded their manufacturing capabilities to meet the various demands of the holography industry.

In this chapter; we have published the most recent technical specifications available from four major manufacturers of silver-halide emulsions. We have also included pertinent commentary from professional A holographers who have had a chance to test and review some of the newly available emulsions.

Introduction

Silver-halide recording materials are commonly coated" onto films or glass plates for use in traditional cameras, in laboratories, and in factories. However, they were not originally designed for holographic applications.

The important difference between photographic and holographic materials is resolving power, usually expressed in lines per millimeter. Whereas photographic films usually cannot resolve more than 50-100 -lines per millimeter, holographic applications require 1,250-2,500 lines per millimeter.

Another difference is sensitivity, which is typically expressed as an ASA number. For example, popular photographic films are rated at ASA 120-400. Exposures are usually measured in hundredths of a second. Silver-halide holographic recording materials are so much less sensitive than standard films that their ASA would only be rated as fractions. Therefore, their sensitivities are usually expressed in micro-joules per square centimeters (or ergs per centimeter squared). Exposure times are typically measured in seconds, or even minutes, depending on the amount of laser light available.

The silver-halide emulsions are coated on either glass plates or film. Glass is preferred for most holographic applications, especially reflection holography, due to its rigidity. The most popular size glass plates are 4" x 5" and 8" x 10", though display holographers often prefer to work with bigger sizes. It is most economical to purchase larger sizes and cut plates to your own specs, if this is feasible.

The same factories that coat glass plates usually offer their emulsions on film, too. A major advantage of film is that it is less expensive than glass, is easier to cut and curve, and is much more suitable for automated reproduction processes - as it is often available on rolls. The main difficulty faced by holographers using low powered CW lasers and film is keeping the film absolutely motionless during the exposure. Although the film can be sandwiched between clear pieces of glass, better results are obtained when the emulsion is left uncovered. Vacuum mounts and various other devices have been designed to accomplish this feat.

Some plates and films are supplied with an antihalation coating on the back, which can be useful when making transmission holograms, as it helps to cut down on unwanted internal reflections. These plates cannot, however, be used to make reflection holograms, so check product codes before ordering.

Current Availability

In response to anticipated demand, several major manufacturers of silver-halide emulsions adapted their existing production techniques and formulations to provide afford- able recording materials for holographic applications. This spurred the growth of the holography industry, which needed a reliable supply of basic materials. These silver-halide emulsions were high quality, ready to use, and had a reasonably long shelf life.

Unfortunately, the world-wide demand for these particular photosensitive materials has leveled off, due in part to the increased use of electronic imaging, especially in the field of Non-Destructive Testing. Since the current combined needs of commercial holographers, educational facilities and hobbyists do not compare to other industrial and mass market customers, some major manufacturers (most notably, Agfa and Ilford) have ceased production entirely. However, the supply situation appears to be gradually improving as other factories expand their production capacity and establish new distribution networks. (To prevent potential supply problems, some commercial holographers have learned to utilize silver-halide recording materials with characteristics similar to holographic films but which are intended for other industrial uses, such as micro-lithography.)

Currently, the main manufacturers of silver-halide holo- graphic recording materials are Slavich (Russia), HRT (Germany) and Eastman Kodak (USA). The first two companies are actively involved in researching and developing emulsions and substrates suitable for professional applications. In contrast, Kodak's product line has remained unchanged for years and its products are still used mainly by hobbyists and schools. Emulsions from Holdor (China) are also being developed and becoming available. We have included all the information we could gather regarding these holographic recording materials in the following chapter.

Editor's Notes: All the aforementioned manufacturers have, or are developing, distribution channels for their products. Check the Business Directory in this book for an updated listing of suppliers.

Some distributors still have stockpiles of the popular Agfa and Ilford product, though production has been suspended for some time. Please refer to the previous editions of Holography MarketPlace for relevant technical specifications regarding these emulsions.

Selecting Silver-Halide Materials

The general rules when selecting a silver-halide material are: match the peak sensitivity of the material as closely as possible to the wavelength emission of the laser being used to expose the material; and select an emulsion with the lowest possible graininess characteristics, and highest possible resolution. Let's examine why.

There are five atoms which, because of their atomic similarity, are called the halides. They are chlorine, bromine, iodine, fluorine and astatine. Silver-halide emulsions are made using either silver chloride, silver bromide, or silver iodide. The other two halides are not used because silver

fluoride is insoluble in water and astatine is radioactive.

A typical silver-halide emulsion is made by adding a solution of silver nitrate to a solution of potassium bromide and gelatin. Silver bromide crystals form in the emulsion. The emulsion is heated for a certain amount of time, which is called the ripening process.

During the ripening process, the grain size increases and the speed of the emulsion is increased. Some doping agents may be added to the emulsion at this time tofoster proper crystal growth. Afterwards, the gelatin is allowed to cool. It is then shredded, and the soluble potassium nitrate is washed out of the emulsion.

The emulsion is heated again, with more gelatin added; then it is cooled and applied to a base. The thickness and hardness of the emulsion is important in holography because emulsions that are too thick tend to deform during development. Emulsions that are too hard can either retard chemical reactions or create vacuoles in the emulsion left by migrating atoms. These vacuoles tend to scatter light.

The Photochemical Reaction

Let's assume the emulsion is made and we now want to expose it to light. It sounds surprising, but a perfectly structured crystal of silver bromide does not react to light in any appreciable way. A crystal with defects, however, does react with light. Fortunately, most silver bromide crystals will have defects which consist of some interstitial (out of order) silver ions displaced in the crystal structure.

The process of the photochemical reaction is not known in exact detail, but it is believed that when light strikes a silver bromide crystal, enough energy is available to remove an electron from an occasional bromide ion. The electron produced is able to migrate through the crystal until it comes in contact with an interstitial silver ion. The silver ion takes the electron and becomes silver metal. Silver atoms formed by this mechanism apparently act as a nucleus for the formation of aggregates of 10 to 500 silver atoms, known as latent images because they are too small to be seen by the naked eye.

After exposure, the emulsion is developed. The developer goes to the site of any silver bromide crystal with a latent image and causes all the silver in that particular silver bromide crystal to be reduced to silver metal and deposited on the already-existing latent image of silver metal. This causes a worm-like grain of silver metal toform which is limited in size by the amount of silver available in the silver bromide crystal. This growth is considerable, amplifying the size of the latent image silver metal by a factor on the order of 10^6.

If the developer is left in contact with the emulsion long enough it eventually attacks all the silver in the emulsion. The speed of development is slow enough, though, that you can use a timer to take the emulsion from the developer just after the latent image, but not the unexposed silver bromide crystals, have been developed. At this point the developer has converted silver ions to silver metal if, and only if, they belong to a silver bromide crystal that was exposed to light.

The emulsion is then placed in a fixer solution which attacks all silver bromide crystals that were not exposed to light. The fixer makes these silver bromide ions soluble and removes them from the emulsion. The result is an emulsion with black spots where light has struck, and clear spots where no light struck.

Three Main Factors to Consider

An ideal silver-halide emulsion depends somewhat on its use but there are three main factors to consider in any emulsion: thickness of emulsion, grain size of silver-halide crystals, and sensitivity (or density of silver-halide crystals) in the emulsion. We can generally state the following: It is agreed that emulsions of more than 10mm are neither practical or theoretically necessary to produce most volume holograms. Thicknesses above this size cause problems in development.

Grain size becomes an important issue in holography because it involves recording fringe patterns that are wavelengths apart. Too large a grain size may create excessive scatter, which may fog or destroy your hologram, and too small a grain size makes the emulsion have no usable sensitivity. It is generally agreed that the ideal grain size is in the range of .01mm to .035 mm.

The ideal exposure would probably be 100-300 mJ/cm² to give a useful density ($D = 2$-3). If exposures are much longer than this, the main attraction of silver-halide emulsion, its speed, comes into question and other emulsions become more attractive.

Slavich Silver-Halide Emulsions

by Alex Cheimets (3 Deep Co.)

The Russian manufacturer "Kompania Slavich" J.S. Co. supports holographers using CW or pulsed lasers with high-resolution emulsions for transmission or reflection holograms. The company's "Micron" plant has produced high-quality recording materials for years and is expanding its product line and distribution channels to meet the demands of users worldwide. Currently only coated glass plates are available, but emulsions on film are expected very soon.

Summary of Emulsion Specifications

<table>
<tr><th colspan="9">SLAVICH PRODUCT SUMMARY</th></tr>
<tr><th>Emulsion Type</th><th>Color Sensitivity</th><th>Application</th><th>Sensitivity Range (nm)</th><th>Spectral Sensitivity (J/m^2)</th><th>Diffr. Efficiency %</th><th>Agfa Equiv.</th><th>Res. l/mm</th></tr>
<tr><td>PFG-01</td><td>Red</td><td>Transmission with pulse laser</td><td>633 nm
694 nm</td><td>1</td><td>45</td><td>8E75</td><td>3,000</td></tr>
<tr><td>PFG-01M</td><td>Red</td><td>Transmission and reflection with pulse laser</td><td>633 nm
694 nm</td><td>1</td><td>45</td><td>8E75HD</td><td>5,000</td></tr>
<tr><td>PFG-03M</td><td>Red</td><td>Reflection holograms</td><td>633 nm</td><td>20</td><td>45</td><td>n/a</td><td>5,000</td></tr>
<tr><td>FPR</td><td>Green</td><td>Transmission with pulse laser</td><td>532 nm</td><td>1</td><td>45</td><td>8E56</td><td>3,000</td></tr>
<tr><td>FPR - M</td><td>Green</td><td>Transmission and reflection with pulse laser</td><td>532 nm</td><td>1</td><td>45</td><td>8E56HD</td><td>5,000</td></tr>
<tr><td>PFG-03C</td><td>Blue
Green
Red</td><td>Color reflection</td><td>457 nm
514 nm
633 nm</td><td>20
30
30</td><td>25
40
40</td><td>n/a</td><td>5,000</td></tr>
</table>

Manufacturer's Specifications for PFG-01 and PFG-01M

PFG-01 and PFG-01M are designed to record holograms

in convergent beams under the Leith method using continuous or pulse laser radiation in the red spectrum range (for instance, helium-neon or ruby laser). These plates are used to:

a) record holographic three-dimensional portraits of people and animals;

b) make artistic holograms for demonstrating them in museums, exhibitions, offices and trading facilities with educational, publicity and designing purposes;

c) register three-dimensional images of moving subjects and rapid processes;

d) manufacture matrices for image recognition systems;

e) perform interferential non-destructive testing of important parts;

f) produce masters for making copies during mass-production of artistic and technical holograms.

<table>
<tr><th colspan="2">PFG-01, PFG-01M
HOLOGRAPHIC PROPERTIES</th></tr>
<tr><th colspan="2">Holographic Sensitivity</th></tr>
<tr><td>For exposure by helium-neon laser 633 nm according to co-directed scheme and processed with bleaching</td><td>$< 0.6 \ J/m^2$</td></tr>
<tr><td>For exposure by ruby laser in free running mode</td><td>$< 1.0 \ J/m^2$</td></tr>
<tr><td>For exposure by ruby laser in monopulse mode</td><td>$< 2.0 \ J/m^2$</td></tr>
<tr><th colspan="2">Diffract. Efficiency at Spatial freq. of 1/1000 mm</th></tr>
<tr><td>For exposure by helium-neon laser 633 nm</td><td>35-40%</td></tr>
<tr><td>For exposure by ruby laser in free running mode (pulse duration is 3×10^{-6} seconds)</td><td>> 45 %</td></tr>
<tr><td>For exposure by ruby laser in monopulse mode (pulse duration is 3×10^{-6} seconds)</td><td>> 40 %</td></tr>
<tr><th colspan="2">Average Diameter of Microcrystals</th></tr>
<tr><td>Average diameter of microcrystals</td><td>60 nm</td></tr>
</table>

Manufacturer's Specifications for FPR, FPR-M

FPR and FPR-M are designed to record holograms in convergent beams under the Leith method using continuous or pulse laser light in the red spectrum range (for instance, helium-neon or ruby laser). These plates are used to:

a) record holographic three-dimensional portraits of people and animals;

b) make artistic holograms for museums, exhibitions, offices and trading facilities with educational, publicity and design purposes;

c) register three-dimensional images of moving subjects and rapid processes;

d) manufacture matrices for image recognition systems;

e) perform interferential non-destructive testing of important parts;

f) produce masters for making copies during mass production of artistic and technical holograms.

FPR AND FPR-M: HOLOGRAPHIC PROPERTIES	
Holographic Sensitivity	
For exposure by helium-neon laser according to co-directed scheme and processed with bleaching	< 0.6 J/m²
For exposure by neodymium laser	< 1.0 J/m²
Diffraction Efficiency at spatial frequency of 1/1000 mm	
For exposure by argon laser	>45 %
For exposure by neodymium laser with pulse duration of 1.10-10c	>45 %
Average Diameter of Microcrystals	
Average Diameter of Microcrystals	60 nm

Manufacturer's Recommended Processing Procedures for PFG-01, PFG-01M, FPR and FPR-M

Formulas recommended here are those most commonly used in Russia. Western holographers have experienced success using Western formulations, some of which are described later in this chapter.

A. Mastering (Manufacture of Phase Transmission Holograms)

MASTERING (MANUFACTURE OF PHASE TRANSMISSION HOLOGRAMS)			
Name And Sequence Of Operation	**Time (Min.)**	**Chemicals**	
1. Exposure		20 - 40 mJ/cm²	
2. Latensification*			
3. Development	2-3	SM-6	
		Ascorbic Acid	18 g
		Sodium Hydroxide	12 g
		Phenidone	6 g
		Sodium Phosphate Dibasic	28.4 g
		Water	1 liter
4. Intermediate wash	2-3	Water	
5. Bleaching	Until clear	PBU-AMIDOL BLEACH	
		Potassium Persulphate	10g
		Citric Acid	50 g
		Cupric Bromide	1 g
		Potassium Bromide	20 g
		Amidol	1 g
		Water	1 liter
6. Washing	10-20	Water	
7. Final washing	1	Water with wetting agent	
8. Slow air-drying			

B. Copying (Manufacture of Phase Reflection Holograms)

(1) For a honey-green color of reconstruction:

COPYING (MANUFACTURE OF PHASE REFLECTION HOLOGRAMS) FOR A HONEY-GREEN COLOR OF RECONSTRUCTION:			
Name And Sequence Of Operations	**Time (Min.)**	**Chemicals**	
1. Exposure		$20 - 30$ mJ/cm^2	
2. Latensification*			
3. Development	2-3	MODIFIED PYROCHROME	
		Part A	
		Pyrogallol	20 g
		Phenidone	2 g
		Sodium Metabisulphite	5 g
		Part B	
		Sodium Carbonate	130g
4. Intermediate wash	2-3	Water	
5. Bleaching	Until clear	PBU-AMIDOL BLEACH	
		Potassium Persulphate	10 g
		Citric Acid	50 g
		Cupric Bromide	1 g
		Potassium Bromide	20 g
		Amidol	1 g
		Water	1 liter
6. Washing	10-20	Water	
7. Final washing	1	Water with wetting agent	
8. Slow air-drying			

(2) For an orange color of reconstruction:

COPYING (MANUFACTURE OF PHASE REFLECTION HOLOGRAMS) FOR AN ORANGE COLOR OF RECONSTRUCTION:			
Name And Sequence Of Operations	**Time (Min.)**	**Chemicals**	
1. Exposure		$20 - 30$ mJ/cm^2	
2. Latensification*			
3. Development	2-3	MODIFIED PYROCHROME	
		Part A	
		Pyrogallol	20 g
		Phenidone	2 g
		Sodium Metabisulphite	5 g
		Part B	
		Sodium Carbonate	130 g
4. Intermediate wash	2-3	Water	
5. Bleaching	Until clear	PBU-AMIDOL BLEACH	
		Potassium Persulphate	10 g
		Citric Acid	50 g
		Cupric Bromide	1 g
		Potassium Bromide	20 g
		Amidol	1 g
		Water	1 liter
6. Water	10-20	Water	
7. Potassium iodide bath	2	Potassium Iodide	18g
		Water	1 liter
8. Washing	1-2	Water	
9. Final washing	1	Water with wetting agent	
10. Slow air-drying			

* Latensification

1. Procedure: Illumination with an ordinary 25-W safe-light lamp with dark green filter from a distance of Im for approximately 0.5 to 4 minutes.

2. The correct time of latensification can be found for each particular situation by reference to the following rule: the fog level of unexposed, latensified and developed holographic material must appear to be marginally higher than when no latensification is applied. Practically, you start with a small unexposed test plate and develop it normally. If the fog level is the same as a non-latensified control plate which is developed in the same fashion, go back and apply longer latensification with another test plate. Repeat this procedure until you notice that your test plate develops marginally darker than the non-latensified control.

3. The optimum time period for latensification may change with emulsion batch, age of the holographic material, environmental conditions and choice of chemistry.

Editor's notes: A further discussion of processing and developing procedures for Slavich recording materials can be found at the 3Deep website: http://www.3deepco.com/ Discussion/disciJrm.htm.

We have recently been informed that some Slavich emulsions are now available on film. Please call your Slavich distributor for further information.

Manufacturer's Specifications for PFG-03M

PFG-03M plates are designed to make reflecting holograms and are used to:

a) copy holographic three-dimensional portraits of people and animals;

b) make artistic holograms for demonstrating under ordinary illumination;

c) register three-dimensional images of moving subjects and rapid processes;

d) manufacture matrices for image recognition systems;

e) do interferential non-destructive testing of machines parts, motors, generators, turbines and other equipment;

f) register information using time-lapse technique;

g) manufacture holographic optical elements (HOEs). (PFG-03M specifications continued on following page)

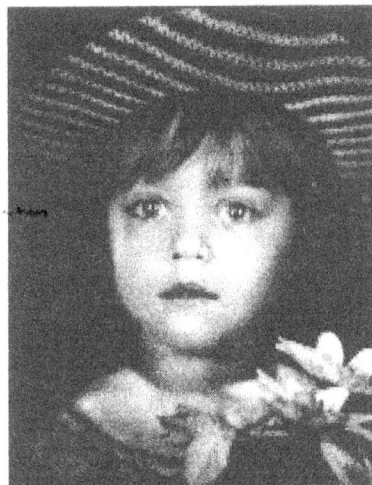

PFG-03M HOLOGRAPHIC PROPERTIES	
Holographic Sensitivity For exposure by helium-neon laser (633 nm)	$< 1.5 - 2.0$ mJ/cm^2
Diffraction Efficiency For exposure by helium-neon laser (633 nm)	$> 45\%$
Resolving Power	> 5000mm^{-1}
Emulsion Layer Thickness	$7.0 +/- 0.5$ mkm
Grain Size	10 nm

Manufacturer's Recommended Processing Procedures for PFG-03M

PFG-03M MANUFACTURER'S RECOMMENDED PROCESSING PROCEDURES			
Name And Sequence Of Operations	**Time (Min.)**	**Chemicals**	
1. Hardening	2-3	Formalin 37%	10 ml
		Potassium Bromide	2 g
		Sodium Carbonate	5 g
		Water	to 1 liter
2. Washing in filtered running water	1-2		
3. Developing in GP-2	10-15	CONCENTRATED GP-2	
		Methylophenydone	0.2g
		Hydroquinone	5 g
		Sodium Sulphite anhydrous	100 g
		Potassium Hydroxide	5 g
		Ammonium Thiocyanate	12 g
		Water	to 1 Liter
		WORKING SOLUTION	
		40ml GP-2 Concentrate + 1 Liter water	
4. Washing in filtered running water	2		
5. Fixing	2	Sodium Thiosulphate (cryst.)	160 g
		Potassium Metabisulphate	40 g
		Water	to 1 liter
6. Washing in filtered running water	5-10		
7. Drying	2	50% Ethyl Alcohol	
	2	75% Ethyl Alcohol	
	2	96% Ethyl Alcohol	

Notes:

1. Drying in 50% and 80% Isopropyl Alcohol is possible.

2. Processing solution temperatures must not exceed 19 $^\circ$ C to avoid emulsion layer deformation.

3. It is admissible a) to process photoplates without fixing, b) to carry out hardening after developing.

PFG-03C HOLOGRAPHIC PROPERTIES	
Holographic Sensitivity	
For exposure by helium Neon laser (633 nm)	< 3.0 mJ/cm^2
For exposure by argon laser (514 nm)	< 3.0 mJ/cm^2
For exposure by argon laser (457 nm)	< 2.0 mJ/cm^2
Diffraction Efficiency	
For exposure by helium Neon laser (633 nm)	> 40 %
For exposure by argon laser (514 nm)	> 40 %
For exposure by argon laser (457 nm)	> 25 %
Resolving Power	> 5000 mm^{-1}
Emulsion Layer Thickness	10 mkm
Grain Size	8 nm

Manufacturer's Recommended Processing Procedures for PFG-03C

PFG-03C MANUFACTURER'S RECOMMENDED PROCESSING PROCEDURES			
Name And Sequence Of Operations	Time (Min.)	Chemicals	
1. Hardening	6	Formalin 37%	10 ml
		Potassium Bromide	2g
		Sodium Carbonate	5g
		Water	to 1 liter
2. Washing in filtered running water	1-2		
3. Developing in VRP	4-5	CONCENTRATED VRP	
		Sodium Sulphite anhydrous	194g
		Hydroquinone	25g
		Potassium Hydroxide	22g
		Methylophenydone	1.5g
		Potassium Bromide	20g
		Potassium Metaborate	140g
		1,2,3-Benzotriazole	0.1
		Distilled Water	to 1 liter
		WORKING SOLUTION	
		1 part of VRP Developer + 6 parts of water	
4. Washing in filtered running water	1-2		
5. Bleaching in PBU-Amidol Bleacher	5-8	Copper Bromide	1g
		Potassium Persulphate	10g
		Citric Acid	50g
		Potassium Bromide	20g
		Distilled Water	to 1 liter
		Amidol	1g
6. Washing	2		
7. Stop-bath	2	Acetic Acid	20g
		Water	to 1 liter
8. Washing	2		
9. Bathing	2	Distilled water with added wetting agent	
10. Drying in normal conditions			

How to Make Better Hologra,ms Than Before; A Detailed Method for Processing the PFG-03M Emulsion Produced by the Slavich Company

by Tung H. Jeong (Professor Emeritus, Center for Photonics Studies, Lake Forest College)

Introduction

Mfa's dominance in the marketplace for silver ha- lide materials for holography had most of us to ignore what has been available to Eastern Europeans for a long time. Now that we are faced with an inevitable change, it is time to learn what the Russian holographers have known for decades.

What follows is not new knowledge. It is a practical step- by-step instruction on how to make good holograms, particularly the Denisyuk variety. It is written with the instructor of a beginners' holography class in mind.

The GP-2 developing procedure has been found to be most suitable for the PFG-03M emulsion. Both the developer and the emulsion are discussed in the book Silver Halide Materials for Holography and Their Processing by Hans Bjelkhagen (Springer-Verlag 1996).

GP-2

The ingredients of GP-2 are:

•	Methylphenidone	0.2 g
•	Hydroquinone	5.0 g
•	Sodium Sulfite (anhydrous)	100g
•	Potassium hydroxide	5.0 g
•	Ammonium thiocyanate	12.0 g
•	Distilled or de-ionized water	1.0 liter

Before mixing or using chemicals, consult a professional chemist or a certified chemistry teacher concerning safety procedures. Observe safety rules at all times.

Components of GP-2 can be purchased separately from chemical suppliers or as a kit. (For instance, the Integraf company sells a processing kit with pre-measured ingredients.)

Dissolve the contents of each container of the five chemicals in one liter of distilled or de-ionized water warmed to 50° - 60° C. When pouring chemicals, prevent contents from escaping into the air.

Keep the mixed solution in a tightly capped bottle and label it as "GP-2 Developer, Stock." At room temperature, this solution should be good for one year.

To make a "working solution", mix 15 ml of the stock solution with 400 ml of distilled or de-ionized water. Label this as "GP-2 Developer, Working Solution." Keep this at room temperature.

Hologram Exposure

Make and develop holograms in a darkened room illuminated by a minimum of green light. A green bulb (such as those from a Christmas tree decoration) can be used underneath a counter or table. Place obstructions around the light so that after'dark adaptation, there is just enough light for the holographer to safely move around. Make certain that there is no direct light on the holographic plate.

Those who are familiar with the Agfa packaging of plates will need to re-learn how to open and close the Slavich box - they are different! Carefully observe the box and think through the procedure before opening it in the dark.

The sensitivity of the PFG-03M emulsion is about 500 $\mu J/cm^2$, which is a factor of 5 to 10 lower than Agfa's 8E75 emulsion. But the PFG-03M has a resolution of 10,000 lines/mm, twice. that of the Agfa material, resulting in brighter and clearer holograms. Since the sensitivity changes based on the history of the emulsion (age, storage conditions, etc.), each batch should be tested for optimum exposure. In general, the exposure time for PFG-03M plates should be four tofive times longer than for Agfa 8E75 plates. The superior hologram you get is well worth it!

As a specific example, suppose we use a helium-neon laser with an output of 5 milliwatts and expand the beam to a diameter of approximately 10 cm. At the center of this beam we wish to make a 2.5" x 2.5" Denisyuk hologram. A good exposure would be about eight seconds. This means that the holography system must be extremely stable and free from vibrations.

Handling the plates by the edges only! Never touch the emulsion (sticky) side, except to check. The emulsion is very soft and has the consistency of Jello™. The sticky (emulsion) side of the plate should face the object.

Developing Procedures (at room temp.)

In addition to the GP-2 developer, you will need a 1-liter bottle of Kodak Rapid Fix™ with hardener, available in most photography stores. The fix should be diluted with distilled water in an 8: 1 ratio (or twice the dilution as recommended). Add a proportionate amount of hardener to the working solution just before use.

In a tray larger than the hologram, fill with enough working solution of GP-2 so that the hologram will be fully submerged. Remember, one liter of stock yield 27 liters of working solution, so don't be stingy. Develop each hologram with a tray of fresh solution.

1. To develop, place the exposed hologram into the solution with the emulsion facing upwards and do not agitate. Cover the tray with something opaque and let develop for 12 minutes.

2. Wash the developed hologram in running water for two minutes.

3. Place the hologram in the fix solution with hardener for two minutes.

4. Wash in running water for two minutes.

5. (Optional) Soak hologram in a dilute (a few drops in a pan of distilled or de-ionized water) Kodak Photoflo™ solution for 30 seconds.

6. Let the hologram dry in a vertical position. Alternatively, use a windshield wiper as a squeegee to wipe both sides of the hologram, then use warm air from a hair drier tofinish it.

For transmission holograms, an image can be viewed (with laser light) while the hologram is still wet. For reflection holograms, a transmission image can be seen while wet, but a sharp image can be seen only when the hologram is thoroughly dried. Finally, spray paint the emulsion side of the hologram with an antique-black paint. The Krylon™ brand available in hardware stores is recommended.

For further information, conract the author at: Center for Photonics Studies, Lake Forest College. Lake Forest. FL 60045 USA Fax.' 847-615-0835 E-mail. jeong@fc.edu

Additional Slavich Processing Recommendations

by Jeffrey Murray (Holography Institute)

Pre-develop Gelatin Hardening bath:

Sensitizes and maintains colors, allows squeegee use.

Distilled water	750 ml
Formaldehyde 37% (Formalin)	10 ml (10.2 g)
Potassium bromide	2 g
Sodium carbonate (anhydrous)	5 g

add distilled water to make 1 L

Processing time: 6 minutes. (Developing times may increase with harder gelatin.)

Fogged Plates:

Plates that are fogged by accidental exposure may be reconstituted by a rehalogenating bleach (such as PBU-Amidol). Try a very weak solution of 1-3% PBU/Amidol in distilled water, 30 seconds rinse.

CWC2 developer / PBU-Amidol bleach:

This process combination is great for all holograms. See next section (HRT processing notes) for formulas. Omitting the Cupric Bromide from the bleach still seems to work.

Desensitizer:

Transmission holograms to be used as masters can be left unbleached to maintain maximum resolution and contrast (expose less for lighter development, - D 1.5), followed by non-rapid fixer, or, for ultimate clarity with less emulsion shrinkage, a few minutes in desensitizer:

0.5 gram Phenosafranine dissolved in methanol / liter water.

The plate appears darker and pinker. If used after colloidal GP-2 development of a reflection hologram, this desensitizer maintains the color, avoiding the greener broadband shift from using fixer.

Colloidal Processing

Russian traditional colloidal developing process: use Russian Holography Developer GP-2, no bleach.

Russian Holography Developer GP-2 for colloidal development of reflection holograms using ultra-fine-grain Slavich emulsions:

Methylphenidone	0.2 grams
Hydroquinone	5 grams
Sodium sulfite (anhydrous)	100 grams

Potassium hydroxide	5 grams
Ammonium thiocyanate*	12 grams
(*or Potassium thiocyanate	24 grams)

Mix into 1 liter water for stock solution.

To use: mix 15 ml stock and 400 ml water.

Develop 12-24 minutes, to opaque (no agitation,).

Rinse thoroughly.

Fix (or desensitize) and rinse. Not rapid fixer.

No bleaching is used; the plate remains red/black.

Alcohol rinse optional. Photo-Flo optional.

Air dry, avoididg heat, or any emulsion stress.

Typically 10X exposure is required over traditional processing, not because of low sensitivity, but because overexposure assures establishment of small grain centers, and the slow, dilute developing, almost counter acted by solvent (thiocyanate), prevents filamentary grain growth. The desired end result with colloidal holographic processing is the tiniest spherical grain - an exposed spot must develop holding exactly the original size and shape.

Western "high-speed-film-processing" utilizes chemical amplification and filamentary grain growth to turn a little exposure light into a big fluffy dark spot.

For more info contact: Holography institute (415) 822-7123

Holographic Recording Technologies GmbH

by Richard Birenheide (HRT)

The BB-Plates are a range of silver halide materials specially designed for holography. We began developing them four years ago, working with Jeff Blyth. Our goal was to find a way to manufacture an emulsion that has low scatter and high diffraction efficiency combined with a sensitivity close to popular Western type materials.

The starting point was a special kind of "Lippman" emulsion developed by Blyth. These emulsions are characterized by a grain size of approximately 15 nm, compared to common Western emulsions measuring 35-40 nm. The smaller the grain size, the less the scatter or milkiness in a correctly exposed and processed hologram. This low grain size, which would be perfect for holography, has the serious disadvantage of having low sensitivity, mainly for three reasons.

First, the number of photons needed to make a grain developable is fairly independent of its size. Therefore, larger grains are more sensitive because they have a larger cross-sectional area to trap the required number of photons. Second, it is necessary for such a low grain size that the formation of the silver-halide crystals is performed under a low concentration of the reaction partners. This leads to a low overall concentration of silver-halide in the final emulsion. Last, most of the common "tricks" used to raise sensitivity lead to grain growth after the silver-halide particles have formed. Although Blyth's method allowed us to make emulsions with quite high silver-halide concentration, the sensitivity and long term stability needed improvement. In time, a way was found to make emulsions which are a compromise with respect to grain size and sensitivity – the BB plates.

The BB-Plates: Characteristics of the Different Emulsions

Emulsions are designed for high diffraction efficiency combined with low scatter in display holography. They are highly suitable for mastering where fringe integrity is a must.

All emulsions are coated with a thickness of approximately 6.5μm onto glass plates.

Long-term storage stability is one year or more.

The sensitivity is approximately 100-150μJ/cm² exposed at 633 nm and developed in a metol/ascorbate developer to a density of 2.5. Please note that this is a very rough estimate only and may vary widely depending on wavelength and development.

In most cases one will get good results by giving two to three times the exposure which was necessary for the analogous Agfa emulsions.

Grain size is approximately 20-25 nm.

a. BB-640

Sensitivity: 633 nm (He-Ne), 647 nm (Kr+). NOT for use at 694 nm (ruby).

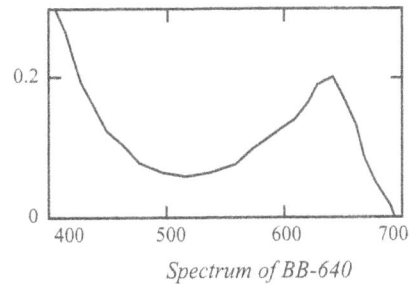

Spectrum of BB-640

Sensitivity region: 580 nm to 650 nm.

Suitable both for reflection and transmission holograms, excellent Denisyuk performance. These plates need approximately three times the exposure of the former Agfa 8E75HD (@633nm).

b. BB-520

Sensitivity: 488 nm (Ar+), 515 nm (Ar+), 532 nm (frequency doubled Nd-YAG).

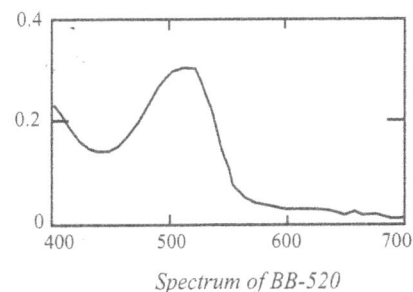

Spectrum of BB-520

Sensitivity region: 480 nm to 540 nm.

Suitable both for reflection and transmission holograms, good Denisyuk performance. S-catter is low enough and grain size small enough to record excellent reflection holograms, although mainly it used for transmission mastering. The exposure is approximately two times the exposure needed for the former Agfa 8E56HD emulsion.

c. BB-PAN

Sensitivity: 488 nm to 647 nm "panchromatic" emulsion. Sensitivity is not the same for all wavelengths. You might find lower sensitivity for a specific wavelength compared to our other emulsions.

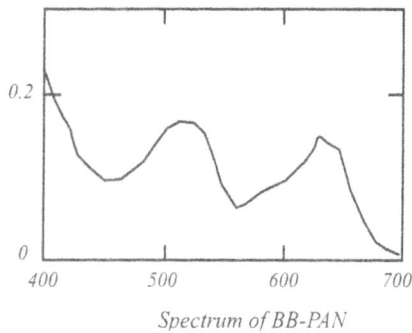

Spectrum of BB-PAN

Sensitivity region: 480 nm to 540 nm and 580 nm to 650 nm.

Suitable both for reflection and transmission holograms, excellent Denisyuk performance for all common wavelengths, specially designed for full color Denisyuk holograms. It is a first generation derivative of BB-640 and BB-520 and contains both dyes. There is a pronounced absorption minimum in the region of 560 nm.

d. BB-4S0

Sensitivity: 442 nm (He-Cd), 458 nm (Ar+); Sensitivity region: 410 nm to 470 nm.

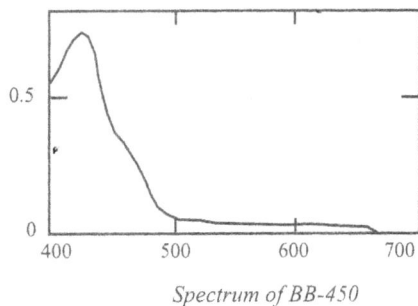

Spectrum of BB-450

This emulsion is specially designed for embossed origination (HI) using a He-Cd laser. It combines the low scatter of all of our emulsions with reasonable sensitivity at the (blue) He-Cd wavelength. Not recommended for reflection or Denisyuk work.

Available Formats

2.5" x 2.5"; 4" x 5"; 8" x 10"; 30 cm x 40 cm; 50 cm x 60 cm. The first two sizes are on 2 mm glass, the other on 3 mm glass. (Please note we can supply 60 cm strips for "rainbow" masters or any other size below 50 cm x 60 cm on request.)

Specialties

We would like to fulfill customer's non-standard requirements even for quite low minimum orders. These specialties include:

• Thicker emulsion layer (bear in mind that this will increase scatter if no compensation is done).

• Thinner emulsion layer.

• Reduced scatter (but with reduced sensitivity). Higher sensitivity (but with increased scatter).

• Higher dye content (increased sensitivity; might lead to problems with reflection holograms)

• Special substrates (must be rigid and flat.)

• Softer emulsions (e.g., In a Denisyuk hologram, this can create a more broadband replay which appears brighter because it uses more of the wavelengths available in a white light source).

Instructions for Use

These are preliminary instructions for the use of BB-plates together with some of their characteristics. These plates are designed for ease of use in common with conventional Western type silver-halide material. The plates have been tested with the common types of developer and bleach such as the "pyrochrome system" or alternatively a developer based on ascorbic acid and either a ferric EDTA bleach or a bleach based on copper sulfate (Blyth's recipe; see below). However, you can try your favorite system, possibly allowing longer development times than you are used to.

This emulsion is hardened quite strongly in order to prevent distortion of the fringes during processing. If you would like to make reflection holograms with a more broadband replay, please ask for plates which are less hardened. Storage conditions for plates should be cool, around 4°C but not below OOC! Please allow the plates to acclimate to the relative humidity and temperature of your lab once you have opened the polyethylene bag before shooting holograms. For making masters (HI) this is of paramount importance. For these, intensity distribution of the reference beam should be as uniform as possible.

Note - If exposure time is too long compared to what you are used to with conventional material, please try to use a higher reference/object beam ratio in multiple beam setups. Since our material has a steep gradation this can provide you a shorter exposure time without losing diffraction efficiency.

*Suggested Processing Formulae**
1. Rehalogenating Process
a. Developer

> 700 cc water (deionized, if available)
>
> 70 g sodium carbonate, anhydrous
>
> 15 g sodium hydroxide*
>
> 4 g metol
>
> 25 g ascorbic acid
>
> Add water to 1,000 cc.

Please allow any constituent to dissolve before adding the next on the list. It is important to limit the time the surface of this developer is exposed to air (oxygen). Therefore the use of two closely fitting plastic dishes is strongly recommended so that one dish floats ("floating lid") on top

(Continued on page 66)

The World Leader in Holographic Security Solutions

Hot Stamping Foils

Product Applications: Transaction Cards, Banknotes, Travellers cheques, Fiscal Stamps, Transit passes, and other documents of value.

Tamper-Apparent™ Labels

Product Applications: Pharmaceutical, Petroleum, Software, Hardware, Audio/Video and licensed products.

Transparent Laminates

Product Applications: Passports, National ID cards, Drivers licenses and other forms of identification.

A.D.V.I.S.E.

The Complete Security Solution.

For more Information Visit Our WEB site at

http://www.abnh.com
or contact us at
1-800-966-ABNH
(1-800-966-2264)

ABNH
AMERICAN BANK NOTE HOLOGRAPHICS, INC.
399 Executive Blvd. Elmsford, N.Y. 10523 (914) 592-2355 Fax: (914) 592-3248

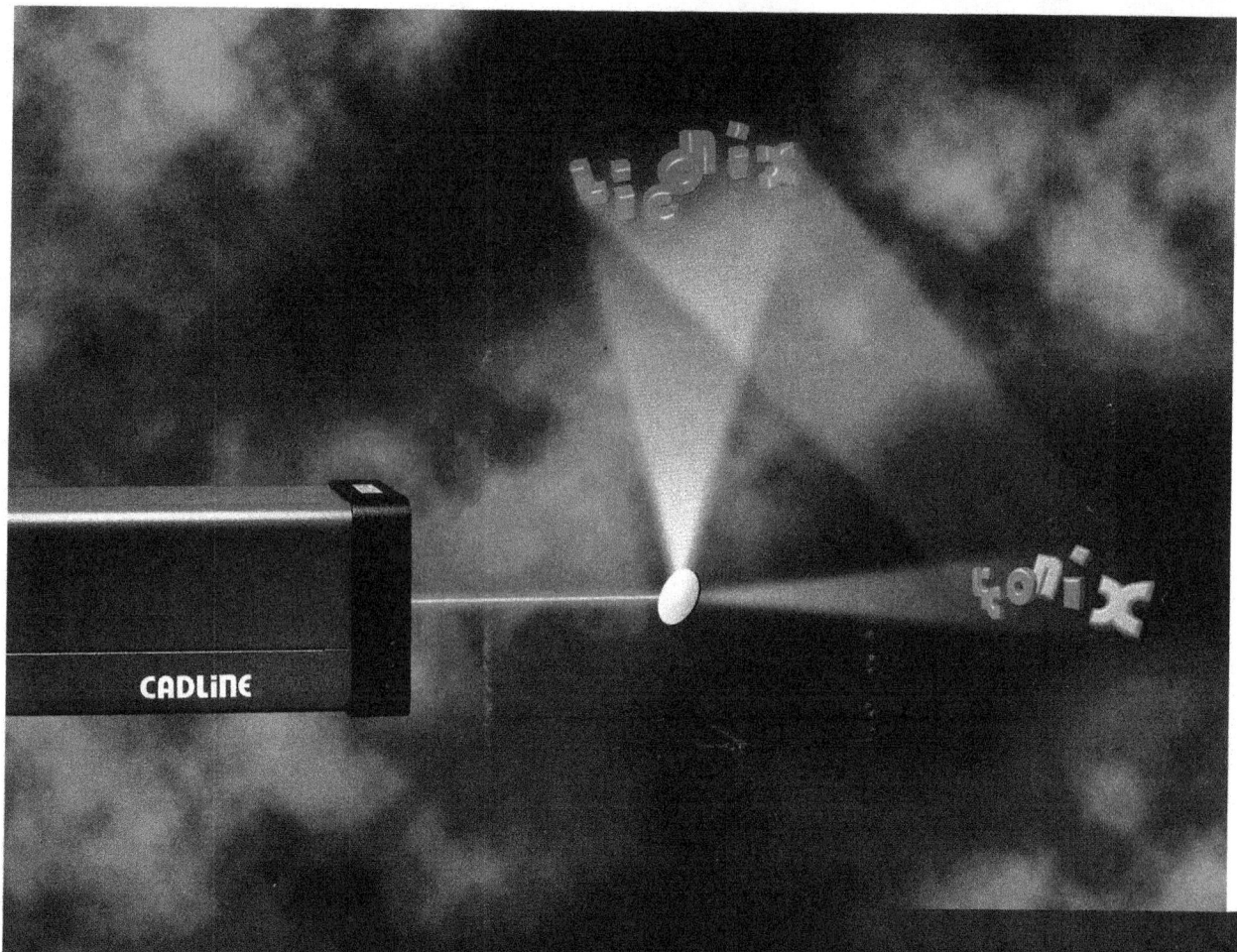

HOLOGRAPHY MARKETPLACE
COLLECTOR'S EDITIONS

This set of limited-edition books contains 50 rare and collectable holograms from the world's major hologram manufacturers as well as pertinent reference material and historically significant information. A must for your bookshelf. Limited Quantities.

192 pages. $19.95

160 pages. $20

192 pages. $20

174 pages. $20

220 pages. $20

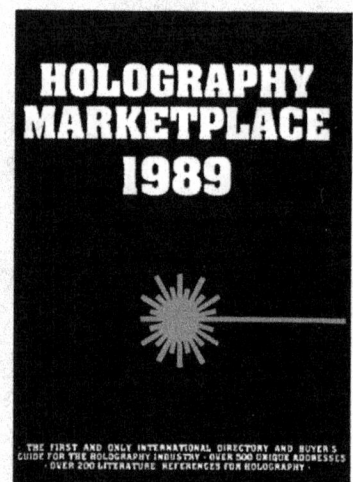

Sold out! 180 pages. Was $100

Special Offer – Editions 2,3,4,5 & 6 for $50!

Delivered to your door

All prices **include** shipping and handling:
in Continental USA...............................$ 65
to Alaska, Hawaii, Canada, Mexico..... $ 75
to all other overseas countries............$100
Books shipped promptly by UPS or Air Mail.

To order

Provide your name and the address where you want books sent. We accept payment by VISA, Mastercard, American Express (include credit card number and exp.date), or check (US$) payable to Ross Books.

Ross Books P.O. Box 4340 Berkeley, California 94704 Toll free in USA 800-367-0930
phone 510-841-2474   fax 510-841-2695   email sales@rossbooks.com

THE SOURCE OF QUALITY HOLOGRAMS

IHMA
INTERNATIONAL HOLOGRAM MANUFACTURERS ASSOCIATION

Holograms now play an essential part in the production of documents, displays, packaging, gifts and other products throughout the world

Obtaining your customised hologram from an IHMA member gives you many advantages

Their work benefits from the most advanced and highest quality production techniques

Members commit themselves to the Association's Code of Practice, offering their clients assured security and reliable business ethics

And only IHMA members can record your hologram on the unique Hologram Image Register – a worldwide database of holograms, which helps reduce the risk of fraudulent manufacture and copyright infringement of holograms.

For further information on the **IHMA IN EUROPE**	*For further information on the* **IHMA IN THE USA**
IHMA	IHMA
Runnymede Malthouse, Runnymede Road, Egham	PO Box 887
Surrey TW20 9BD England	Englewood CO 80151 USA
Phone: +44 (0)1784 497008 Fax: +44 (0)1784 497001	Phone: 800 741 6552
E-mail: Ian_Lancaster@CompuServe.Com	E-mail: ReconnUSA@aol.com

Photographs of full-color holograms produced by holographer Hans Bjelkhagen on Slavich's panchromatic silver-halide emulsion. Refer to Chapters 3 and 4 for more details.

A photograph of a full-color hologram produced by holographer Larry Lieberman on DuPont's panchromatic photopolymer film. Refer to Chapters 3 and 5 for more details.

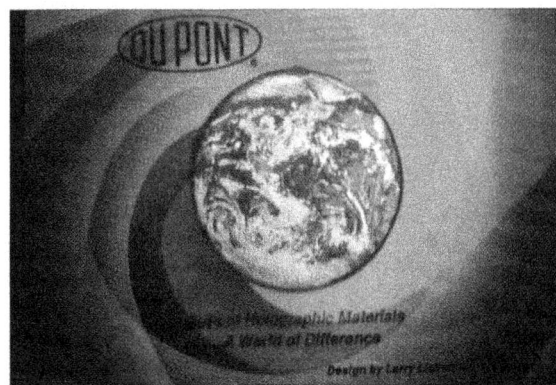

- TWO LAYERS
- 3 D IMAGE
- 600 DPI DOT MATRIX
- MICRO - TEXT
- 200 DPI DOT MATRIX

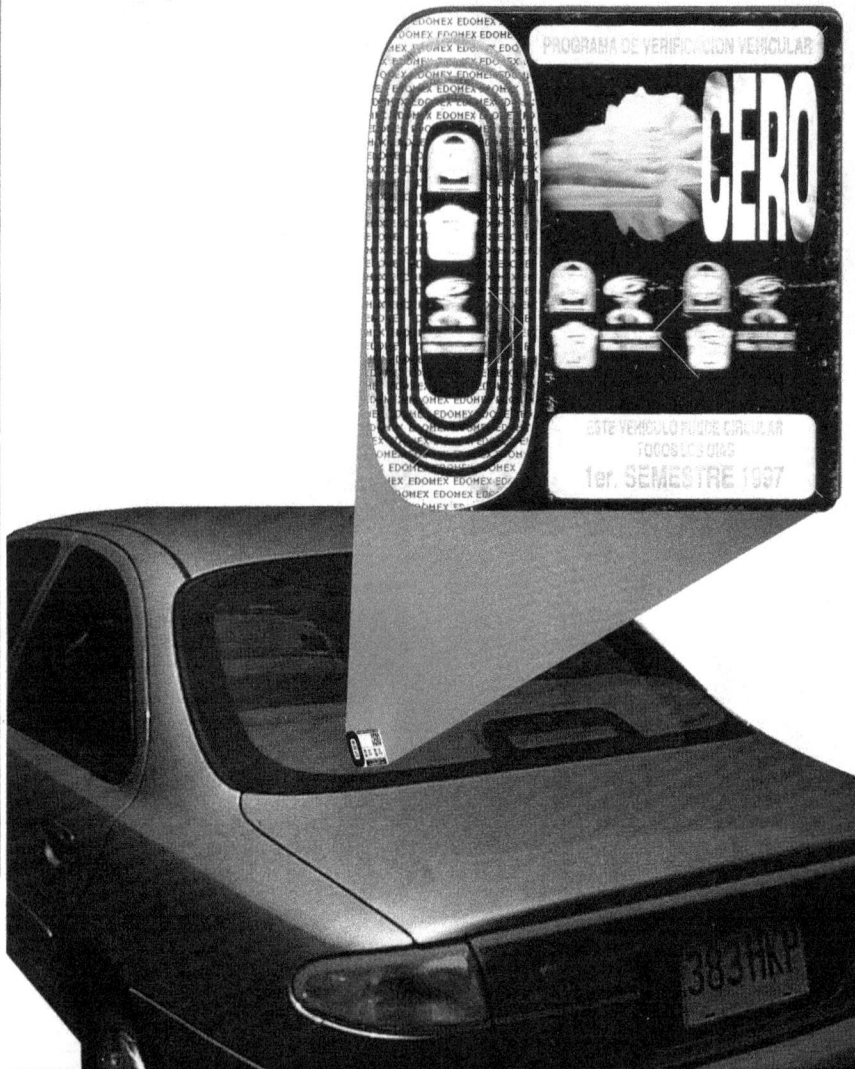

Embossed production, Die-cutting and kiss-cutting, Hot-melt, and hot-stamping adhesives, Computerized art department, Photoresist production, 2D/3D, 3D, Stereograms origination, and Holomatrix™.

Our embossed materials are: polyester, polypropylene, PVC, paper, and Hot-stamping foil. Our embossed widths are between 6" to 43". Among our specialties we can have added protection to your holograms adding bar codes, sequential numbers, etc.

HOLOGRAMAS S.A. DE C.V.

We are a group of vertically integrated companies, something that let us do every process in-house.

PINO 343, LOCAL 3, COL. ATLAMPA. C.P. 06450, MEXICO D.F., MEXICO. Email: holomex@holomex.com.mx
Home Page: http//www.holomex.com.mx DIALING CODE TO MEXICO CITY 525 + NUMBER
541 - 1791, 541 - 3413, 541 - 1696, 541 - 1506, 541 - 9046 FAX 547 - 4084, 541 - 3374.

of the developer and the only time the whole surface is exposed to air is when the development progress is being monitored. Develop to a density of at least 2.5.

b. Stop bath

> 1,000 cc water (deionized, if available)
>
> 5 g sodium hydrogen sulfate* crystals
>
> Rinse briefly in cold water and then bleach.

c. Bleach

> 700 cc water (deionized, if available)
>
> 35 g copper sulfate (pentahydrate)*
>
> 100 g potassium bromide
>
> 5 g sodium hydrogen sulfate crystals*
>
> Add water to 1,000 cc.

Bleach until the hologram is clear. Rinse in cold water. Normally, unless you are making reflection master holograms, this bleach should always be followed by an "anti-print out" bath such as:

d. Anti printout

> 700 cc water (deionized, if available)
>
> 3 g potassium dichromate*
>
> 6 g sodium hydrogen sulfate*
>
> Add water to 1000 cc.

Treat for about one minute with agitation.

To rinse off, just give a brief dip in a bath of deionized water free of any trace of developer. The brevity of the rinse is necessary to actually leave some of the anti-print- out solution within the body of the hologram. Please note that dichromate is a toxic material and treated unprotected holograms should not be handled by young children.

This "anti-printout" has the additional advantage that it will break down any residual dye present in the emulsion from sensitization. However, with the "pyrochrome" system (see below) that bleach will serve as a dye remover.

2. Pyrochrome Process (van Renesse)**
a. Developer

> Solution A
>
> 700 cc water (deionized)
>
> 15 g pyrogallol
>
> 5 g metol
>
> Add water to 1,000 cc.

> Solution B
>
> 700 cc water (deionized)
>
> 30 g sodium carbonate, anhydrous

> 7 g sodium hydroxide (may be replaced by
>
> 10 g potassium hydroxide)
>
> Add water to 1,000 cc.

Mix 1 part of solution A and 1 part of solution B immediately before use. The developer lasts for many hours if the "floating lid" (see above) trick is used. Develop to a density of about 2.0 (although a higher density might give a better result). This may take longer than you are used to.

After developing, the hologram should be well rinsed in tap water and then rinsed in deionized water before being placed in bleach bath.

b. Bleach

> 700 cc water (deionized!)
>
> 5 g potassium dichromate
>
> 15 g sodium hydrogen sulfate crystals
>
> Add water to 1000 cc.

Bleach until hologram is clear. Before washing with tap water the hologram should be rinsed thoroughly with deionized water. This is in order to prevent silver ions in solution precipitating in the emulsion with residual chloride ions in the tap water; this would lead to increased scatter in the final hologram.

This process is, in comparison to the rehalogenating bleach system, quite sensitive to overexposure. If your results with respect to diffraction efficiency are poor although developed density is high, please check this out.

* Note - HRT Holographic Recording Technologies is not liable for any damage caused by the use of these recipes. Please note that some of substances given in the formulae are toxic and/or corrosive. Please handle these substances according the safety regulations which apply to your country.

** Graham Saxby, Manual of Practical Holography, Focal Press, 1991

Manufacturer's Comments From Users

Some difficulties in the usage of BB-Plates for transmission holograms in conjunction with the ferric-EDTA bleach have been reported. Please use the copper sulfate bleach until further notice. Dated: 22nd January, 1997

A very useful remark comes from Mike Medora, England. The sensitivity can be increased remarkably by a pretreatment with a 3% TEA bath. Plates should be used within 24 hours after pretreatment. Dated: 12th September, 1997

Recommendations for Working With HRT Red Sensitive - BB-640
Plus a Developer-Bleach Combo for Most Silver-Halide Emulsions

by Jeffrey Murray (Holography Institute)

These may be the best plates ever made - very clear with no scatter, and they produce very bright, quality images. Here are a few procedures I have recently tested:

Recommendations

Plates should be presoaked, squeegeed, and dried before exposure, for much increased sensitivity.

1. Presoak-

Water and a few drops Photoflo is a minimal presoak solution. You can also add:

- Triethanolamine (sensitizes better, and swells the emulsion, giving an eventual color shift. Start at 2% solution, up to 20% for blue - be careful, the gooey thick solution is hard to control and the triethanolamine must not contaminate holograms at any other stage.)

- Developer will work as a sensitizer, perhaps by starting the process as soon as the exposure occurs. It gives no color shift.

2. Exposures -

Should be 3-4 mJ /sq.cm. (Example: a ten milli-watt HeNe laser, expanded to light a 4" x 5" plate, exposed 30-40 seconds).

Low power lasers (1 mW) should use very small plates.

3. Develop-

5 minutes using CWC2.

4. Rinse

5. Bleach-

5 minutes using PBU/amidol 6. Rinse

Note - These plates will also work well with Russian GP-@ colloidal development. Use more exposure time.

CWC2 Developer and PBU-Amidol Bleach

(also works for other types of silver-halide emulsions)

CWC2 Two-Part Developer:

PART A solution*	
Warmed distilled water	500 ml.
(Pyro)Catechol	10 grams
L-Ascorbic Acid (Vitamin C)	5 grams
Sodium Sulfite (anhydrous)	5 grams
Urea	30 grams

PART B solution*	
Warmed distilled water	500 ml.
Sodium Carbonate	30 grams

*Wait at least 30 minutes for chemical activation.

Part A is good for one month, Part B indefinitely

Mix equal parts. A & B to activate developer just a minute or two before use. Use just enough developer to cover one hologram. The mixed solution is active for 20 minutes. Discard it after one use to assure each hologram has optimum development.

Develop plate for at least FIVE minutes @ 68° F with constant agitation when using low powered lasers (also when using Slavich PFG-03M plates). Two minutes for Agfa plates.

Rinse in distilled water.

View a green safelight through the rinsed plate to judge density - some variation is OK.

Adjust exposure/developing time to achieve a final developed density of:

D 1.5-2: medium gray (for an unbleached transmission hologram)

D 2-3.5: very dark (reflection holograms).

D 4: appears opaque.

Do not use fixer if it will be bleached! (Reflection holograms are usually bleached.)

Notes on developed density - This stage is where you figure if exposure, ratio, gleam spots, beam centering, even illumination, and overall light levels and their recorded patterns are OK for the next shot as well, or need adjustment. After the plate is bleached clear, these clues are gone. Although dark, wet, and hard to see, observation of different gray levels is important, hopefully understanding what caused each visible pattern.

PBU-Amidol re-halogenating Bleach (Phillips-Bjelkhagen Ultimate):

Potassium Persulfate	10 grams
Sodium Bisulfate (or Citric Acid)	10 grams
Potassium Bromide	20 grams
Cupric Bromide	1 gram
Amidol (add last!)	1 gram

Mix one at a time, in sequence, into 500 ml warmed distilled water, then add another 500 ml distilled water to make 1 liter.

Wait at least 30 minutes for chemical activation.

Bleach unfixed plate for 3-4 minutes @ 68° F. until clear + 1 minute. Rehalogenating (and image brightening) continues after clearing.

Rinse, rinse, rinse in distilled water.

Air dry with a low-heat blower or drying cabinet for approximately 15 minutes - not too fast, not too slow.

An acetic acid rinse after bleaching may help reduce print-out (the emulsion will darken a bit after you run out in the daylight to see your image). I prefer to avoid intense sunlight until aged a few days.

Re-bleaching later will partially clear a darkened plate and give some immunity to further printout.

Bleach can be re-used a few times, and is usually good for two weeks - the red color will fade to clear, indicating exhaustion. Beware sediment as it ages - do NOT attempt to re-mix before each use! Decant and do not dump dregs out onto emulsion. Bleach will leave permanent purple stains on everything! Handle carefully.

Many thanks to Cooke and Ward, Hans Bjelkhagen, Nick Phillips and Ed Wesly for the many trials it took to allai!! the basic formulation.

Silver-Halide Plates from HOLDOR

by Wai-Min Liu (Control Optics) and Chuck Paxton (Photon Cantina)

Manufacturer's Specifications

Wavelengths: 6943, 6470, 6328, 5320, 5145, 4880, 4416.

Resolution: 3,000-10,000 lp/mm

Sensitivity: About 2.5 mj/cm^2

Emulsion thickness: 7-15 m

Storage time: > 1 year (room temp)

Size: 4"x5", 8"x10"and soon.

Processing: CWC2 Developer or Pyrogallol Developer

Applications: Transmission holograms, reflection holograms, embossed holograms and holographic interferometry.

We have been testing these new plates for the last couple of months. The following information should give anyone interested in using these plates a good starting point.

The Emulsion

The emulsion is available in red, green, blue, and panchromatic. We found the panchromatic to be the most sensitive. When compared to Agfa, we found the sensitivity less than that of Agfa 8E75 by a factor of about five. This means the stability of the set-up becomes very important to handle the long exposure times required. One thing we noticed is that the emulsion is very hard and holds up well to processing without the need to harden before processing, which is necessary with the Slavich plates.

Pre-sensitization

We found it important to pre-sensitize prior to exposure with a 2% solution of triethanolomine for two minutei When using a He-Ne laser this will give image playback in the yellow-green region. This also helps to reduce the exposure time by 50%.

Exposure

The laser we used is Spectra Physics 125 currently producing between 30-40 mW. Our set-up is single beam reflection type and our exposure time was 10 seconds using the panchromatic plates with a 2 minute pre-sensitizing step.

Processing

We tried several types of developers and bleaches and found the best to be a modified CW-C2 developer and PBU- quinol bleach. The developing time was between three to five minutes , which gave a plate density of about 70% dark. Bleaching time was two to three minutes.

Results

We found the image resolution to be excellent and image brightness to equal that of Agfa 8E75. We feel that this emulsion should be used for single beam or contact copy-type holograms. For split-beam holograms, I would recommend the use of a fringe locker to handle the long exposure times.

Processing Steps

1. Pre-sensitize emulsion with a 2% solution of tri-ethanolomine two minutes, then dry plate.
2. Expose plate.
3. Develop plate for three minutes (70% dark).
4. Rinse plate for three minutes.
5. Bleach plate for three minutes.
6. Rinse plate for five minutes.
7. Photoflo for one minute.
8. Dry plate.

MODIFIED CWC2 DEVELOPER	
Part A	
Distilled Water	400 ml
Catechol	10 g
Ascorbic acid	5 g
Sodium sulfite (anhydrous)	5 g
Add distilled water to make 500 ml	
Part B	
Distilled Water	400 ml
Sodium carbonate (anhydrous)	30 g
Add distilled water to make 500 ml	

Take A:B = 1:1 compounded for use.

PBU QUINOL BLEACH	
Distilled Water	750 ml
Cupric Bromide	1 g
Potassium Persulfate	10 g
Citric Acid	50 g
Potassium Bromide	20 g
Add distilled water to make 1 liter	

After the above chemicals have been mixed, add 2 g hydroquinone (quinol) for PBU-quinol bleach. The bleach Clffibemefro after six hours.

Comments

I am impressed with the improvement in the overall quality of the Chinese emulsion from the time we started working with it. Working with Dr. Wai-Min Liu at Control Optics, we were able to communicate problems as we came across them directly to the manufacturer. The first problem was the coating not covering all the way to the edge of the plate. This problem has been addressed. The next is the overall sensitivity. Dr. Liu and Prof. Weiben Yuan (HOLDOR), feel that the sensitivity of the red sensitive plates can be increased by another 25%. We plan to continue working with these plates to refine the process and update anyone interested.

KODAK Products for Holography

(reprinted with permission from Eastman Kodak Company)

Red Sensitive Holographic Emulsions

For hologram recordings made with Helium-Neon lasers (633 run) and Krypton lasers (647 nm).

1. KODAK High Speed Holographic Glass Plates, Type 131-01, Type 131-02

(To place orders and obtain current pricing and availability information, call 1-800-823-4474.)

131-01 (with back side antihalation; dye and polymer backing)

Size	Catalog #
2" x 2"	#1723485
2.5" x 3.5"	#1729656
4" x 5"	#1233139
8" x 10"	#1297431(possibly n/a)

131-02 (unbacked)

Size	Catalog #
2" x 2"	#1240308
2.5" x 3.5"	#1729656
4" x 5"	#1231547
8" x 10"	#1241082

2. KODAK High Speed Holographic Film SO-253

(To place orders and obtain current pricing and availability information, call 1-800-822-9442.)

SO 253 - (backed sheet film)

Emulsion coated 9 microns on a clear ESTARTMFilm Base. A dyed gelatin pelloid on the base side provides antihalation protection.

Size	Catalog #
4" x5"	#1772672

Note - Rolls of film in larger sizes are available as special order item. Contact your KODAK distributor for availability and pricing.

General Characteristics of Red Sensitive Holographic Emulsions

High speed, high contrast.

Resolving power - 1250 lines + per mm. Microfine grain size

Energy required - approximately 5-10 ergs per sq. cm at 633 nm.

Negative working.

Glass plate is 0.04 inch thickness. SO-253 polyester film base is 0.004 inch thickness.

Processing - Process in KODAK Developer D-19™.

Additional information

This film provides extraordinary speed when exposed with helium neon or krypton lasers. At the same time, its micro-fine grain structure and other emulsion characteristics combine to yield high diffraction efficiency and low noise upon reconstruction of holograms recorded at spatial frequencies as high as 1500 cycles / run. It is recommended primarily for holographic interferometry and micrography, and it is particularly useful for general holographic procedures with low power HeNe lasers.

Recommended Processing Technique

• Developing - When exposed with HeNe lasers or the red line from Krypton lasers and processed for six minutes in D-19 developer at 68° F (20° C), exposures of 5 ergs/cm sq. should be sufficient to achieve maximum reconstruction brightness. (As with other holographic materials, an increase in development time, to 8-10 minutes, will result in higher speed and diffraction efficiency at the expense of reduced exposure latitude and playback signal-to-noise ratio.)

Following development for the indicated times, processing is continued with the following steps, all at 65 - 70° F (18.5 - 21° C):

• **Rinse** in running water or KODAK Indicator Stop Bath or KODAK Stop Bath SB-1™ with agitation for 10-30 seconds.

• **Fix** using KODAK Fixer or KODAK Fixer F-5™ with agitation for 5-10 minutes.

• **Wash** with moderate agitation for one minute.

• **Rinse** in a solution of KODAK Hypo Clearing Agent™ with agitation for 4 minutes. (This rinse contributes to washing equivalent to the criterion for archival keeping as described in ANSI Standard PH 4.8 - 1971.)

• **Wash** with moderate agitation for three minutes.

• **Rinse** in a solution of three parts methanol and one

part water with agitation for five minutes. (The methanol rinse is required to remove a high level of residual sensitizing dye from the emulsion. The dye is distinctly blue in appearance and would greatly reduce reconstruction brightness when operating with a red-emitting laser.).

• **Wash** with moderate agitation for five minutes. Use a wash water flow rate sufficient for one change of water every five minutes.

• **Dry** in a dust free atmosphere. Drying marks can be minimized by treating the film in KODAK Photo-Flo Solution (prepared as directed on the bottle label) after washing. The use of Photo-Flo solution will promote uniform drying of film surfaces. For best results, dry film slowly at room temperature.

Latent Image Decay - Like most films with extremely fine grains, these emulsions exhibit significant latent image fading during the hours just following exposure. It is good practice in determining an optimum exposure level for a given holographic setup to process as soon after exposure as possible, provided that the elapsed time can be maintained for all subsequent operations with the same setup.

Storage - Unexposed film should be stored in a cool place (700 F or lower) in the original sealed package. Prevent condensation, which may result in spotting, ferrotyping, or sticking. In addition, thermal expansion during exposure will result in smearing of holographic fringes.

Safelight - Total darkness is recommended when handling this film. Green safelights at very subdued levels may be tolerable.

Note - This film can also be exposed efficiently with Helium-Cadmium (442 run), Argon (515 run), and frequency doubled Nd:Yag (532 run) lasers. For exposures in the blue or green, 25 - 40 ergs/cm² will be required. Some reduction in holographic performance is to be expected at progressively shorter wavelengths as a result of Rayleigh scattering in the emulsion during exposure, but this is characteristic of all silver-halide holographic materials.

Blue / Green Sensitive Holographic Emulsions

Suitable for hologram recordings with Nd: Yag laser (532 nrn), or other blue lasers below 550 ~,.

1. KODAK High Plate, Type A

(To place orders and obtain current pricing and availability information, call 1-800-823-4474.)

Type 1A

Size	Catalog #
12" x 16"	# 8943904

2. KODAK High Resolution Film SO-343

(To place orders and obtain current pricing and availability information, call 1-800-248-3022.)

SO-343

Size	Catalog
4" x 5"	# 1929132
8" x 10"	# 1929074
20" x 24"	# 1666965

General Characteristics of Blue / Green Sensitive Holographic Emulsions

Extremely slow speed. Extremely high contrast.

Resolving power of 2,000 per mm.

Energy requirement approximately 1,000 ergs per cm² at 532 nm.

Negative working.

Glass plate thickness 0.04-0.06 inch. Film thickness is 0.007 inch.

Processing - Process in KODAK Developer D-19 or KODAK HPR Developer.

Additional information

"Microfine grain structure. Thin emulsion. Frequently used for pattern generation of fine reticles, preparation of printed circuit board artwork, television shadow masks and other masks for microelectronics."

Safelight - red

For further information, contact the KODAK Information Center ph: 1-800-242-2424 ext. 19.

5

Other Recording Materials

This chapter will discuss the following holographic recording materials:

- *photopolymer films;*
- *dichromate gelatin;*
- *photoresist (solutions and pre-coated plates);*
- *and photo-thermoplastic films.*

If you are reading this chapter you are probably a practicing holographer or someone interested in investigating recording materials other than silver-halide emulsions. In either case, you are likely to know that the primary supplier of silver-halide emulsions (Agfa) has recently ceased production. As the last chapter states, other suppliers of silver-halide emulsions have begun to increase their production and distribution in order to meet the existing demand for these materials.

However, many holographers and hologram production facilities are unaffected by this change of suppliers as they rarely use these silver-halide emulsions to record the type of holograms that they sell. Many users prefer the visual attributes or require the physical properties that holograms made on other recording materials offer. In fact, it is probably safe to say that most high-volume industrial and commercial hologram manufacturers do not use silver-halide emulsions at all, unless their replication systems copy silver-halide masters.

For instance, clients that need relatively deep-image, highly-detailed, easily viewable display/holograms for security, advertising, signage, or giftware products are increasingly choosing photopolymer film to make their holograms on. Industrial users alsofind them ideal for HOE applications. They combine the imaging qualities of silver-halide glass plates and films with the high-speed, high-volume capability of embossed hologram production.

Over the past decade, photopolymer recording materials have become quite cost effective to use. Per square inch, the unexposed photopolymer cost much less than silver-based films. In addition, with the proper laser it is possible to do both mastering and replication on photopolymer, which can lower manufacturing expenses. Most importantly, since exposure, processing and finishing procedures can all be automated using high-speed equipment, replication costs are significantly reduced - which should further increase the appeal of these holograms to all end-users.

Another recording material, dichromate gelatin (DCG), is undergoing a slight renaissance, as holographers all over the world have begun to research photosensitive materials that they can make themselves using easily obtainable ingredients, thus alleviating all potential supply problems. Although the giftware market has slowed drastically since the days when a considerable number of retailers carried jewelry and w.atches made with DCG holograms, this material could become more widely used by artists and other display holographers equipped with the proper lasers. We encourage any holographers with the patience and know-how to experiment with DCG.

Due to the fact that embossed holography requires a surface relief hologram to replicate, photoresist is now probably the most widely used recording/mastering material. As many readers know, the use of embossed holograms has skyrocketed since the packaging and security industries have integrated the technology into their production lines. Although most hologram manufacturing plants use the same photoresist formulations that they always did, commercial users have pushed suppliers to provide larger precoated plates, while still retaining quality. Some facilities have even begun to coat their own plates in custom sizes. Instructions for doing this are included in this chapter.

It is important to mention photo-thermoplastic recording materials in this chapter, too. These films, along with the "instant-hologram" cameras that employ them, offer benefits that other recording materials and systems can't. Industrial holographers doing Non-Destructive Testing might be most interested in following developments in this field. However, educators, researchers and even embossed hologram manufacturers could find photo-thermoplastic systems extremely valuable in their work as well.

Photopolymer Recording Materials

PhotoPolymer is a plastic compound that reacts to light. It is formulated so that certain wavelengths of light create specific molecular changes in the material, thereby making it useful as an optical recording medium. These changes in the chemical and physical structure of the film cause changes in the refractive index of the material, which is a useful method for recording a holographic interference pattern.

Chemical processing is not required to produce an image, as is the case with conventional silver-halide materials. Instead, holograms are created "real time" in the photopolymer emulsion during the primary exposure to laser light and "fixed" using a subsequent exposure to ultraviolet light. A simple baking process follows.

Photopolymers have been formulated 'to produce crisp, bright holographic images. Diffraction efficiencies greater than 95% (for transmission) and to 99% (for reflection) holograms have been measured.[1,2] This means these holograms can be viewed under the less-than-ideal-lighting conditions that are present in many commercial environments, which is of paramount concern to many potential users. Under proper illumination, photopolymer materialsare capable of displaying high-quality images with great image depth. Photopolymer compounds can also be used for related industrial applications, such as optical processing and optical storage.

Photopolymer holograms do cost more than embossed. holograms to mass produce due to the fact that the former process utilizes photosensitive materials and optical replication rather than foils and mechanical stamping presses. At present, typical production runs cost 5 to 15 cents/sq. inch for finished product. However, no Gther holographic technology is more cost effective for customers that require high impact three dimensional images that can be seen under "normal" lighting conditions.

Advantages

There are several advantages of using photopolymer film to record and reproduce display holograms:

• The material is well suited for a variety of commercial applications. Being a flexible plastic film, the photopolymer material is more easily integrated into existing manufacturing, finishing, and distribution systems than other types of reflection holograms. Most notably, photopolymers are less expensive to manufacture than silver-halide emulsions and much easier to handle than dichromates. The material itself is lightweight, durable and has a long shelf life. It is convenient to store and ship. Finished holograms can easily be cut to shape and attached to other paper or plastic products.

• You can record high-quality, fully-dimensional deep-image holograms. Photopolymer films that are expressly designed for display holography record "volume phase" reflection holograms (i.e., the image is recorded throughout the entire thickness of the emulsion). This results in a high density of information being recorded, which in turn allows for fully-dimensional, highly detailed images to be captured. Images with a very wide viewing angle (up to 180 degrees) and considerable depth (up to several feet) have been recorded and reproduced.

In contrast, they're more widely-used embossed hologram manufacturing process creates microscopic surface relief patterns on a plastic film or on a sheet of foil. The process results in relatively shallow images measured in inches (or less in most cases). In addition, the techniques used to make most embossed holograms eliminate vertical parallax and limit horizontal viewing angle to approximately 60 degrees.

• The overall color tone, "weight" and "look" of images produced on photopolymer film are much more realistic than those produced by embossing. Photopolymer holograms look more like highly detailed three-dimensional black and white photographs. Most have a green, gold or orange color tone and a solid black background, instead of the ever-changing prismatic colors and reflective metallic background typical of embossed holograms. Multi-color and true color photopolymer holograms may be available in the year ahead.

The actual color of the holographic image will be similar to that of the laser light used to make the recording. Typically an argon ion laser is used to expose the photopolymer film (green 514.5 nm) which results in a green tone image. Some hologram reproduction facilities are experimenting with the new Diode Pumped Solid State lasers which also emit green (532 nm). Methods have been developed to further adjust color. The photopolymer emulsion itself is transparent; however, it is usually attached to a dark backing material for higher contrast.

1. Stevenson, et. al. SPIE. Vol. 233. p 60-70.

2. Gambogi, et. al. SPIE. Vol. 1555. p256-267.

Two Major Manufacturers

Currently, there are only two major manufacturers of photopolymer recording materials for the holography industry: Polaroid Corporation and E.J. duPont de Nemours and Company (DuPont). Sofar; the companies have employed contrasting marketing strategies regarding their hologram products and related technologies. However, both mainly deal with high volume users, rather than casual holographers. The following articles will profile these two important suppliers of photopolymer holograms and photopolymer recording materials. (Refer to the Polaroid ad insert and the Krystal Holographics ad insert in this book to see actual holograms produced on these materials).

Polaroid Photopolymer Products for Holography

Polaroid has been involved with the research and development of holographic imaging for over 30 years. Using its proprietary photopolymer material DMP-128, Polaroid was the first company to mass produce photopolymer holograms for commercial applications.

(Editor's note - See the First Edition of Holography MarketPlace for a sample of one of the world's first publicly available photopolymer holograms; a 4 " x 4 " hologram which depicts the interior of a Polaroid camera. A true collector's item! Editions Three and Four also have very noteworthy Polaroid photopolymer holograms on their front covers.)

However, in an effort to maintain the security of its authentication customers, Polaroid's (unexposed) recording material is not for sale. Instead, the company offers a full range of custom-design, modeling, origination and reproduction services for its customers. Mass production is done "in-house" at Polaroid. Clients receive finished holograms.

Mirage™ Display Holograms

Polaroid continues to produce its Mirage™ line of display holograms for large volume commercial applications including product security/authentication, advertising premiums, stickers, trading cards, phone cards and wall decor. (See the Polaroid advertising insert for a sample of a photopolymer hologram produced for a phone card). Sizes range up to 9 inches by 14 inches. Holograms can be converted in accordance with the client's needs. Some stock images may be available.

The company, in conjunction with Red Beam and Holos Gallery, was the first to produce relatively large (4 inch by S inch) display holograms on photopolymer for the giftware market [in 1989], which popularized the material and set quality standards. Although Polaroid only offers monochromatic image reproduction, the company experimented with multi-color photopolymers at one time.

Imagix™ Holographic Reflectors

Polaroid's industrial division continues to develop applications for its Imagix™ Holographic Reflectors. These Holographic Optical Elements (HOEs) are specially designed to improve the brightness and contrast of reflective and transflective LCDs by concentrating the emerging light from an LCD screen at a narrow viewing angle, while simultaneously redirecting glare. The company claims screen brightness is increased by two or three times using their product. Manufacturers of Personal Desktop Assistants (PDAs), pagers and watches (such as Motorola and Timex) have already integrated Polaroid's Imagix holograms into their products. Additional HOE applications using Polaroid's proprietary materials and processes are expected.

For further information, contact: Polaroid, Holographic Products Division. Phone 800-237-5519.

DuPont Photopolymer Products for Holography

DuPont manufactures photopolymer recording films and replication equipment which it sells directly to authorized replication facilities. Currently there is only a small number of such operations worldwide; however, they are capable of high volume production (for instance, Doug Miller of Krystal Holographics International reports production capacity at his facility of over 1 million units a day), DuPont continues to provide select holographers materials for research and development work.

DuPont's line of holographic photopolymer film features OmniDex®706, a blue-green reflection material, currently sold in SOOfoot by 12 inch rolls. R&D sample films are also available: HRF-700 (reflection), HRF-600 (transmission) and HRF-150 (transmission).

In addition, DuPont has developed a full color photopolymer film which is sensitive to multiple wavelengths. This panchromatic film can be exposed with a white light laser or a multi-laser setup to create multi-color and/or realistically colored images. See the picture in the "color page insert" section of this book for a photograph of a full color photopolymer image produced in 1997 by holographer Larry Lieberman in conjunction with DuPont. As you can see, all colors are well reproduced. The material could certainly have many applications in the security, packaging, advertising and giftware industries.

[Editor's note: See the article in this book by Hans Bjelkhagen (Chapter3, page44) for an overview of color reflection holography.]

Ease of Use

One of the attributes of DuPont's materials which makes them so appealing to commercial holographers is the ease of post-exposure processing; i.e., the steps which "fix" an image to make it permanent. Unlike conventional silverhalide emulsions that require complex chemical reactions to occur during the developing process, DuPont's photopolymer materials are "developed" using ultraviolet light and heat.

Exposure to ultraviolet light fixes the image, then heat processing (in a forced air convection oven) increases the brightness of the image. This "dry" developing technique lends itself to automated mass production. Holograms made on silver-halide materials are much more cumbersome to replicate, which has severely impeded their widespread application.

OmniDex® 706 - A Detailed Look

OmniDex® 706 Holographic Recording Film is a photopolymer film that records volume-phase reflection holograms. It can record H1 holograms directly, or it can record H2- holograms by contact-copying from a master hologram. Masters produced on silver halide materials, dichromate or even on another piece of photopolymer have been successfully copied.

The 706 film is comprised of three layers: a Mylar®polyester film base, a middle layer of photosensitive photopolymer and a polyvinyl chloride (PVC) cover sheet. (See table 5.1 and figure 5.1.) The middle layer is a mix of polymeric binders, monomers, an initiator system, plasticizers and dyes sensitive to visible light. Photopolymerization and diffusion of the monomers is responsible for image formation.

The product is sold with a matching amount of OmniDex® CTF-7S, a color tuning film which is laminated to the recording material during the processing procedure to enhance image color and brightness. The color tuning film is also comprised of three layers.

Manufacturer's Specs

Film performance is characterized by the sensitivity of the unexposed film (absorption and film speed) and the brightness of the resulting holographic image (reflection efficiency, bandwidth and integrated efficiency). The brightness of the image is dependent on the magnitude of the variation of refractive index of the periodic microfeatures recorded in the film (i.e., index modulation).

Photosensitivity

As shown in figure S.2, OmniDex®706 film is sensitive to blue-green lines of an argon-ion laser (458-S28 run). Upon processing the dye absorption diminishes as shown by the dotted line.

Film Speed

Film speed is defined as the minimum amount of laser exposure necessary to produce a saturated image in terms of its brightness. Experimentally, film speed is determined by making a series of holographic exposures at a fixed laser intensity with various exposure times. The exposure energy (ml/cm²) is simply calculated by multiplying laser intensity (mW/cm²) by exposure time (sec.). Film speed curves are shown in figure 5.3. Brightness of the hologram initially increases with exposure energy of about 25 ml/ cm².

Mass Production Process

High volume reproduction utilizes a master hologram, a hologram copy setup, a roll of OmniDex®706, a roll of OmniDex®CTF-7S, an ultraviolet source, an oven, lamination machinery and transport equipment. Here is the basic procedure used to mass produce these photopolymer holograms:

1. Contact copies are recorded from a master hologram in a step-and-repeat process. DuPont's replicating system employs a web transport mechanism in conjunction with a laser and optics. The master hologram is loaded into a frame and OmniDex®706 film is contacted to it (this procedure calls for the direct lamination of the film onto a glass plate or another piece of photopolymer. Contact DuPont for lamination instructions).

2. A flood or scanning laser beam copies the image. Laser exposure of 40-S0 ml/cm² at 514.5 run is recommended.

OMNIDEX® 706 FILM STRUCTURE		
Layer	Material	Thickness
Cover sheet	PVC	60.9 μm
Photopolymer	compound	20.0μm
Base Sheet	200D Mylar®	50.8μm
OMNIDEX® CTF-75 FILM STRUCTURE		
Layer	Material	Thickness
Cover Sheet	Polypropylene	17.8μm
Photopolymer		24.6 μm
Base Sheet	200D Mylar®	50.8μm

Table 5.1. OmniDex®706. CTF- 75 film structure

3. The film is advanced another frame and the process is

4. Ultraviolet exposure is done in line (100 ml/cm² at 300- 366 nm).

5. A roll of exposed film and a roll of color tuning film are loaded on a DuPont OmniDex®laminator. The machine removes the cover sheets from the film and the color tuning film while laminating them together at 6 ft.lmin. (It is very important that the twofilms laminated together are free of tension mismatch and wrinkles. Also, removing the cover sheet geuerates a great deal of static electricity. This causes any suspended particulate in the air to be attracted to the film. This is one of the main causes of defective product. To avoid excessive wastage, manufacturing must be done in a dust free "clean" room).

6. The roll is fed through a DuPont OmniDex® scroll oven at 1400 C for eight minutes. Both the temperature and the time used in the baking processing influences the color and the brightness of the hologram. If color tuning film is not used, heating the hologram is still recommended to increase brightness. However, it is advisable to cover the bare photopolymer with a sheet of protective material such as Mylar (not PVC) so the film is not damaged in the

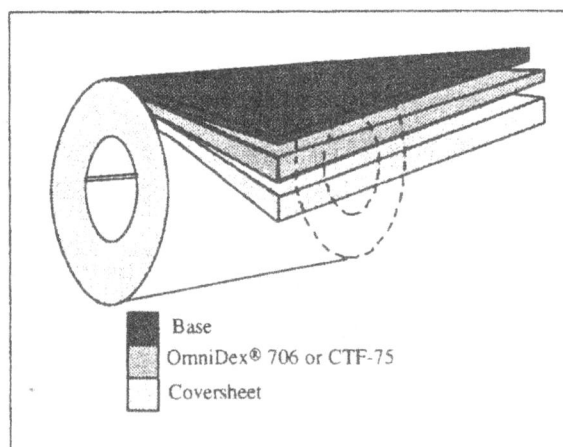

Base
OmniDex® 706 or CTF-75
Coversheet

Figure 5.1 OmniDex® 706 , CTF-75 Film structure

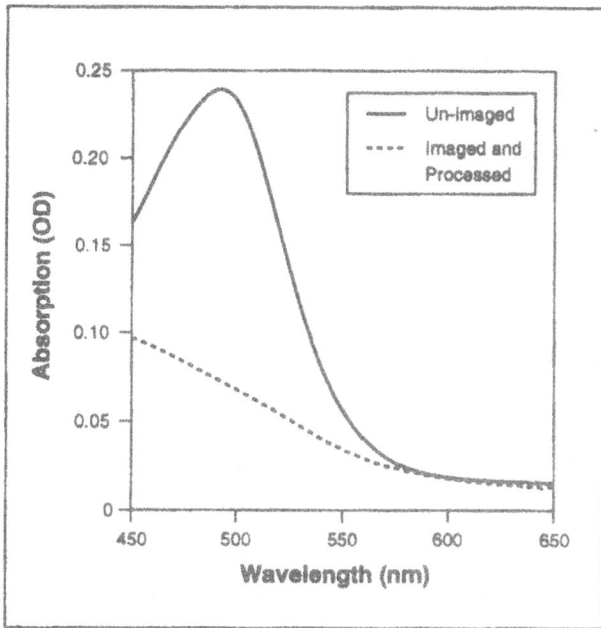

Figure 5.2. Absorbtion spectra of OmniDex® 706film
(dolled line UV cured and baked).

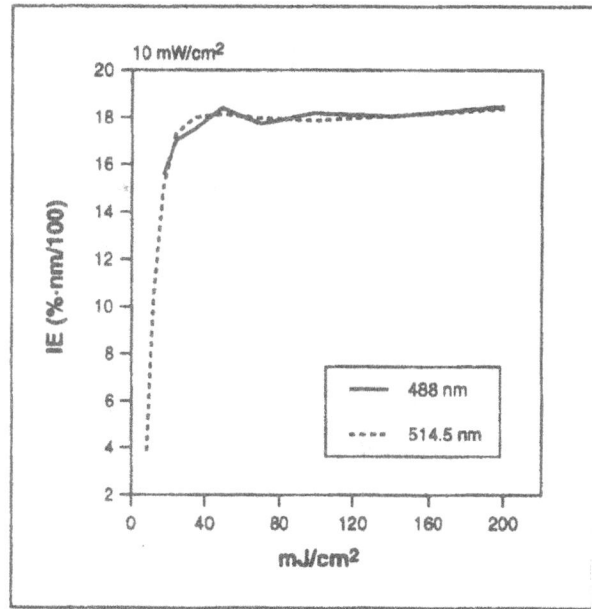

Figure 5.3. Film speed curves of OmniDex 706film
at 514.5 nm and 488 nm.

oven (as it is softened by the heat) or exposed to air (which would cause a blue shift).

7. The finished roll still has polyester base sheets attached (they were laminated face to face) that can also be removed. Different adhesives materials or protective coatings can be applied depending on the requirements of the user, followed by die-cutting or trimming.

Those interested in replication should contact DuPont. Some replication facilities are mass producing photopolymer with custom-built equipment, although the basic process used is the same as the one DuPont recommends. Company spokespersons are available for consultation.

For further information contact: DuPont Holographic Materials. Ph: 1-800-542-5467.

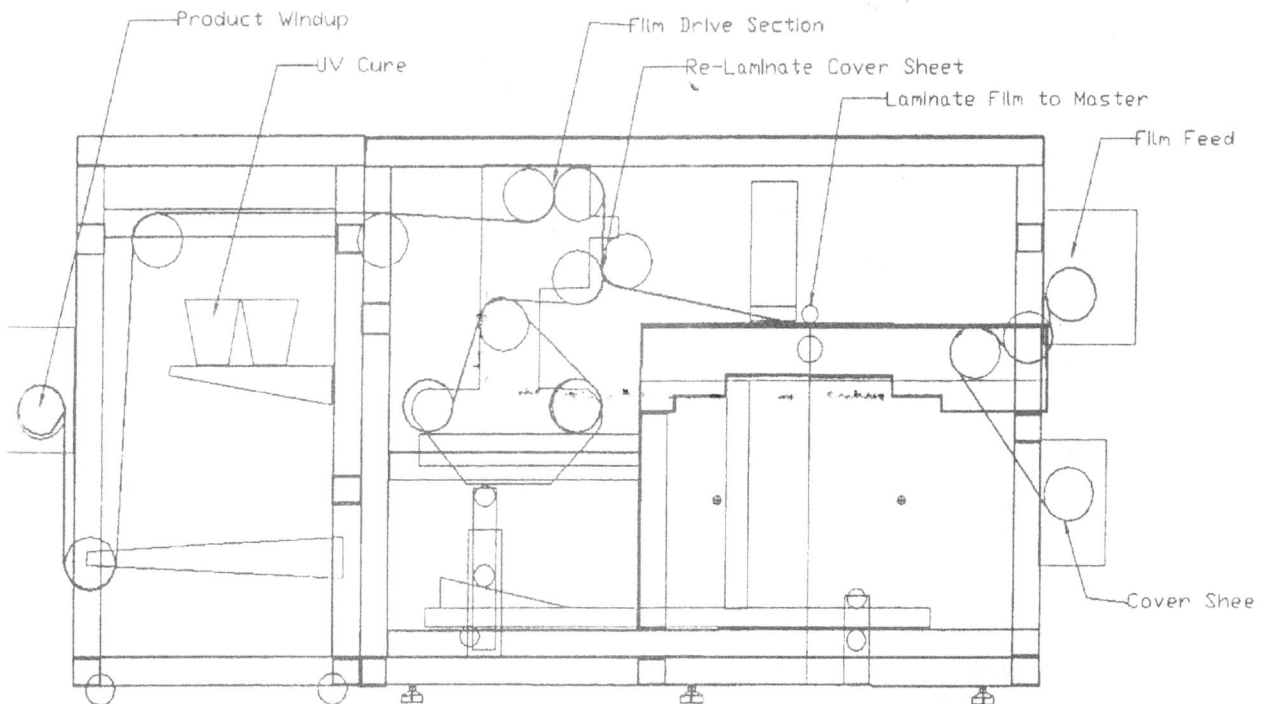

Figure 5.4. Schematic of DuPont's Direct Lamination Replicator Machine.

Dichromated Gelatin (DeG)

(information provided by Lasart Ltd. and Holocrafts)

Dichromated Gelatin (DCG) has the highest index of refraction of any emulsion used in holography. Therefore, holograms recorded on this emulsion create the brightest and most easily viewable images under a variety of lighting conditions. This makes it an ideal material to use for commercial displays, art and giftware. In addition, DCG produces little scatter in blue light, making it valuable material to use when manufacturing precision optical components, such as HOEs.

Making DCG Plates

This emulsion consists of Ammonium Dichromate or Potassium Dichromate, gelatin and water. Dichromates are available through chemical supply houses. Gelatin is easily obtained from gelatin manufacturers, by the barrel. Recording plates can be made by coating a piece of glass with a uniform thickness of the liquid emulsion using standard application methods, such as spin coating. To obtain the best results, this should take place in a dust free environment.

DCG is one of the easiest emulsions to work with and high quality holograms can be consistently-produced manufacturing variables are identified and controlled. The major variables to be aware of are the concentration of dichromate used in the emulsion, the "hardness" of the gelatin used in the emulsion, and the temperature and amount of humidity that the emulsion experiences once it is coated onto the recording plate.

Pre-coated DeG Plates

Russian emulsion maker Slavich lists a dichromated gelatin coated plate PFG-04 in its latest Product catalog. The company literature states that these plates "are designed to record holograms in contrary beams by the Denisyuk method using continuous laser emission in the blue and green spectrum (for instance using helium-cadmium, argon or neodymium lasers) PFG-04 keeps its holographic properties [up to] half a year before exposure. Such long [shelf] life is achieved due to special manufacturing methods." Additional information can be obtained from your local Slavich distributor. *

Mastering and Processing

Mastering setups depend on the size and complexity of the particular job. A single beam "Denisyuk" type of setup is most commonly employed when recording small objects, such as miniature models. Due to the ease of preparing a shot, mastering charges for dichromate production runs are often comparable to, or even lower than, mastering charges for other types of holograms.

The laser power necessary to expose DCG emulsions varies according to the formula used, but a rule of thumb is that the more dichromate in the emulsion, the shorter the exposure. DCG is blue green sensitive -- the shorter the wavelength, the more sensitive the emulsion becomes. Most holographers working with DCG use 5 watt Argon lasers, but small holograms can be made with smaller Argon (40 mW) or Helium Cadmium lasers.

Color control is somewhat limited in DCG since exposures are made using the shorter wavelengths of light. Some holographers shoot wide-band, which results in a white or silver tone image. Others shoot narrowband, which often produces a gold ad,bronze colored image. However, some laboratories have;perfected multi-color production techniques by selecting certain wavelengths and painting the subject matter to match.

Processing is quite simple - starting with a fixer, then a water rinse, then drying with isopropyl alcohol. After processing, one must isolate the emulsion from atmospheric moisture and direct contact with water. This is done by laminating another piece of glass over the emulsion and sealing the sandwich with an appropriate glue. Otherwise, the emulsion will dissolve and the holographic image will disappear.

Mass Production

It is very cost effective to produce short runs of DCGs. Large runs are more expensive due to the amount of time consuming hand labor used throughout the process, especially in the sealing and finishing stages. Many production facilities mass produce DCG holograms by repeatedly "remastering" each shot using the original model or subject matter. Higher quality results can usually be obtained by copying from a master hologram.

The most popular finished product is glass discs (they can be more easily sealed) which are used as watchfaces and as jewelry items. Other companies produce ready-to-frame plates for wall decor and executive gifts. In the past, manufacturers have attempted to mass produce DCGs by laminating them in plastic, but holograms sealed in this way tend to fade quickly.

Other Considerations

The major drawback to using DCGs for a wider range of commercial applications is the fact that they are thick and fragile, since they are sealed in glass. This usually makes them impractical to use in product packaging and publishing. DCG holograms display more image depth than embossed holograms, but less than those produced on silver- halide.

Editor's request: We are unaware of any holographers using pre-coated DCG plates in their work. If you have successfully used these products, please notify us.

Photoresist Recording Materials

Most commercial holographic applications utilize embossed holograms. A crucial step in the production of embossed holograms is transforming the microscopic optical recording into a useful production tool. To accomplish this, holographers record their images onto a high-resolution photosensitive material called photoresist. Once an image is recorded onto a photoresist emulsion, it can be then be processed in a manner suitable for mechanical mass replication.

Although most holographers shooting holograms on photoresist materials prefer to use pre-coated, ready-made recording plates, some holographers might want to (or need to) coat and process their own plates. The following article outlines the procedure for making plates "from scratch" using Shipley photoresist emulsion. Shipley is the main manufacturer of these emulsions and sells its product in liquid form mainly to high volume users. Although it is mostly used by the microelectronic and semiconductor industries, their photosensitive formula has proven suitable for holographic recordings.

After the aforementioned information from Shipley, we will provide a description of the pre-made products available from Towne Technologies - a major U.S. manufacturer and distributor of photoresist recording plates. Towne is not the world's only supplier of these products; however, it has extensive experience serving the holographic industry and its products have been used successfully by many embossed hologram production facilities.

Shipley Products for Holography

(reprinted with permission of Shipley Co. Inc.)

Coating Your Own Plates

The following instructions cover the use of MICROPOS-IT® S1800 Series Photo Resist for holographers interested in coating and processing their own plates. Microposit S1800 Series Photo Resist is a positive working photoresist system optimized to satisfy industry requirements in advanced optical lithography' and related holographic applications. Microposit S1800 Series Photo Resist is a replacement for 1300/1400 type resists providing an alternative to cellosolve acetate as the casting solvent.

Step 1 - Dehydration Bake to Prime Substrate

To obtain maximum process reliability, bake all substrates immediately prior to coating at 200°C for 30 minutes. Cool to 18°-25°C ambient before coating.

For maximum resist adhesion to all semiconductor surfaces, vapor phase priming with Microposit Primer is recommended. For liquid phase priming use Microposit Primer Type P.

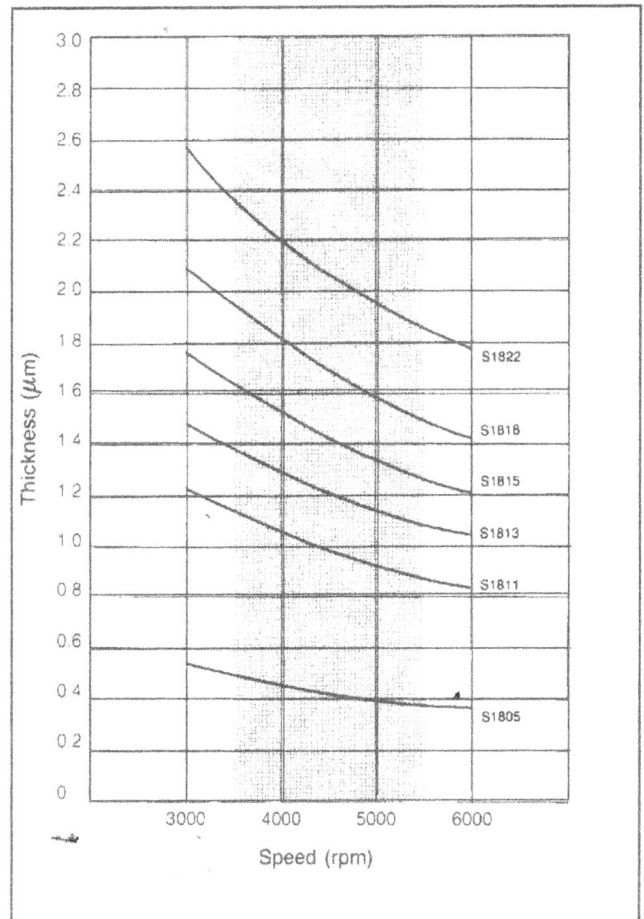

Figure 5.5 is a graph showing coating thickness for Sl 800 products from 3000 to 600 rpm. The shaded area is optimum spin range.

Step 2 - Spin Coat

Microposit S1800 Series Photo Resist is a resist designed to produce low-defect coatings. The improved coating characteristics include enhancements to thickness uniformity for optimized critical dimension control across a wafer as well as wafer-to-wafer.

Select the appropriate S1800 Series Photo Resist to give the optimum coating thickness at the optimum spin speed. The chart above (figure 1) shows typical coating thickness vs. spin speed for the standard Microposit S1800 Series Photo Resists. Spin speeds between 3500 and 5500 rpm are recommended for maximum coating uniformity.

Use the following parameters to obtain maximum resist coating uniformity:

Dispense -	Static.
Spread static -	2 seconds recommended.
Or spread dynamic -	500 rpm, 2 seconds maximum.
Ramp -	Maximum acceleration.

Spin-   3,500-5,500 rpm.

Spin time -   25 seconds minimum,

Microposit S1800 Series Photo Resists are manufactured to give reproducible coating thicknesses. The nominal thickness your process produces may vary slightly from our published nominal thickness due to process and ambient variables. Reproducibility will be achieved when processing parameters are held constant.

Product Type	Approximate Viscosity (cSt)	Resist Thickness (μm)			
		3000 rpm	4000 rpm	5000 rpm	6000 rpm
S1805	5	0.54	0.46	0.41	0.36
S1811	15	1.23	1.06	0.93	0.83
S1813	20	1.51	1.28	1.14	1.02
S1815	26	1.78	1.53	1.34	1.22
S1818	35	2.10	1.82	1.59	1.45
S1822	48	2.58	2.20	1.95	1.77

STANDARD MICROPOSIT S1800 SERIES PHOTO RESISTS

Table 5.2 (above) relates specific SI 800 SERIES products to viscosity and corresponding film thickness at 3,000 to 6,000 rpm. The standard product name is derived from the coating thickness obtained at 4,000 rpm spin speed. This should be used as a guide to selection of a particular product.

Step 3 - Edge Bead Removal
Microposit EBR-10 is recommended for removing resist buildup occurring at the edge of wafers during spinning. The formulation has been specified to be free of Cellosolve acetate, acetone, and xylene. EBR-10 can be used with most coating equipment designed to include, an edge bead removal process.

Step 4 - Soft Bake
The following baking parameters are optimal:

Oven - Forced air convection (do not use nitrogen)

Temp. - 90-100° C

Time - 30 minutes (after recovery to operating temperature)

Cool - To ambient

In-line track baking equipment should be adjusted (speed/temperature) to yield photospeed, contrast and unexposed resist loss through developing equivalent to or better than those obtained using the above forced air convection conditions.

Step 5 - Expose Plate
Microposit S1800 Series Photo Resists can be exposed with light sources in the spectral output range of 350-450 nm contained in commercially available exposure equipment.

Standing Waves and Exposure Requirements:
The resist film thickness on a reflecting surface will determine the intensity and location of the standing waves in the film . Exposure time is minimized and line size control in the photoresist will be optimum in a resist film thickness that is an odd multiple of one quarter the exposing wavelength. By choosing this thickness a valid comparison of relative exposure energy required versus wavelength can be made. The exposure energy threshold (Et) is the minimum dose required to remove 100% of the photoresist in a large open area on a wafer. Contact the manufacturer for additional data regarding exposure times.

Step 6 - Develop
The following developers are designed for use with Microposit S1800 Series Photo Resists:

Ready To Use Developers: Microposit 352, Microposit 353, Microposit 354, Microposit 355, Microposit MF-312 Developer CD-27, Microposit MF-314, Microposit MF-316, Microposit 452, Microposit 453, Microposit 454, Microposit 455, Microposit Developer CD-30

Custom solutions and ready to use developer formulations are manufactured to tight analytical and functional specifications, and are therefore recommended for maximum process control. These developers will also avoid costly production line downtime caused by in-house dilution errors.

Note - Contrast data is available from the manufacturer.

Step 7 - Hard Bake
A hard bake is recommended for all photo levels to optimize process reliability in wet etching and maximize selectivity in dry processing steps. The following baking parameters, or equivalent, are recommended:

Oven -   Forced air convection

Temperature -   100-110° C

Time -   30 minutes

For maximum thermal stability, blanket deep UV exposure or plasma treatment prior to hardbake is recommended. Either of these procedures will assure image integrity through plasma etching, high dose implanting and other high temperature process steps.

Where taper etching is desired, using dry processing, the thermal properties of S1800 allow for the appropriate resist profile adjustment through controlled hard bake.

Step 8 - Etch / Ion Implant
Microposit S1800 Series Photo Resist can be used with all common semiconductor wet etchants as well as with plasma and ion implant processes.

Step 9 - Strip
Microposit S1800 Series Photo Resists Can Be Removed

Using Microposit Remover 140, Microposit Remover 11121a, Microposit Remover 1165, or oxygen plasma. Refer to the individual remover data sheets for specific processing instructions, specifications, and other product information.

Properties as Delivered

Microposit S1800 Series Photo Resist is filtered to 0.2 μm absolute. Each container is date coded.

Microposit S1800 Series Photo Resist Properties (F-or S1805, S1811, S1813, S1815, S1818, S1822):

- Film thickness reproducibility for:

    S1805, S1811, S1813, S1815

    +200 A with respect to a ref.

    S1818   +300 A with respect to a ref.

    S1822   +500 A with respect to a ref.

- Water content   0.5 % max.
- Index of refraction   1.64 @ 6328 A

    1.68 @4360 A

- Na content   < 1 ppm
- Fe content   < 1 ppm
- Type of solution   solvent base

    propylene glycol

    monomethyl ether acetate

- Flash Point (closed cup)

    approximately 46° C

- TLV rating (Rating is for propylene glycolmonomethyl-ether)   100 ppm

Handling Precautions

CAUTION: Note that this solution is combustible and harmful if swallowed. Use adequate ventilation, avoid breathing vapors, keep away from heat, sparks, or open flame. Avoid contact with skin and eyes. Handle with care. Wear chemical goggles, rubber gloves and protective clothing.

Toxicological and Health Information

Ethylene glycol monoethyl ether acetate (also known as 2 ethoxyethyl acetate or Celloso ve-R-acetate) is used as a diluent solvent for most conventional positive photoresists. The solvent used in Microposit S1800 Series Photo Resist is propylene glycol monomethyl ether acetate. It has been demonstrated in toxicological studies reported in the NIOSH Current Intelligence Bulletin 9, (512183) that the propylene glycol derivatives contained in Microposit S1800 Series Photo Resist do not demonstrate the adverse blood effects and reproductive effects that the ethylene glycol derived ether acetates do. These significantly lower health risks are reflected in the most current American Conference of Governmental Industrial Hygienists (ACGIH) TLV values which lists: TLV ethylene glycol monoethyl ether acetate 5 ppm propylene glycol monomethyl ether - 100 ppm. Material Safety Data Sheet is available upon request from Shipley.

Equipment

Microposit S1800 Series Photo Resist is compatible with most commercially-available photoresist processing equipment. Recommended compatible materials include stainless steel, glass, ceramic, unfilled polypropylene, high density polyethylene, polytetrafluoroethylene, or equivalent materials.

Disposal

Microposit S1800 Series Photo Resist should be disposed of according to Shipley Waste Treatment Procedure WT 78-13 (include with other solvent wastes). Contact your Shipley Technical Sales Representative for details.

Storage

Store in dry area at 50°-70°F (10°-21°C) in closed original containers away from light, oxidants, heat, sparks and open flame.

Towne Products for Holography

Towne Technologies of Somerville, NJ (USA) is a producer of photoresist plates for use in holography and other fields. Towne supplied the following description of how its plates are manufactured.

The materials that are used to produce the large Iron-Oxide coated holographic plates are purchased to specification requirements of the microelectronic, semiconductor and printed circuit board industries for which it was originally designed.

For example, an optical grade polished (both sides) soda lime, float glass substrate 24" x 32" x 0.190" (609.6 x 812.8 x 4.83 mm) has a flatness tolerance of 150×10^{-6} inches per linear inch (flat to within 150 microinches per linear inch). Before it is acceptable for FeO_2 coating, each piece of glass is cleaned and surface-polished to ensure that the slightest surface imperfections and even micro-dust particles are removed.

The pure Iron-Pentacarbonyl used has a controlled specific gravity of 1.44-1.47 @ 20°C and its deposition is carried out in Class 100 clean room conditions. After the FeO2 deposition, 100 Angstroms thick, the plate is inspected for integrity of coating. Pinholes are marked, and when the 24 inch by 32 inch plate is cut into final working plates, the pinholes are avoided.

The plates are dried in a thermostatically-controlled Class 100 environment, then cleaned and inspected again. From that time to the deposition of the photosensitive coating, nothing is permitted to contact the surface of the plate. The Microposit-S-1400-30 highly sensitive photoresist is used for coating because it is specifically formulated to be striation-free. On plates up to 15 inch by 15inch, (381 mm square) the photoresist is applied by a spinning process to a final standard thickness of 1.5 +/- 10% micrometers subsequent to a 0.2 micrometer filtration process.

The success of the iron-oxide coating is owed primarily to two inherent characteristics of FeO_2 coating, e.g., the iron oxide coating effectively absorbs any laser light that may be transmitted through the photosensitive coating. This virtually eliminates light backscatter and the possibility of damage to the primary image. Second, and possibly more important, the iron-oxide coating greatly increases the adhesive quality of the photosensitive coating, thus ensuring the integrity of the imaging and electroplating processes to follow.

Sal LoSardo, the sales representative from Towne, says that of the people who buy their plates, about 30% use them for holography, and "the number is growing." A particular area of expansion has been the Far East and China.

Towne has not needed to make any specific modifications to its plates to accommodate holographers, except for the need to install two "oversize" spinners. Plates for the electronic industry are usually dip coated when the size is above 7 inch by 7 inch. Holographers require the smoother finish of spin coated plates. Prior to 1984, Towne could only spin coat up to 7 inches by 7 inches , but now they can spin up to 15 inch square or 18 inch octagonal plates.

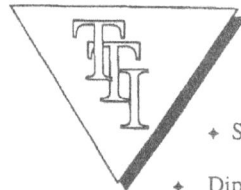

Photo-thermoplastic Films and Instant Holography Systems

Contributing author Alex Chaihorsky (Ultra Res)

I magine looking at a fully developed hologram seconds after it is exposed, while at the same time eliminating all darkroom processing. And if you did not like the resulting hologram, you could erase it with a flip of a switch, re-shoot and quickly see a new image.

This method of "instant" and "dry" holography is currently available using photo-thermoplastic recording materials in conjunction with specially designed "holo-camera" systems. Applications exist for interferometry / NDT, prototyping, display holography and classroom presentations.

Photo-thermoplastics for holographic applications were developed by researchers at several major corporate laboratories (such as Honeywell, IBM and Xerox) over two decades ago. However, related commercial products were not available until the 1980's when the Newport Corporation manufactured a complete film/camera system designed for interferometry.

Using a three-layer sandwich of relatively inexpensive materials, surface-relief "phase" holograms are created quickly using electricity, heat and, of course, laser light. No chemical processing of the recording material is required. Photo-thermoplastics are unique among holographic recording materials in that they can be erased and reused. The material is especially suited for industrial interferometry (its optical properties are ideal for recording and producing viewable interference patterns) and are suitable for some display holography applications as well.

How Does It Work?

The typical photo-thermoplastic recording material is composed of three thin layers:

Lower - a transparent support material very thinly coated with an electrically conductive material (such as titanium indium oxide);

Middle - a layer of photoconductive 'material; and

Upper - a deformable plastic (thermoplastic) with a softening temperature of about 70-100° C.

The hologram formation cycle can be broken into four parts:

1. Just before exposure the aforementioned composite is electrically charged to sensitize it. Upper surfaces now have a positive charge. *Note - Before this step takes place, the film is not affected by light like other photographic emulsions, so darkroom handling is never necessary!*

2. During the exposure, the laser light affects the middle photoconductive layer in a predictable way - exposed areas are charged negatively and the holographic interference pattern is converted from an illuminance pattern into an electrical one.

3. The thermoplastic layer is heated. As it softens, it distorts in accordance with the electrical charge distribution, creating a surface relief pattern that mimics the original interference pattern. This occurs because positive charges on the surface film are attracted to the negative charge underneath.

4. As the thermoplastic cools rapidly, it hardens, and the hologram is permanently recorded. However, additional heating can "erase" the material and make it ready for another recording cycle.

Available Film and Camera Systems

Ultra Res of Reno, Nevada (USA) is very actively involved in the research and development of photo-thermoplastic films and associated hardware. The company has developed an "instant hologram" system that can be custom modified for a client's needs. Its basic holo-camera system sells for just under $10,000. Options for real time interferometry can be added for an additional charge. Recording film for the device is 35-mm. It costs several dollars each and is reusable.

In contrast, a similar system that was marketed by the Newport Corp. (but discontinued several years ago) sold for approximately $15,000. Its reusable recording plates cost $461 each and were smaller (30-mm) than the Ultra Res films. In addition, the size and complexity of the New- port system made it more difficult to use and transport (it required the use of pressurized nitrogen, for instance).

A French manufacturer, MD Diffusion sells a system called Holodata which is primarily designed for interferometry. Options include a remote control and compatible video equipment. The film comes in 70-mm wide reels 25 meters long (the "holo-camera" makes holograms 50 x 70 mm.)

Ultra Res System Specs

Ultra Res's photo-thermoplastic recording material is named "T-Film™". The company reports that the material is sensitive from 500 to 700 nm and its sensitivity reaches 1 erg/cm.square (10^{-7} J/cm.square). According to the manufacturer, the relatively high sensitivity allows for hologram formation times measured in seconds, which significantly reduces vibration isolation precautions. Under proper conditions, the actual "exposure" itself measures mere fractions of a second. More sensitive films are under development.

Company president Alex Chaihorsky explains that, "Our panchromatic recording material allows the system to be integrated into a variety of holography set-ups. In addition, eliminating the wet film develop-and-fix cycle saves time, money, precious lab space and expensive silver recovery procedures. Since our recording materials are erasable and reusable, our system does not consume or waste anything except electricity."

According to Chaihorsky, Ultra Res's system works by creating a 0.1 micron deep "ripple" on the surface of the photo-thermoplastic film as a result of electrostatic attraction between positive charges (deposited on the surface by a Corona discharge) and negative charges (that are induced in the underlying photoconductive layer by the illuminating light photons.) The ripple is a recording of an interference pattern.

To better explain how the system is actually used, here are some instructions from the Ultra Res user's manual:

1. Place the camera where the recording plate would be in a standard transmission hologram set-up.

2. Insert a frame (a holder and a piece of T-Film™) inside a slot in the camera and close the heater door.

3. Set controls on the unit. These are:

- Plate temperature (usually remains unchanged).
- Corona voltage (needed to sensitize the film).
- Process time (6-8 seconds).

4. Push the start button on the controller, which initiates the sensitizing charge. Once sensitized, hologram formation proceetis.

5. Several seconds later a loud click will announce that the process is finished and heater door will automatically open. The finished hologram can be viewed in or out of the unit.

If the results are unsatisfactory, the flip of a switch and closing the heater door will start the erasing the plate. Thirty seconds later it will be ready for use again. (Ultra Res also sells a bulk eraser for the T-Film™, which completely erases three frames at a time.)

Who Could Benefit From Using This "Instant Hologram" Technology?

Display Holographers: Holographers can use the "instant" holo-camera systems as a quick and affordable tool for setup verification and fast prototyping for clients. Security and embossing specialists doing low runs could save mastering and plating charges. According to Chaihorsky, holographic patterns recorded on T-Film™ can be stamped into certain materials right out of the camera, creating an "instant embossing".

For instance, the relief pattern could be easily pressed into Mylar to generate samples for clients interested in larger - runs of conventional embossed holograms. In fact, if the surface relieflayer of the Ultra Res film is properly plated,

it could even be used in place of a photoresist embossing shim .

Interferometrists: Since the hologram can be viewed while it remains in the camera, there is no need to wait for photo-processing, which consumes valuable time during testing sessions. More importantly, there is no need to reposition and realign the hologram in the viewing device for "real time" interferometry - a common dilemma. Once a hologram is recorded, the operator (or video camera) can look through it while the tested materials are re-stressed. Double-exposure holograms for other NDT applications are easily produced.

Teachers: Educators can use instant holo-camera systems to make teaching and learning holography easier since isolation problems are minimized and darkroom processing is eliminated. Students can see results immediately. This helps teachers with limited classroom time and resources.

Researchers: Holographic Memory is another field where users are finding new applications for holographic photo-thermoplastics.

Some Limitations

Systems employing photo-thermoplastic films only produce small laser viewable transmission holograms. Since they are surface relief holograms, image depth and viewing angle are limited. Resolution is relatively low compared to other types of recording materials; up to 1,200 line pair/mm (with best overall performance being at 1,000 line pair/ mm) .

Film size and camera size have been dictated by market considerations; larger formats can be manufactured. Since film is inherently less rigid than a glass plate, Ultra Res sells one version of its film that is stretched and glued in the handing frame to combat potential film stability problems. In addition, MD Diffusion's catalog reports that thermoplastic coated onto glass plates is 30-100 times less sensitive than film , as thicker coatings are needed. Therefore, they also prefer to use film in their systems.

The Ultra-Res INSTANT HOLOGRAPHY Camera

* Instant (3-5 second) hologram - No chemical development neede
* Very high sensitivity: up to 0.1 μJ/cm^2 (10^{-7} J/cm^2) in 400-700 n
* Inexpensive thermoplastic film media. Erasable and reusable.
* The hologram is recorded as 0.1micron-high ripple on the surface of the film.
* Ready for embossing on softened plastic.
* Applications: Holography, interferometry, setup verification, HOEs phase filters, embossing, security.

Ultra-Res Corporation

1395 Greg St. - Suite 107
Reno, NV 89431
Phone (702) 355-1177   Fax: (702) 359-6273
Email: alex@acds.com
Web page: http//www.acds.com/UR/ultra-res.html

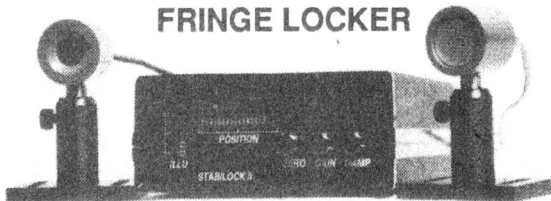

6

Optics and Equipment

Many of the newly introduced silver-halide recording materials are several times less sensitive than the Agfa emulsions they are intended to replace. This results in longer exposure times being used, which greatly increases the probability of unwanted vibration during the mastering and copying processes.

This chapter describes a system whereby unwanted motion can be detected and compensated for, so that high quality holograms can be recorded in spite of longer exposure times. This feedback-based system can also help holographers using other recording materials in a wide variety of research or production situations.

Fringe Stabilizers

by Jeff Odhner (Odhner Holographics) ...

In order to make a good hologram, every optical component that is directly involved in the holographic recording process must be immobilized while the emulsion is being exposed. To be more accurate, it is the interference pattern which makes up the hologram (and represents the optical path difference between the reference and object beams) that must be held perfectly still. The dark and light lines on this interference pattern are referred to as fringes. Fringe movement on the holographic recording material is the primary cause of failure to produce a hologram.

No matter how good the recording setup is; no matter how expensive the components are; there will always be some amount of fringe movement during the exposure time when using a CW laser. Anything that causes a change in the effective optical path length between the reference and object beam will cause the fringes to move. Culprits include: air currents, temperature gradients, thermal drift, vibrations, laser frequency shifts, scene motion, and component settling. In practice, many of these variables are uncontrollable past a certain point. Due to the fact that recent supply problems have forced holographers to switch to less sensitive emulsions (which means longer exposure times) the problem offringe movement becomes even more important to solve. One solution that has proven effective is called **active fringe stabilization.**

What is a Fringe Stabilizer?

A fringe stabilizer senses a movement of the dark and light lines (fringes) in the interference pattern that makes up the hologram and then quickly reacts by changing the effective optical path length of either the reference or object beam to compensate for that fringe movement. If this optical path difference is compensated for fast enough, the fringes will effectively remain stationary. A good fringe locker, when properly set up, can stabilize to within 1/100th of a fringe for exposure times exceeding an hour, although a tenth of a fringe stability is usually more than sufficient.

How Does a Fringe Stabilizer Work?

A fringe stabilizer works on the principle of a closed loop feedback and control system. A sensor detects the fringe movement and converts this into a proportional voltage (fringe error). This voltage is fed into a control system which changes the effective optical path length of the reference or object beam until the fringe error (movement) disappears. (See figure 6.1.)

A fringe stabilizer consists of three basic components:

1. A Phase Detector which senses phase difference changes between the reference beam and the object beam (manifested by fringes).

2. Electronics to generate a control signal which is an error voltage proportional to the fringe movement.

3. A Phase Modulator which changes the relative phase of the two beams in proportion to the error voltage.

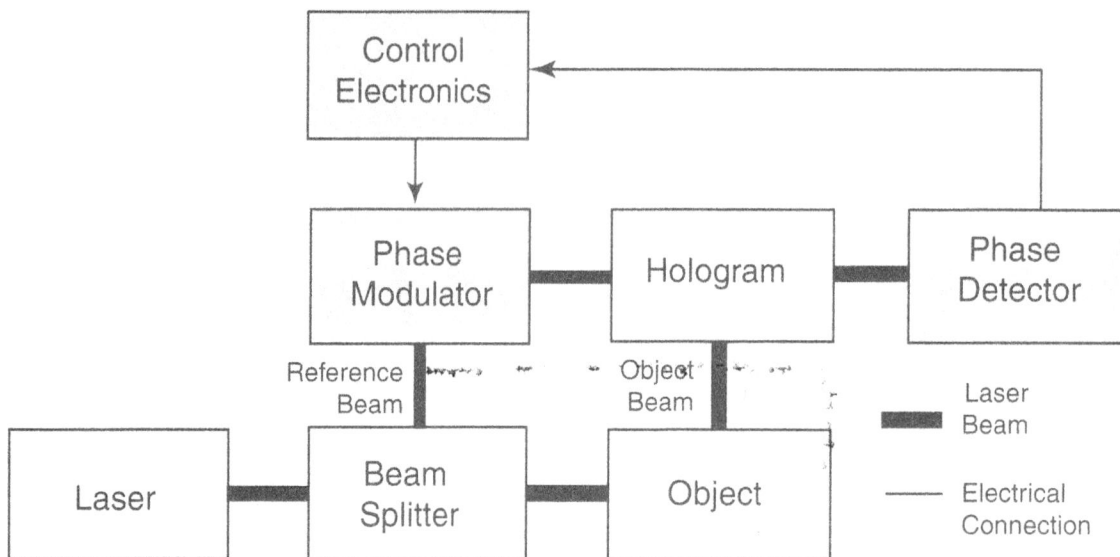

Figure 6.1. Basic components of a Fringe Stabilizer and how they integrate into a hologram recording setup.

Phase is not easy to measure directly, but the fringe movement seen by two interfering laser beams is directly proportional to the phase difference between the two beams. In other words, if anything moves in the optical paths, the fringes move. For visible laser light, the phase detector is typically a pair of ultra-low noise silicon detectors which look directly at the fringe pattern. When the fringes make the slightest movement across the detectors, the voltage difference between the detectors changes. This voltage difference gets amplified by the control electronics and then that signal drives a phase modulator which is placed in either the reference beam or the object beam path. The phase modulator changes the phase difference between the two paths until the phase difference between the two paths is equal

to zero. This will "lock" the fringes so well that fringe movement is imperceptible even over long periods of time.

When Should a Fringe Stabilizer be Used?

Try setting up a Michaelson interferometer on your optical table with the longest legs possible and look at the fringe movement over several minutes. If these fringes are drifting over what would be the required exposure time, then it is worth the trouble to set up a fringe stabilizer in one of the legs of your hologram setup. (See figure 6.2.)

Different Types of Phase Modulators
Piezo-electric

The most popular phase modulator is a piezo-electric (PZT) ceramic driven mirror. When voltage is applied to the PZT ceramic, it provides a pure lateral movement to the mirror. This mirror is incorporated into the optical path and thus changes the path difference. From a safety point of view, it is preferable to find a PZT ceramic which can move +/- several microns with a low voltage applied as opposed to some PZT crystals which require high voltage to move +/- several microns.

Speaker

A mirror glued to a speaker has the advantage of being inexpensive. However, speakers are sensitive to acoustic vibrations, and can have some angular displacement which can throw off optical alignment.

Angle Plate

If a plate of thick glass is moved by a galvanometer in the optical path, the distance that a laser beam travels as it passes through a thick glass plate will change as the angle of the glass is varied. This has the effect of changing the optical path length of the leg it is introduced into and thus can be used to control the fringe pattern. The advantage to this is simplicity. The disadvantage is a large lateral beam displacement, low frequency response due to the mass of the glass, and the expense of a galvanometer. The large lateral displacement can be solved by

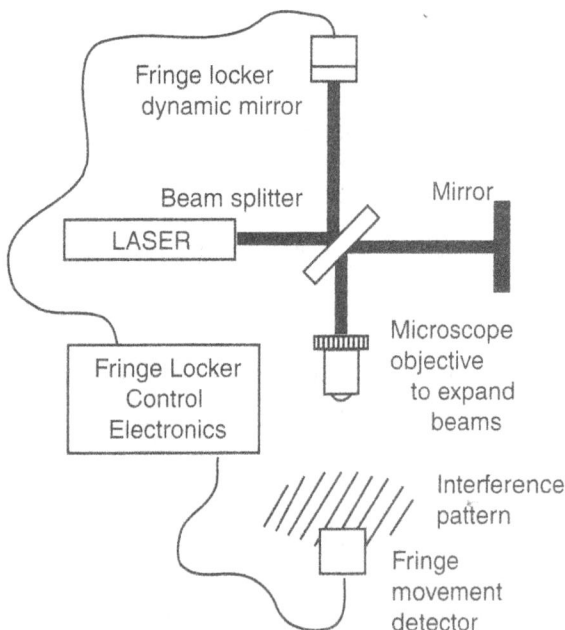

Figure 6.2. How to incorporate a fringe locker into a Michaelson interferometer.

using another glass plate of equal thickness placed in the path at the opposite angle, but this also adds the expense of another galvanometer.

Magneto-restrictive

These are devices which can move a mirror with applied voltage but are expensive and not commonly used.

Crystal Phase Modulators

Certain crystals will change their index of refraction with applied voltage. If this index change is high enough, or the crystal long enough, it can sufficiently affect the optical path length enough to control the fringe pattern. Unfortunately, these crystals tend to be expensive and the potential required is generally high voltage.

How to Set Up a Fringe Stabilizer

The phase modulator (i.e., piezoelectrically driven mirror) is often put in the reference beam leg of the holographic setup replacing one of the mirrors before the beam has expanded. The phase detector is placed to the side or in back of the light sensitive recording media where the beams are recombined and a fringe pattern is present.

When the fringe stabilizer is first turned on, an intensity difference will be detected by the two detectors in the phase detector and the mirror will move. The direction it moves is not important but that it moves until the detectors are looking at equal intensity points on a pair of fringes. Next the gain and damping are adjusted to strengthen the fringe "lock".

The fringes sampled directly can be quite small so an alternative is to generate wider fringes at the film plane. This is done by Richard Rallison of Ralcon in the following manner: "I place a beam slitter behind or to the side of the film holder and orient it so that one side of the input beam is reflected to be collinear with the other beam. A small cross,. scratched into the surface of the splitter helps align the two beams on a screen placed on a distant separate table. Ev-

erything is adjusted to get the best contrast and size. The locker will dofine with only a 1 mm fringe, but I like to have a few fringes blown up to several inches wide on a cross hatched white screen so I can watch even the tiniest quiver or catch the fluttering mode or a mode hop during the exposure."

It is interesting to note that multiple fringe stabilizers can be set up in the same holographic setup to control multiple reference and object beams pairs.

Low Light Level Applications

Holographic fringes are on the order of a micron. When a microscope objective is used to expand the fringes to roughly the size of the fringe locker detector spacing, the intensity decreases as $m \cdot 2$ (m = magnification).

Using a cylindrical lens to magnify the fringes improves the situation somewhat as the intensity will decrease as only m^{-1}.

But for low light level applications, a moire technique can be used to get secondary fringes. A second hologram (H2)is made just behind or off to the side' of the hologram of interest (H1), developed and then placed back in the same position. (See figure 6.3.)

A similarly sized focusing lens placed behind the hologram in place of the microscope objective to focus the secondary fringes can provide an intensity gain on the fringe locker detector of several orders of magnitude. [2] The presumption is that there is at least enough stability to make a dim hologram of the original setup to be used as the phase control grating. This is an especially useful technique not only for low light levels, but also when using wavelengths which are the edge of the silicon photo detector response (exposing photoresist with UV).

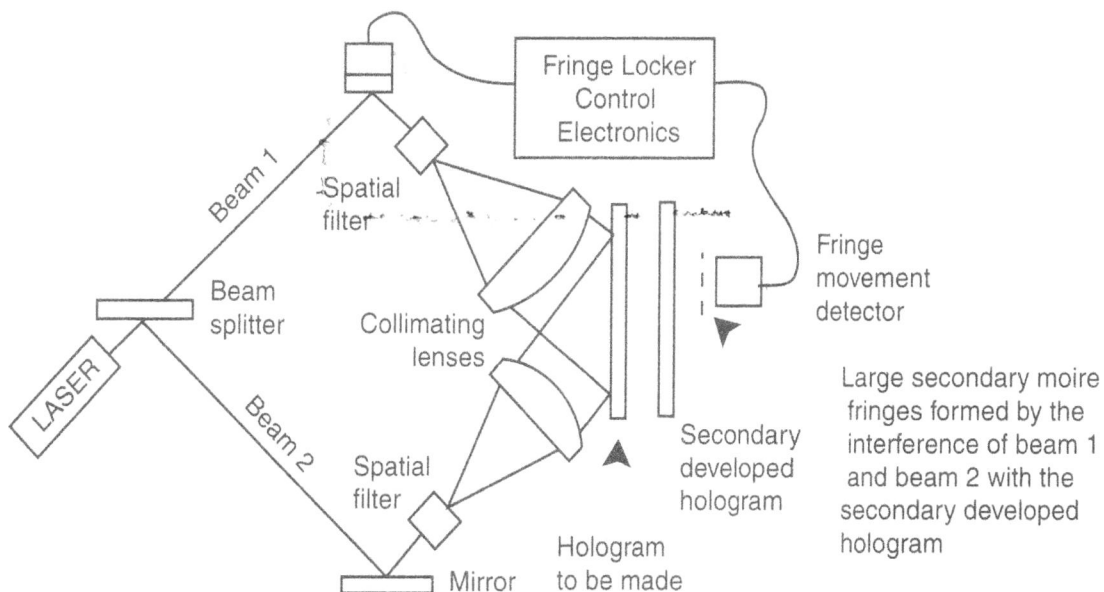

Figure 6.3. How to use a secondary hologram to increase fringe spacing in a diffraction grating setup.

Gain Relationship Between Detector Spacing and Fringe Width

The relationship between detector spacing and fringe width is important to understand for successful use of the fringe locker. Assume for a moment that the fringe intensity varies sinusoidally. If the spacing of the fringes is equal to the spacing of the detectors, then as the fringes move across the detectors, there would be no difference in the detected intensities. Each detector would register either bright or dim light intensity. Since the most common fringe stabilizer uses a detector pair and these work differentially, there would be no detectable error signal as the fringes move. Now consider the case where the f~spacing is - twice the detector spacing. In this case, as the fringes move, one detector would be in a bright band while the other would be in a dark band and the maximum differential signal would be produced. Thus, the proper setup of the fringe size makes a large difference in the total system gain. In actuality, the assumption of a sinusoidal intensity variation is somewhat in error and a differential signal is always.generated. Also, fringe locking is possible even if-the fringe width is as much as an order of magnitude wider than the detector spacing.

Intensity and Contrast Variations

There is an additional gain relationship in the optical system caused by intensity variation and contrast. If where the light recombines destructively the result is zero energy, then where it combines constructively the amplitude is dependent on the optical energy launched into the system. The brighter the light, the greater the optical gain of the system. Also, in real systems there is a contrast factor. The . greater the difference in intensity between the light and dark regions of the fringe, the greater the optical gain.

Modes of Operation

This system gain generates the feedback signal that controls the mirror movement. It is a high gain system and falls into the category of a classical gain stability system. This classical system has three possible modes of operation: under-damped, over-damped, or critically-damped.

Under-Damped

Under-damped is a condition where the system gain is so large at the system goes unstable. This is similar to squelching in audio system. When a fringe locker is under-damped, an audible tone is often heard, but a more reliable detennination is tb observe the fringes . If fringes are visible when the fringe locker is off but blur beyond recognition when the fringe locker is on, then an under damped condition is in present.

Over-Damped

Over-damped is a condition where there is insufficient system gain to effect the desired operation. This is viewed as sloppy stabilization and weak locking and typically the fringe will move across the detectors as much as 30 degrees and destabilizing forces of 10% of the lock range will cause a break lock.

Critically-Damped

Critically-damped is the desired condition. When properly set up, the fringes in this mode will be crisp and clear, and slow destabilizing forces are easily compensated for.

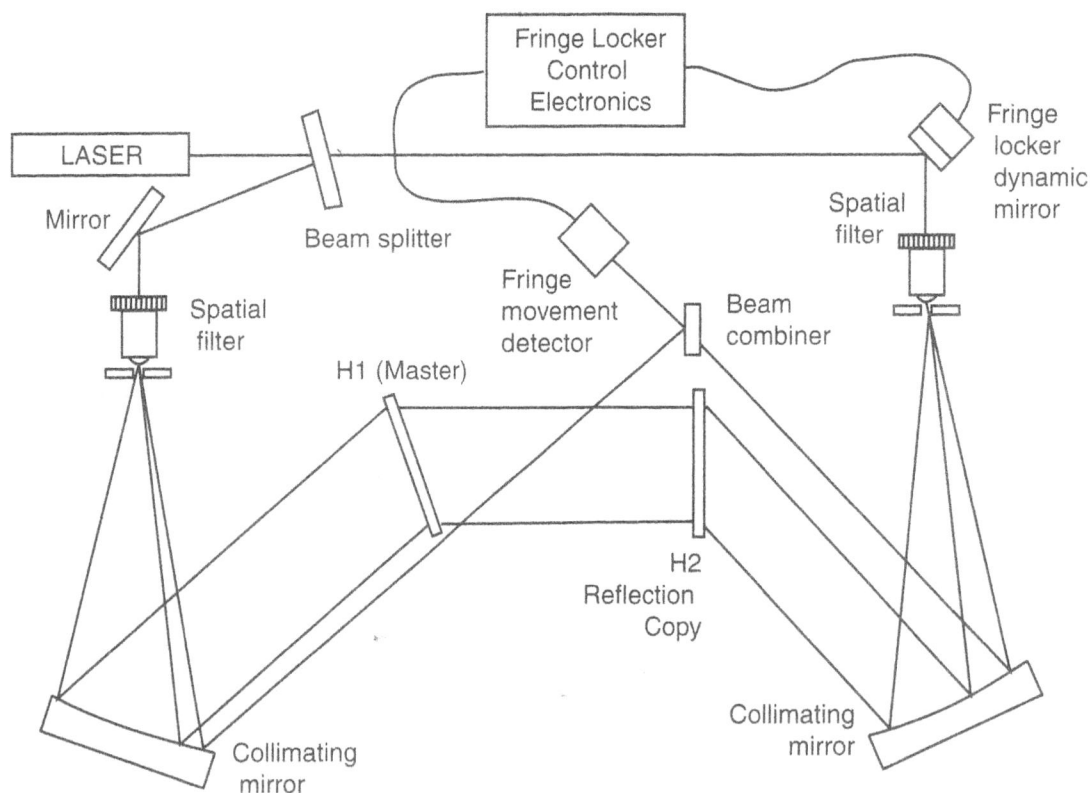

Figure 6.4. Using a fringe locker to make reflection transfer holograms.

Unstable

Fringe locking systems which do not have adjustable gain and adjustable damping do not have good locking characteristics. Even those systems that do require an experienced operator to achieve best results. The best way to acquire these skills is to experiment.

Applications

Fringe lockers are applicable anywhere it is desirable to have a stationary interference pattern. They have been used in atmospheric interferometers to study air turbulence, LU-PIs (laser unequal pather interferometers), and vibration monitoring. But the most common applications are in the area of holography. The following is a sampling of some of the more common applications of fringe lockers.

Reflection Hologram Copies

Production transfer copies will get consistently better results through electronic fringe stabilization and holograms made with materials such as DCG (dichromated gelatin) which require long exposure times can be impossible to make without a fringe stabilizer.[3] A transfer copy set up is shown in figure 6.4.

Phase Locking Two Lasers Together

Phase locking two lasers together was suggested by Neumann and Rose and demonstrated by McGrew.s The high reflectance mirror on one laser is replaced with the fringe stabilizer mirror. Movement of this mirror controls the laser frequency. The output of the two lasers is then combined with a beam splitter toform an interference pattern which is sampled by the fringe locker detector. See figure 6.5. Initially the fringes race across the screen with a speed proportional to the frequency difference, but this can be slowed by electronically moving the fringe locker mirror until they momentarily slow to the point where the fringe locker can take over. At that point the lasers are mutually coherent. In practice, this would allow a long reference beam path with a short object beam path or other unusual geometries.

Reduce the Effect of Laser Wavelength Shifts

Shifts in the laser wavelength correspond tofringe movement. These slight frequency changes can be partially compensated for with a fringe stabilizer.6 (see figure 6.6.) If one of the lasers is carefully adjusted, so that the frequencies are matched, the interference pattern will momentarily stop moving and then the fringe locker can take over.

Holography in the Classroom

When a portable holography table is taken into different classrooms, the conditions for exposure are unpredictable and there often is not time to trouble shoot the setup with multiple exposures. One chance is typically all that is given. But temperature variations and component heating changes can be compensated for with a fringe stabilizer to ensure a successful holographic exposure.

Conclusion

An active fringe stabilizer compensates for thermal drift in any interferometer and can solve stability problems

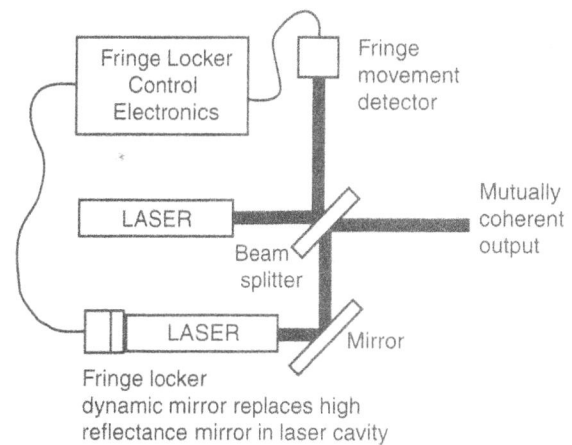

Figure 6.5. How to use a fringe locker to lock two lasers together to get a single coherent beam.

which cannot be solved in any other way. Compensation is over a limited range and there is some effort involved in incorporating fringe locker in a setup, but the improvements in the finished product can be dramatic. Certain types of holograms can not be made without active fringe stabilization. A fringe stabilizer should be considered whenever working with any laser interference pattern.

References:

1. Graham Saxby, "Practical Holography", Appendix 9 'Fringe Stabilization', Prentice Hall, N.J, pp. 423-432 (198d)

2. David R. MacQuigg, "Hologram fringe stabilization method", Applied Optics 16, pp. 291-2 (1977).

3. Richard Rallison, "Fringe Locking", Holographies International No. 8 Summer 1990.

4. Neumann and Rose, "Improvement of Recorded Holographic Fringes by Feedback Control", Applied Optics 6, 1097-1104 (1967).

5. Steve P McGrew, "An inexpensive fringe stabilizer for long exposure holography", International Symposium on Displav Holography 1, pp. 189-193(1982).

6. Jefferson E. Odhner, "A system for measuring the phase difference between two points of a propagating laser beam", paper presented at the first International Meeting for the Propagation of Waves Through Random Media in Seattle, Washington in the Fall of 1992.

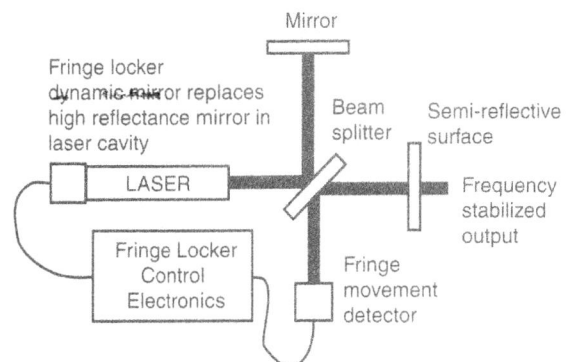

Figure 6.6. How to use a fringe locker to stabilize the output frequency of a laser.

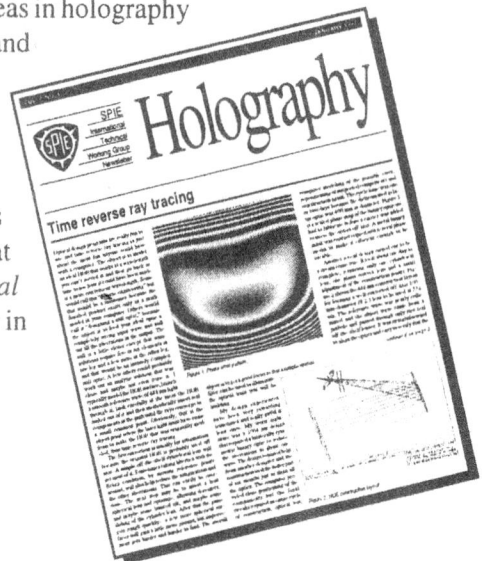

7

Laser Fundamentals

This chapter could be titled "from photons to etalons." It begins with a review of the fundamentals of electromagnetic radiation and then proceeds to a detailed explanation of how the lasers that are most commonly used in holography (CW lasers) work.

Laser Fundamentals

To sufficiently understand the operation of lasers, their many advantages and their necessity in the production of holograms, one must furst comprehend certain properties of our physical world.

The entire universe consists of only two things: **matter and energy**. Matter is all things that have physical substance; energy is the mover, or potential mover, of physical substance. Matter is the stuff we see, smell and feel. It has mass and occupies space.

Energy, on the other hand, is more abstract. It is most often invisible, though sometimes not. Yet, it is everywhere. It lurks in the crevices of every moltcule and sweeps the skies with its magnificence. A master of transformation, energy facilely converts itself from one of its many forms to another, all without sacrifice.

Energy is the driving force behind all forms of motion: the motion of our car, the motion of planets, the motion of atoms. Nothing moves without it. Matter, without energy, is reduced to a dark, frozen lump of nothingness. In a dynamic universe, matter both possesses energy and is affected by it.

Energy not only changes form; it is also easily passed from one object to another. Interestingly enough, no matter how many times it transforms or transfers, the amount of energy involved in any given transaction never changes. The **law of conservation of energy**, one of the most im-

portant laws in the universe, dictates that energy is never created or destroyed; it can only be transferred to another object or converted into a different form of energy.

Because the amount of energy in the universe remains fixed, phrases such as "energy shortage" and "depleted energy" are misnomers. You can not lose energy, nor can you be in short supply. The amount of energy in our environment is so great that it is beyond our comprehension. The discomforts in past decades from "energy shortages" were created only by our inability to either convert energy to a usable form or distribute usable energy to where it was needed.

Energy is measured in **joules**, in honor of the British sci- entist James Joule. One joule is roughly the amount of en- ergy required to lift an apple from your kneecap over your head.

A glass of apple cider has 502,092 joules (equal to 120 calories-the **calorie** is another unit of energy often used when referring to the content of food) of food energy. A gallon of gasoline has over 200 million joules of energy.

The process of applying energy to matter is called **work** (also measured in joules). Work is the mechanism that transfers energy through a system. It is produced by applying a force on an object such that motion occurs over a distance. For example, when you pick up a book, you have performed work on the book. The heavier the book, the greater the force that is necessary to raise it therefore, the more work done. The farther you raise the book, the greater the distance in which the force must be applied. Again, more work is accomplished.

Energy is formally defined as the ability to do work. It can be classified in two categories: stored energy and motion energy. Stored energy is more commonly called **potential energy**. When raising a book in the air, work must be performed on it. While suspended in air, however, the book has the ability to perform work in the opposite direction-courtesy of the earth's gravitational field. This "stored" energy may be released simply by dropping the book.

The potential energy an object possesses due to its position in a gravitational field is called, premetably enough: **gravitational potential energy** (or **GPE**). Any object raised above the earth's surface has gravitation potential energy. Water behind a dam has a significant amount of GPE that may be converted (as mandated by the law of conservation of energy) into electrical energy.

Other forms of potential energy are also commonplace. A coiled spring is the good example of **mechanical potential energy.** By performing work on the head of a Jack-in-the-Box, one can push it into the box. With the lid secured, the box has stored energy (mechanical potential energy): hence, the ability to do work. By unlatching the lid, the stored energy is released and work is performed in the reverse direction on our friend Jack.

ue for electric charges. Electric charges that are positive attract those that are negative, and vice-versa. Equally, two electric charges of the same type (both positive or both negative) repel each other. The attractions and repulsion of electric charges are caused by invisible **electric fields** produced by each charge. An electric field permeates the territory around each charge, affecting all other charges that occupy its space. The larger the charge, the more influence its electric field exerts on its occupants. We encounter electric fields to varying degrees throughout a typical day. We witness them when we use our dryer, for it is electric fields that cause static cling.

To separate two opposite charges or unite two like charges requires work. Like the Jack-in-the-Box, when work is applied to bring like charges together (or to separate opposite charges), **electric potential energy** is created. Remove whatever constraint that holds the stored energy (in the Jack-in-the-Box the constraint was the lid, in electric charges it is usually a non-conductive material like air or plastic) and work is performed in the opposite direction.

A familiar device utilizing electric potential energy is the battery. The stored energy in a battery can be released by placing a conductive path between the positive and negative terminals. The performance of battery is stated in terms of **voltage** (also called **electric potential**, abbreviated with a V and measured in **Volts**). Voltage, an important element in laser operation, is the ratio of electric potential energy (**EPE**, not to be confused with electric potential) to the amount of charge (abbreviated in equations with a **q**, measured in **coulombs**). In equation form:

$$V = (EPE)/q$$

Another common source of voltage is the generator. Generators are machines that convert various types of energy into electric potential energy. Generators in dams convert gravitational potential energy from elevated water. The energy is transported to homes and businesses, readily available for those who wish to do a little work. Since generators produce higher voltages than batteries, they are used to supply power to all gas lasers and most others.

Matter is the greatest repository of energy. Atoms arranged together have binding electrical forces (called **bonds**) that act much like infinitesimal coiled springs. When bonds are broken, stored energy is released. This stored energy in molecules is called **chemical potential energy**. The gas we pump into our cars and the food nourishing our bodies are two common forms of chemical potential energy being utilized in our lives. Forces holding the nucleus of an atom together store an astounding amount of **nuclear potential energy**, as witnessed on July 16, 1945, when the Manhattan Project unveiled the atomic bomb.

Matter in motion possesses **kinetic energy**. An object will gain kinetic energy when work is done to it. An object will lose kinetic energy when work is done against it. The amount of kinetic energy an object gains or loses is exactly the same as the amount of work done on or against it.

For example, the engine of a train converts chemical potential energy into kinetic energy and performs work on the train. The train will move; it now has the amount of kinetic energy equal to the net gain of work done on it. If you turn off the engine, the train eventually stops, even if it is riding on a perfectly level set of tracks. This is because friction (between the wheels and the track; between the wheels and their axles; and between the air molecules and the front of the train) is performing work against the train.

When a train in motion hits a stationary object in its path, work is performed on the object. The object will move. Some of the train's kinetic energy is transferred to the stationary object. If one removes all sources working on and against a moving train on perfectly flat tracks-engine, friction and objects in its path-the kinetic energy of the train will never change. The train will continue to move forever at a constant velocity.

Other kinds of motion energy include heat, sound and electromagnetic radiation.

Heat occurs from the motion of molecules. The faster the molecules move, the more heat generated. A common source of heat is friction. In our previous example, friction performed work against the train. The kinetic energy of the train transformed itself to frictional heat in its wheels, axles and tracks (to a lesser degree, the air molecules). The train eventually stopped because its kinetic energy was entirely transformed into frictional heat. In most energy exchanges in nature, heat is part of the transaction.

Sound is another form of motion energy that occurs when a disturbance in a medium (commonly air) produces molecules to vibrate back and forth creating "sound waves." Each molecule receives the wave, vibrates back and forth and returns to its original position, but not before imposing a similar disturbance on its neighbor. The neighboring molecule repeats the same maneuver, as does each successor, thus creating a chain of disturbances that allows sound energy to propagate through the medium. Eventually: the sound waves hit an eardrum, causing it to vibrate. The vibrating eardrum creates signals to the brain that enable us to "hear" the sound energy.

One of the most important forms of motion energy is **electromagnetic radiation**. It exists everywhere throughout the universe and comes in many forms. Radio and television waves can be transmitted hundreds of miles through the air enabling music, images and conversation to magically appear in our homes. Microwaves, used in radar and modem cooking devices, ensure safe travel and a fast meal. Infrared radiation warms our skin and other vital regions of the universe. Visible light, the only form of electromagnetic radiation that we can see, enables our world to have definition and beauty. Ultraviolet radiation bums our skin and cures our plastics. X-rays help doctors diagnose problems in our bodies while gamma rays are found in many forms of radioactive decay.

Although we perceive and apply them differently, all forms of electromagnetic radiation are essentially the same phenomenon. Only the amount of energy per fundamental unit distinguishes a microwave from a beam of light.

The fundamental unit of electromagnetic radiation is called the **photon**, an infinitesimally small "packet" of energy. Radio waves have relatively low energy per photon. Microwaves have more energy per photon than radio waves but not as much as infrared radiation. A photon of visible light has more energy than a photon of infrared radiation but less than ultraviolet radiation. X-rays and gamma rays carry the most energy of all.

Electromagnetic radiation is created by accelerating or decelerating an electric charge. The greater the acceleration (or deceleration) of an electric charge, the more energy it will produce. It would take a much greater deceleration of an electric charge to create an x-ray photon than a microwave photon. Electric charges that are stationary, or those moving at a constant velocity, do not create electromagnetic radiation.

An electron, the most fundamental unit of negative charge, is the most common vehicle for creating electromagnetic radiation. For example, a radio station produces radio waves by accelerating electrons up and down a transmission antenna in a process called **oscillation**. An antenna is limited in its ability to rapidly accelerate and decelerate electrons, however. This is why antennas do not create visible light. Electron activity in atoms is the most prolific manufacturer of visible light. As explained later, this activity will be the basis from which lasers are created.

Properties of Electromagnetic Waves

Water waves make a good model for the study of electromagnetic waves because they are commonplace, exhibit comparable properties, and move slowly enough to be carefully observed. There are a few profound differences between the two (for example, water waves must propagate *in a medium* - where electromagnetic waves need not), but not enough to impugn our comparison.

A wave is created by a disturbance in a medium (for electromagnetic waves, a disturbance may be created in empty space). In a swimming pool, a swimmer resting in the shallow end of the pool slaps his hand on the water. The disturbance creates a wave that moves from the shallow end to the deep end (for this example and all fictitious pools in this section, allow the edges to absorb all waves that hit it, thus eliminating the effect of reflected waves). The wave has a "crest" (high point) and a "trough" (low point).

A closer look at the wave would reveal that the water molecules do not travel with the wave. In fact, if you measured the net movement of all the water molecules due to the wave, it would total zero. One may ask, "If the water molecules aren't moving in the direction of the wave, what is?" The answer is energy.

If the swimmer slaps his hand many times in regular intervals, a wave with a series of crests and troughs is created. Each pair of one crest and one trough is called a cycle due to its tendency to repeat. If the intervals are fast, the crests and troughs (or cycles) will appear to be closer together. If a sunbather sitting halfway between the deep and shallow end of the pool had a watch, she could count the number of cycles that pass by her each second. This value, the number of cycles per second, is called the **frequency** (for space economy, we use the letter **f** in equations) of a wave. The unit of one cycle per second is more commonly called a **Hertz** (abbreviated **Hz**) in commemoration of the German physicist Heinrich Hertz. Because the frequencies of electromagnetic waves are quite high, larger units such as **megahertz** (one million hertz, abbreviated **MHz**) and **gigahertz** (one billion hertz, abbreviated **GHz**) are commonly used.

The velocity (v-measured in meters per second) of a wave is directly related to the medium in which the wave is travelling. If the pool was drained and filled with molasses, the velocity of the waves would be less than those moving in water. For the remainder of this section, it will be assumed that all waves generated in our fictitious pool are traveling at identical velocities.

The distance between two successive crests (or two troughs) on a wave is called the **wavelength** (l) which is measured in meters or subunits of meters. The most common subunit for wavelength measurement of electromagnetic waves is the **nanometer** (abbreviated **nm**) which is one-billionth (1×10^{-9}) of a meter.

When the swimmer slaps the water with slow intervals, the distances between the crests and troughs of the waves are large, hence the wavelength is long. The sunbather times very long cycles per second on such waves. The frequency is small. However, when the swimmer rapidly slaps the water, the crest and troughs seem to bunch together. The wavelengths are short. The sunbather counts many cycles per second. Waves with higher frequencies, therefore, have shorter wavelengths and waves with lower frequencies have longer wavelengths. The relationship between the two can be summarized in the equation:

$f = v/l$ or $l = v/f$

It is important to remember that, as long as you adjust their numerical values per the above equation, frequency and wavelength are interchangeable. In the study of light it is common to use either term.

The swimmer may also notice that when he slaps the water with more force, the crests become taller and the troughs become deeper. The height of the crests (or in many cases, the depth of the trough) is called the amplitude of the wave. If the swimmer continues producing waves, one long, unbroken string of crests and troughs will span the entire length of the pool. This is called continuous wave transmission (often called just CW transmission).

But, the swimmer could decide to produce one wave, rest (thus saving his energy), and then produce another wave - followed by another rest period. By saving his energy between waves, the swimmer could slap the water harder and produce waves of greater amplitude. This is called pulse transmission. Lasers transmit in a similar manner; they are either continuous wave or pulse.

In a V-shaped swimming pool with two shallow ends converging to one deep end, two swimmers at rest (swimmer A and swimmer B, equal distance from the deep end) start slapping the water at exactly the same time and with exactly the same intervals. Not only would both waves (wave A and wave B) have the same frequency but, as they passed the sunbather, all the crests of wave A would pass exactly at the same time as all the crests of wave B. Similarly, all the troughs of wave A would pass at exactly the same time as the troughs of wave B. The two waves are said to be in phase.

Two waves being in phase or out of phase refers to a comparison of the two waves at a given point. In the example above, the given point is the exact spot where the two waves pass the sunbather. Two waves can be out of phase at a given point for a variety of reasons. The disturbances could have started at different times. The frequencies of the two waves could be different, or the distance travelled to the given point (called the path length) could be more for one wave than the other. If the two waves were in different mediums, they could have different velocities.

Two waves having identical frequencies and velocities that are out of phase, will be so to the same degree at all points. For example, two waves at 60 Hz that are 80° out of phase at the sunbather will be 80° out of phase at the deep end of the pool and all points in between.

Two waves with equal frequencies and velocities starting at the same time can be out of phase if their path lengths are different, e.g., if swimmer A was slightly further from the sunbather than swimmer B. Holographers use this fact to create their holograms.

When wave A and wave B meet at the deep end, they will join together and this phenomenon is called interference. How they interfere depends on what degree the two waves are in or out of phase (called their phase relation-

ship). If wave A and wave B are in phase, the crests of the two waves will meet and combine to form one large crest for each cycle whose amplitude is the sum of the two individual crests. The troughs of wave A and wave B would also combine to form one large trough in each cycle whose amplitude is the sum of the two individual troughs. This is called constructive interference. (See diagram A.)

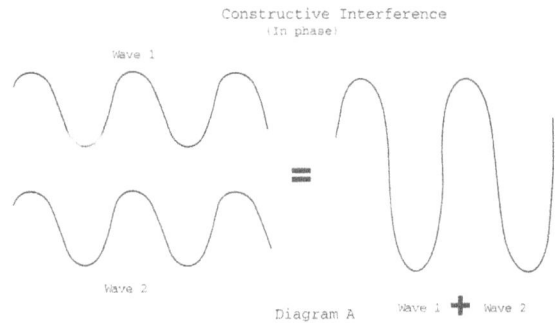

Diagram A

If the two waves are 180° out of phase, the crests of wave A would merge with the troughs of wave B (and vice versa), cancelling each other's amplitudes. The result is a combined wave with little or no amplitude. This is called destructive interference. (See diagram B.)

Because the phase relationship of two waves can change from one point to another, the two waves can be in phase when they pass the sunbather, but out of phase when they hit the deep end of the pool. The ability of the two waves to stay in phase while they travel the length of the pool is called coherence. Waves that stay in phase for a long time are said to be very coherent.

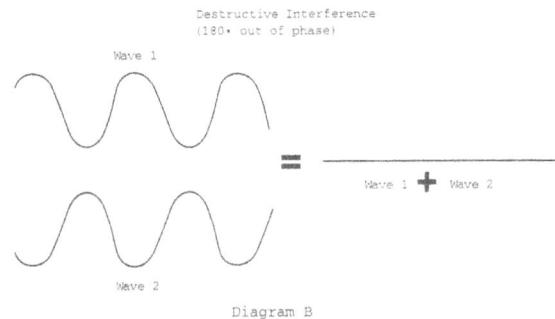

Diagram B

Suppose the two swimmers agreed to create continual constructive interference at the deep end of the pool. They would start slapping the pool at the same time while trying to maintain exactly the same frequency. If both swimmers have extremely good timing, they can keep the two waves coherent for a long period. But the swimmers, like other producers of waves (lasers, for example) aren't perfect. Their frequency may be slightly off.

The difference in frequency of any source (source is a common term for a device or system that produces waves) is called its bandwidth (abbreviated f, and measured in Hertz). The coherence of a source can be determined by measuring how long the waves stay in phase. Be it swimmers or a laser, the distance that the source can guarantee

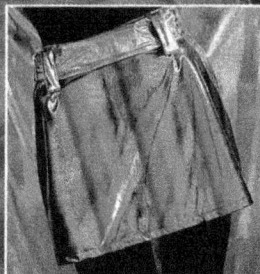

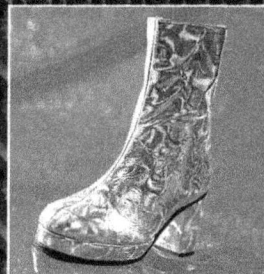

Lock Out Mode-hops Drift and Jitter

With Z-Lok and J-Lok

Lock in the ultimate in single-frequency performance with a BeamLok® ion laser equipped with Z-Lok® and J-Lok®.

This is the only commercial system to actively stabilize the three important parameters for single-frequency operation: mode hops, long-term frequency drift and short-term frequency jitter.

Mode-hops, drift, and jitter are all stabilized with the same compact PZT device which is integrated into every BeamLok ion laser.

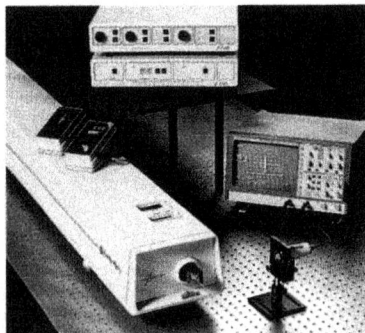

The Ultimate in Single-Frequency Performance.

❏ Mode-hop free operation
❏ Drift ≤ 30 MHz/°C
❏ Jitter ≤ 2 MHz

So, if you're performing holography, interferometry, LDV studies, or other applications that require the most stable single-frequency performance available, specify a BeamLok ion laser with Z-Lok and J-Lok. And lock in stability.

In North America, call your local Spectra-Physics office, or dial 1-800-775-5273.

S Spectra-Physics
Spectra-Physics Lasers

Call: 1-800-SPL-LASER (775-5273)   Web: http://www.splasers.com   E-mail: sales@splasers.com

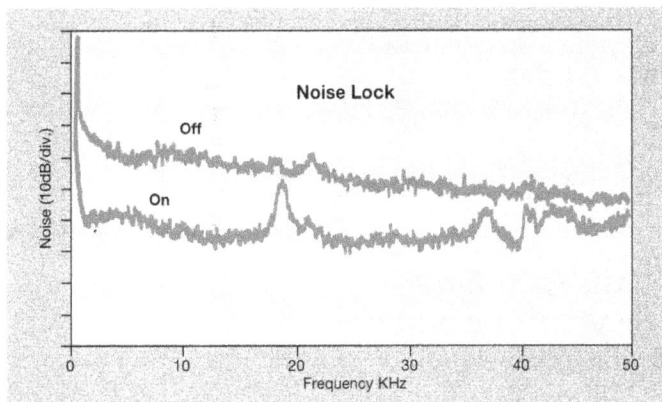

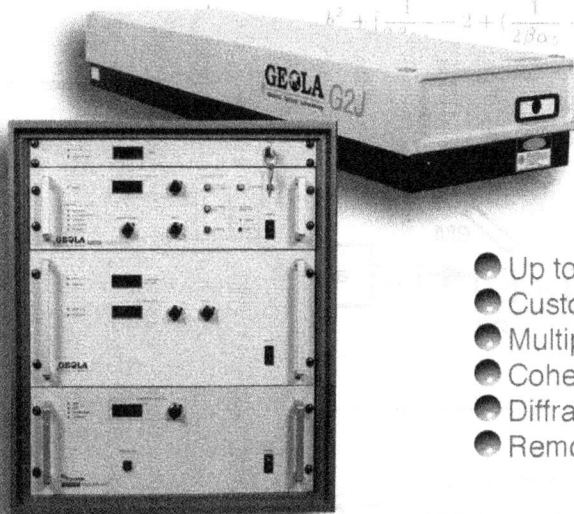

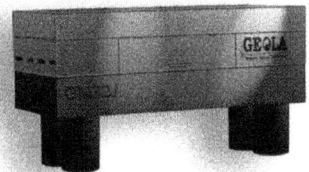

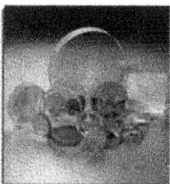

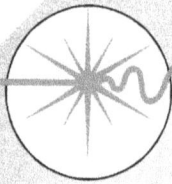

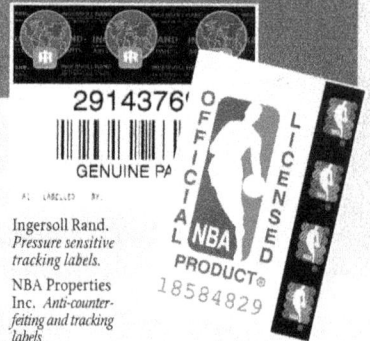

the waves will stay in phase is called its coherence length (abbreviated **L**; it is measured in meters or subunits of meters). The coherence length of a source is of great importance to holographers. It is directly related to bandwidth by the equation:

$$L = v/Df$$

Lasers are used to produce holograms primarily because no other source offers enough coherence.

Although electromagnetic waves exhibit exactly the same wave properties as the water waves described above, there are some notable differences between the two. Water waves propagate in two dimensions on a plane. Electromagnetic waves tend to propagate in three dimensions. As mentioned earlier, electromagnetic waves can propagate with or without a medium. Electromagnetic waves move extremely fast; water waves move relatively slowly. The interference of a water wave is determined by its amplitude with electromagnetic waves, it is a function of its intensity.

In empty space, electromagnetic waves move at a velocity of 300 million meters per second (3 x10^{8}m/S, also known as the **speed of light**– it is abbreviated in equations with the letter c). In his 1905 paper on special relativity, Albert Einstein correctly defined the speed of light as the absolute fastest velocity possible– a cosmic speed limit, so to speak. In air, the velocity of electromagnetic waves isjust slightly less than "c". In most applications where electromagnetic waves are travelling through air, it is acceptable to use "c" as the velocity of the wave. Therefore, for electromagnetic waves travelling in free space or air:

$$f = c/l \text{ or } l = c/f$$

and

$$L = c/Df$$

Electromagnetic waves are a union of electric and magnetic fields that are at right angles (90°) to both each other and the direction of their movement. (See diagram C.) When electromagnetic waves propagate, there is an infinite amount of directions in which they can travel. A laser is designed to channel waves such that their propagation is substantially in one direction.

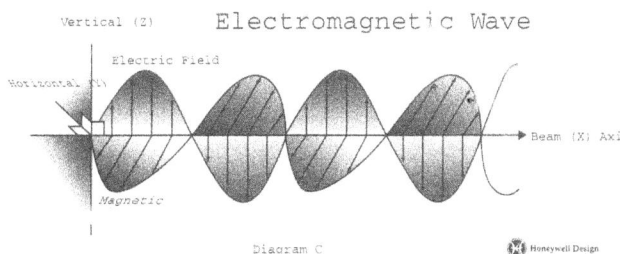

Diagram C   Honeywell Design

This unidirectional propagation of electromagnetic radiation is generally referred to as a **laser beam.**

Even while moving in a common direction, each wave can have its electric field and magnetic field oriented differently. The electric field can point straight down on the first wave, sideways on the second. There is an infinite amount of directions the electric field (or magnetic field, which stays exactly at a right angle to the electric field) can be pointing on the beam "axis" for each wave moving in the same direction.

Many properties of waves are more consistent if their electric and magnetic fields are properly aligned. The ability of electromagnetic waves to be aligned in the same orientation on the beam axis is called **polarization**. The human eye is not sensitive to polarization and cannot distinguish between polarized or unpolarized light waves. Some insects, like bees, are more sensitive and use polarization to determine direction. In holography, where consistency of the source's waves is critical, polarization is essential.

Because a reference point is needed in defining polarization, the electrical field is used to identify the position. If the electric field is travelling directly on the xz plane, the wave is defined as **vertically polarized**. How close the beam is to being polarized in the vertical position can be described by its **polarization ratio**. A laser beam with a 100:1 polarization ratio is very close to being polarized in the vertical plane - a 500:1 ratio is closer still.

Polarization can be achieved by several means, including reflection, transmission, scattering and birefringence. **Birefringence** is a phenomenon that occurs in certain crystals such as calcite and other materials. Such materials limit the absorption of waves to those with specific electric field orientations. Sunglasses use this effect, reducing the amount of glare received by the eyes. Crystals with maximum birefringence allow only one orientation to be absorbed and transmitted. The beam exiting the crystal is polarized.

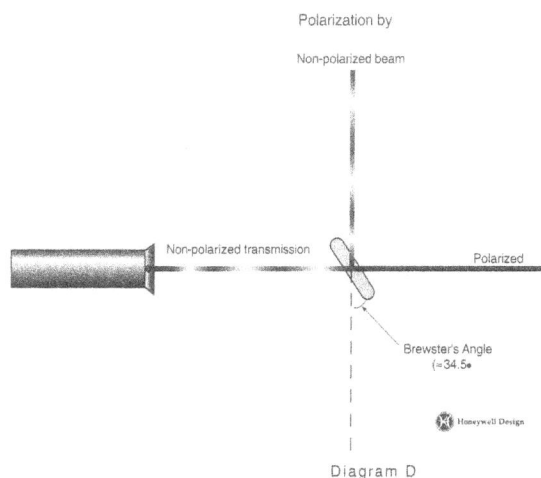

Diagram D

Scattering occurs when an atom deflects a photon away from it. Scattering of electromagnetic radiation in the earth's atmosphere produces partial polarization . Bees use this type of polarization for navigation. Polarization by scattering is not an effective source for most photonic applications. For lasers, polarization by transmission is more relevant.

If an unpolarized beam strikes a non-conductive, transparent target at a specific angle (called **Brewster's** angle; see diagram D), a polarized beam will pass through the target. This process is called polarization by transmission. Waves in the beam that do not have the selected electric field orientation are reflected from the target. Brewster's angle varies for different wavelengths and materials, but for most lasers it is in the neighborhood of 34.5 degrees.

Particle Properties of Light

On October 19, 1900, Max Planck introduced a concept that was to revolutionize science. In an effort to resolve conflicts between scientific theory and experimental evidence, Planck suggested that **energy** (abbreviated Tn equations with the letter **E**) is not continuous, but instead comes in discrete little "packets" called **quanta**. Further, an "energy packet" of light was directly related to its frequency by the equation:

$$E = h/f$$

where h is defined as **Planck's Constant** and is equal to 6.63×10^{-34} Joule Seconds.

Although the mathematics seemed to work, and it did resolve current conflicts between theory and experimental data, the implications of Planck's hypothesis were rather hard to accept. Clearly, light exhibited wave properties such as **diffraction** (bending of a wave around a comer), inter- ference and polarization. Waves are inherently contiritous and not discrete. Frequency, for example, used in Planck's equation is a wave phenomenon. Yet the energy in the same equation describes light as discrete packages of h/f. How could light possibly consist of particles and demonstrate properties of waves?

The numerous experiments and the profound mathematics that followed are extremely significant and detailed. The final result of two decades of scientific fervor was a new definition of the laws of physics now known as **quantum mechanics.**

At the core of quantum theory is the concept of **duality** which states that light, electromagnetic radiation, energy and even matter is both a wave and a particle. Electromagnetic radiation itself is composed of minute "wave packets" called **photons** that demonstrate properties of both continuous waves and discrete particles.

In terms of wavelengths, the energy of one photon is expressed as:

$$E = hlc/l \text{ or } l = hlc/E$$

As stated earlier, all electromagnetic radiation is the same phenomenon. Only the energy per photon is different. The fundamental unit of electromagnetic radiation is the photon. One can classify all forms of electromagnetic radiation by the wavelength (or frequency) of the photon. Because the wavelength involved in the common forms of electromagnetic radiation is small, it is usually measured in nanometers (1 x 10-9 meters, abbreviated **nm**). In visible light, the wavelength of a photon determines its color. Red had the longest wavelength (740-622 nm), followed by orange, yellow, green (577-490 nm), blue (489-430 nm) and

violet (429-390 nm). White light is a mixture of all colors. Photons with wavelengths greater than 740 nm produce infrared (below red) radiation. Photons with wavelengths less than 390 nm create ultraviolet (beyond violet) radiation.

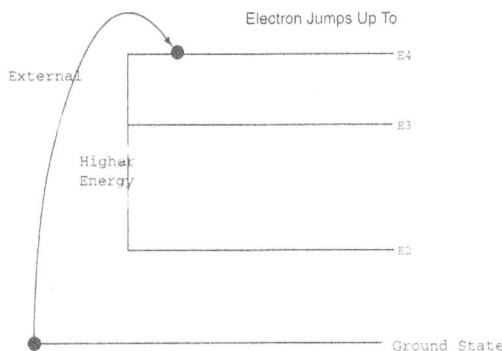

Diagram E

Photons can be created by transferring energy to an atom. In an atom, electrons reside in various positions and energy states. If tlit atom is stable, the electrons are defined as being in their **ground** (lowest level of energy) state. When external energy is transferred to the atom, the electrons get "excited" and respond by jumping up to higher, or excited **energy states.** (See diagram E.)

Quantum mechanics dictate that an electron cannot reside between two energy states. It has to jump up all the way to a higher state or not jump at all. The electron will stay in the excited state for a very short period and then spontaneously drop back down to a lower energy state.

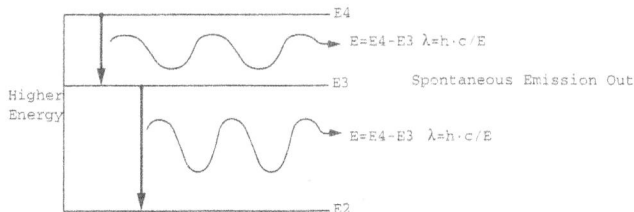

Diagram F

When the electron drops one or more energy steps (also called **transitions**), it releases the amount of energy difference between the two states (see diagram F) in the form of a photon. The wavelength of the photon is determined by the amount of energy released-the energy difference of the two energy states - through the formula $l = hlc/E$. If the energy released is small, the wavelength of the photon will be large and vice versa. This process is called **spontaneous emission.**

There are two principal mechanisms that enable energy to be transferred to an atom: absorption and collision. **Absorption** occurs when a photon bumps into an atom. If the photon's energy (determined by the wavelength of the photon using $E = hc/l$) matches the energy difference between a lower energy state where an electron resides and a higher energy state (called an **energy band gap**), the atom will absorb the photon. The photon energy is transferred to the atom, kicking the electron up to the higher energy state. If the photon's energy does not match any of the energy differences in two excited states of the atom, scattering occurs, redirecting the photon without otherwise altering it.

Collision occurs when a moving particle (an electron, ion, atom or molecule) smashes into the atom with the proper amount of momentum. Some or all of the particle's kinetic energy is transferred to the atom, again raising the electrons of the atom to a higher energy state.

Laser Theory

Albert Einstein was the first to recognize the significance of Planck's concept of quantized energy. He used Planck's $E = h/f$ equation to derive his explanation of the photoelectric effect in 1905, for which he later received the Nobel Prize in Physics. In 1916, Einstein predicted another phenomenon now known as **stimulated emission** - the basis for all laser technology. The same year, Einstein also released his most prized work, the general theory of relativity. The theory of stimulated emission went unnoticed until the late 1940's.

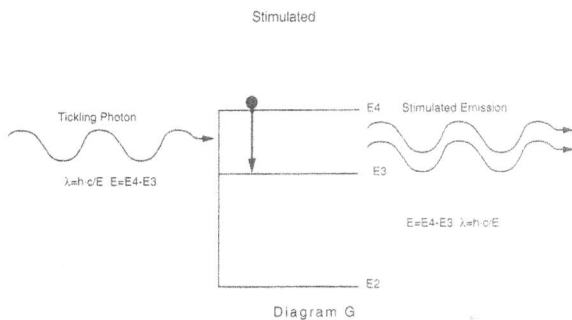

Diagram G

An excited electron in a higher energy state will spontaneously drop to a lower energy state and emit a photon. The wavelength of that photon is determined by the difference of the energy between the two energy between the two energy states $(l = hc/E)$. If, before the electron drops to the lower energy state, a photon with a wavelength identical to that which is about be produced by spontaneous emission passes by the exited atom, it will stimulate the electron to drop. This stimulation (also called "tickling" see diagram G) will force an emission of a photon that is identical to one passing. This process is stimulated emission. Both photons have the same wavelength (and therefore the same frequency), are in phase, coherent, and travelling in the same direction. It is important to note that the "tickling" photon is not absorbed by the atom; it must only pass closely.

In an environment with many identical excited atoms, photons can multiply rapidly through stimulated emission. Two photons quickly become four which quickly become eight. Since atoms and photons are extremely small, eight photons can become billions in a reasonably short distance. This process of multiplying photons is called amplification; it is the essence of a laser. The term LASER itself is an acronym for Light Amplification from Stimulated Emission of (electromagnetic) Radiation. Billions of photons, with the same wavelength, in phase, coherent and travelling in the same direction can be a very useful tool.

For lasing to occur, it is essential that there are more atoms with electrons in a higher energy state than those at the lower energy state-a condition called a population inversion. An atom with its electrons in lower energy states will absorb a passing photon instead of duplicating it. If there is more absorption than stimulated emission, amplification will not occur. How can a baker increase his inventory if he has five hungry children and only four cookies in the oven? A population inversion assures continuous multiplication of photons.

To create a laser (see diagram H), many atoms of one type must be contained in a given space. The atoms used for lasing are called the active medium and are housed in a container. Since the energy states of the active medium create most or all of the stimulated emissions, it is the active medium that determines the possible wavelengths produced by the laser ($1 = hc/E$). Solid state lasers have active mediums made from matter that is solid at room temperature. Neodymium and chromium ions (an ion isjust an atom with either an excess or deficit of electrons) are common active mediums used in solid state lasers. The active medium in liquid lasers are, of course, liquid and those in gas lasers are gas. Common active mediums used in gas lasers are neon, argon and ionized cadmium.

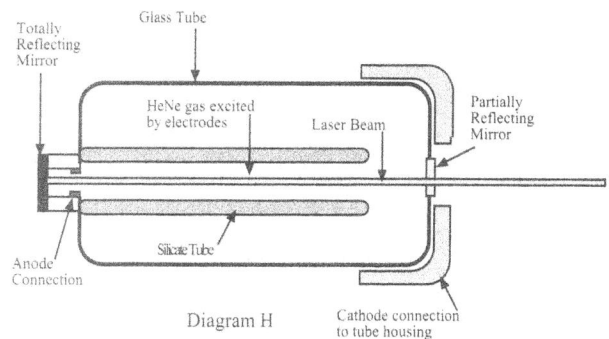

Diagram H

A pump transfers energy to the active medium, raising their electrons to an excited (higher energy) state. Brisk and continuous pumping will create and maintain a population inversion. As stated earlier, energy can be transferred by either absorption or collisions. In solid state lasers, an optical pump produces a flood of photons with energies that are easily absorbed by the active medium. Common optical pumps are flash lamps, laser diode arrays and other gas, liquid and solid state lasers.

The vast majority of gas lasers use electron collisions to pump energy to the active medium. Normally, electrons will not flow through a neutral gas. If a large voltage (from 1,000 to 20,000 Volts, depending on the gas) is applied to the gas, the gas will "break down" and allow a **discharge** of electrons to rush through it. This initial voltage is called a **spike**, and is applied to the gas through two electrodes (an **anode** and a **cathode**) on opposite ends of the container. Once electron current is flowing through the tube, the voltage is automatically reduced (by 90 to 4,000 Volts, depending on the gas). The discharge of electrons and other charged particles collide with the atoms in the active medium, enabling energy to be transferred. Both the spike voltage and the operating voltage necessary to operate the laser are furnished by a power supply, usually an external box that transforms either 117V AC or 220V AC to the required voltage.

Many active mediums are not efficient at receiving energy from the pump. In such cases, the active medium must be combined with a transfer medium-a substance that compensates by efficiently collecting energy from the pump and then passing it on to the active medium. Solid state transfer mediums should be good absorbers; gas transfer mediums require the ability to efficiently receive energy from collisions.

The transfer medium must be compatible with the active medium in two ways. First, it must have some common higher energy states that provide a channel to efficiently pass energy to the active medium. Second, it must be chemically compatible, allowing peaceful coexistence with the active medium as well as the other components inside the tube.

There are typically 5-20 transfer medium atoms for every active medium atom. Inert helium makes a good transfer medium and is used in helium-neon (**HeNe**, pronounced Hee-Nee) and helium-cadmium (**HeCd**, pronounced Hee-Cad) gas lasers. Common transfer mediums in solid state lasers are yttrium atrium garnet (called "**YAG**"), yttrium lithium fluoride (**YLF**, pronounced "Yelf'), sapphire and ruby.

The container that holds the active medium (and transfer medium, if applicable) must protect it from elements that may interfere with the lasing process. In solid state lasers, the crystalline transfer medium encompasses the active medium atoms, forming a strong durable solid structure that serves as the container. Because the transfer media in solid state lasers house and transfer energy to the active medium, they are called **hosts**.

For gas lasers, the container is almost always a long cylindrical **tube**. Tubes are made of various materials; ceramic and glass are the most common. Inside the tube is an equally-long, yet very narrow « 3 millimeters) **bore** that allows lasing to occur in a straight, usually horizontal path.

During lasing, a wide variety of spontaneous and stimulated emissions occur throughout the tube with photons propagating in every conceivable direction. All emissions except those travelling down the bore exit from all sides of the tube with limited or no amplification. Those photons travelling down the length of the bore continue to multiply through stimulated emission. By making the bore long enough, one could continue the lasing process until adequate amplification was achieved. At this point, a billion or so photons exit the bore in the form of a laser beam.

Unfortunately, in order to achieve sufficient amplification, the bore would have to be forty feet long. A forty foot laser would be extremely awkward to both transport and operate. A shorter bore is needed to make the device more practical.

By passing photons back and forth (called **optical feedback**) several times through a shorter bore, adequate amplification can be attained. This is accomplished simply by placing mirrors on both sides of the bore. If both mirrors are 100% reflective, extremely high amplification is achieved. However, no photons exit the laser (photon exiting the laser is called **transmission**). This, of course, has no value whatsoever.

However, if one mirror could reflect some of the photons back into the bore while allowing the remainder to exit, both amplification and transmission could occur. Such a mirror found on lasers is called the **output coupler**, or **OC**. The second mirror is known as the **high-reflector**, or **HR**. A perfect high-reflector will provide 100% reflection. In actual lasers, there is a small amount of light transmitted from the HR, referred to as **leakage**.

Power is the amount of energy a source produces each second. It is measured in units of Joules per second, more commonly called **Watts** - in honor of the British scientist James Watt. The power in a laser reflects the amount of photons per second exiting the laser.

Laser designers strive to maximize the power produced by the laser. However, if the OC allows too many photons to exit, the number of photons returning to the bore may not be adequate to provide significant amplification. This, in turn, limits the amount of transmission. Therefore, the amount of transmission and reflection provided by the OC must be properly balanced to compliment both the amplification process in the laser **cavity** (area between the two mirrors) and the transmission from the cavity. In most lasers, output couplers allow 1-3% of the photons to be transmitted.

It is common to have mUltiple wavelengths lasing inside of the cavity. Because there are a variety of paths of energy states for which an excited electron can travel back to the ground, there are a variety of energies (and therefore, wavelengths) that can be emitted throughout its descent. The electron can also bounce down two steps at a time-maybe three-each time emitting photons of higher energy and lower wavelengths.

Certain transitions are more dominant, however. The more spontaneous emissions produced at a given wavelength, the greater probability of stimulated emission. The more stimulated emissions, the more amplification. The most dominant wavelength in a given laser is called the **primary wavelength**. The second most dominant wavelength is called the **secondary wavelength**.

Photons with undesirable wavelengths can be eliminated from the cavity by putting special thin film coatings on the mirrors. Such coatings only allow a specific range of wave- lengths (for example 430-460 nm) to reflect back into the cavity. Thus photons with undesirable wavelengths will not multiply.

The lasing process described above is continuous wave transmission. Lasers that produce this kind of transmission are called, expectably, **continuous wave lasers** (or just **CW lasers**). Lasers that provide pulse transmission are called **pulse lasers** by some and **pulsed lasers** by others.

A continuous wave laser can be converted to a high energy pulse laser by installing a Q-switch in the laser cavity. **Q-switches** are devices that enable the active or transfer medium in the cavity to collect maximum pump energy before beginning the process of stimulated emissions. The Q in Q-switches is an abbreviation for "quality factor" to represent the quality of a feedback system. In the "low Q mode", a Q-switch limits optical feedback in the cavity for a very short period. If optical feedback is blocked, continuous stimulated emission will not occur.

During this period, the active medium continues to collect energy from either the pump or the transfer medium. A high percentage of electrons are elevated to higher energy states creating a large population inversion. When the Q-switch opens (high Q mode), a rapid and powerful episode of stimulated emission occurs in the cavity until the Q-switch is again closed. The active medium begins receiving energy, and again its electrons begin to elevate in preparation for the next high Q mode.

The repetitive bursts of lasing in the cavity result in a string of powerful energy pulses departing from the output coupler. The average length of time of a pulse is referred to as its **pulse length** (measured in seconds and subunits of seconds). The number of pulses per second is called its **repetition rate** (or more commonly, **rep rate**, measured in Hertz).

There are four types of Q-switches: chemical, electro-optical, acousto-optical and mechanical. The first three are common and found in a variety of applications. Mechanical Q-switches are seldom used because they tend to be slow, noisy and produce unwanted vibrations.

The laser tube and mirrors are held by a mechanical support structure almost unanimously referred to as a **resonator**. The title, nevertheless, is wrong. A resonator is system consisting of a laser cavity and mirrors that enables rapid bi-directional optical feedback. It oscillates. Most physicists will readily admit that the term is incorrect; however, there is no other term available other than "mechanical support structure". Conforming to the majority, I will use the term "resonator" in this paper with the knowledge it is incorrect. The resonator has the task of keeping the bore straight and aligned with the mirrors. This is not an easy assignment. The lasing process produces an ample amount of heat. The heat creates thermal expansion, which tends to shift the mechanisms that hold the mirrors (**mirror mounts**) and bore. The bore itself, being long and quite narrow, is extremely susceptible to thermal distortions.

When the laser cools down, the components of the laser tend to contract. Even the best resonators will sometime fail to keep the bore and mirrors in line. Because it is easier to align the mirrors than the bore, alignment devices are placed on the mirror mounts.

Such devices, called **tilt plates** or **xy plates**, enable the operator to change the positioning of the mirrors either sideways (horizontal, or "x" position) or up/down (vertical, or "y" position) without changing the z position (frontwards and backwards-this would change the cavity length which can only hurt the laser's performance-see section V). Proper adjustment of the tilt plates enables maximum amplification inside the cavity, producing maximum laser power. The resonator must also provide mechanical protection from the routine bumps and bruises that may occur.

The tilt plates and mirror mounts are secured in the resonator by three or four resonator rods, which span the length of the laser. Resonator rods are the backbone of the resonator. They give mechanical support to the entire laser head and the hardware that holds the tube.

On large lasers, the resonator rods are made from carbon graphite and are generally one to two inches in diameter. Carbon graphite has a very low **coefficient of expansion**, which means it will have minimal movement (expansions and contractions) when temperatures fluctuate. In smaller lasers, invar is generally used. Invar has a larger coefficient of expansion than carbon graphite, but provides equal strength at one-fourth the thickness.

In polarized lasers, **Brewster windows** are attached to the ends of the bore. The windows, non-conductive and transparent panels placed at Brewster's angle, seal the bore and polarize the photons that pass through it. One of three methods may be used for sealing the windows to the bore: epoxy, frit and optical contact. Sealing the bore and securing the Brewster window by epoxy was one of the first methods used on gas lasers. A space-grade epoxy glue is evenly distributed on a quartz **Brewster stub** and meticulously fastened on the bore. Frit sealing involves heating glass between the Brewster stub and the bore.

Perhaps the most effective method of sealing the windows, yet hardest to do properly, is optical contact. Optical contact requires a precise mechanical fit between the Brewster stub and the bore. The bore is heated, microscopically melting the Brewster stub directly to the bore.

In low-powered lasers, such as air-cooled ion lasers, HeNe and HeCd lasers, the Brewster windows are almost always made from fused silica. Fused silica is preferred due to its low absorption of electromagnetic radiation, which enables the highest possible transmission. In higher powered lasers, such as large frame ion lasers, fused silica is susceptible to solarization. **Solarization** occurs when an excessive amount of the transmitted photon energy is absorbed by a Brewster window, changing its optical properties. Two of the more common effects of solarization are thermal lensing and color centering.

Thermal lensing is caused when photon energy absorbed by the Brewster window is converted into heat. Heat circulating in the window changes its optical properties. It also warms the air surrounding the Brewster window, distorting the optical properties of the air. Both the window and the surrounding air act as a randomly shifting lens that causes a slight variation the direction of the beam. When the shifted beam hits the mirror, it may not reflect precisely down the center of the bore. Part of the beam may "clip" upon entering the bore, causing a significant reduction in power.

The effects of thermal lensing are very similar to those of a misaligned mirror. It is common for an unaware operator to try to correct the malfunction by adjusting the tilt plates. Often, it will work-temporarily.

Unfortunately, heat energy is not stationary. The distortions in the lens and its surrounding air can change, causing the beam to shift again. Or, if the operator turns the laser off, thermal lensing may not occur again until hours after restarting it. All previous adjustment of the tilt plates are no longer valid. The laser is now legitimately misaligned.

Thermal lensing is extremely frustrating if not detected. An operator who finds the large frame ion laser constantly out of alignment may find it necessary to inspect the Brewster windows.

Color centering is the result of extreme solarization. In color centering, the Brewster window absorbs enough energy from the laser beam to change its molecular structure. When this occurs, the Brewster window will lose its transparentness.

Because most of the energy of a beam is in its core, the molecular restructuring is generally restricted to the center of the Brewster window. Photons will no longer pass through the damaged region, producing a beam that has no light in its center. This "donut" shaped beam is unusable in most applications. To help reduce the effects of solarization, thermal lensing and color centering, Brewster windows on large frame ion lasers are generally made from crystal quartz.

Additional Properties of Lasers

Any light source that delivers exactly one wavelength is said to be **monochromatic**.

Ideally, stimulated emissions from a group of identical fuel atoms should produce very distinct wavelengths (or frequencies) lasing within the cavity; for example, a HeCd laser would produce amplification at 325.0 and 441.6 nm. By using proper coatings on the mirrors, the 325.0 nm wavelength can be removed from the lasing process, thus producing a monochromatic beam at 441.6 nm. In actuality, however, this is not the case.

In the same manner the two swimmers discussed earlier produced slightly different wave frequencies, the lasing inside of the cavity produces minute variations of photon frequencies (or wavelengths) in the transmitted beam defined as the laser's **bandwidth** (Df).

Several factors contribute to variation of photon frequencies inside the cavity, including the motion of an atom at the time it emits a photon.

Variation of photon wavelengths in a laser, called its **linewidth** (Dl, measured in nanometers) is exactly the same phenomenon as a laser's bandwidth (Df). Quantitatively, however, the two will not have the same numerical values. Linewidth can be numerically converted to bandwidth (and vice versa) by the following equations:

$Dl/l = Df/f$ or $Dl = Dfl/f$ or $Df = Dlf/l$

Often, it is helpful to eliminate frequency entirely from our equations. This may be done by inserting the expression $f=c/l$ (see section II) into the above equations and applying some basic algebra. The results are:

$Dl = Dfl^2/c$ and $Df = Dlc/l^2$

Coherence length (L) can also be expressed in terms of wavelength and lidewidth:

$L = c/Df = l2/Dl$

Waves moving back and forth in a confined region create constructive and destructive interference similar to that in Section 2. If the frequency of the waves matches the resonant frequency of the region, constructive interference will occur throughout the length of the region. A set of non-moving waves, complete with crests and troughs, form in the region. These "standing" waves now dominate.

By changing the length of the region (or the frequency of the waves), the two-frequencies no longer match. Destructive interference is introduced, and the standing waves disappear. If you continue to change the length, you will find other discrete distances (called **harmonics**) that will enable the frequency of the region to match those of the waves. Destructive interference is again replaced by constructive interference, and the standing waves reappear.

By increasing the length of the region, you enable more standing waves to exist within its boundaries. By decreasing the length of the region, fewer standing waves exist.

A laser has waves (photons are waves) travelling back and forth in a confined region (the laser cavity). Because many frequencies exist within the lasers bandwidth (Dt), some of them will match the resonant frequency of the structure. Standing waves will form within the cavity. Only those frequencies creating standing waves will continue to lase. These frequencies are called **longitudinal modes**.

Within the bandwidth are a set of distinct longitudinal modes spaced equally apart. There no frequencies lasing in between them. The distance between each mode is called the **longitudinal mode spacing** (abbreviated m in terms of frequency, measured in Hertz-see Diagram I).

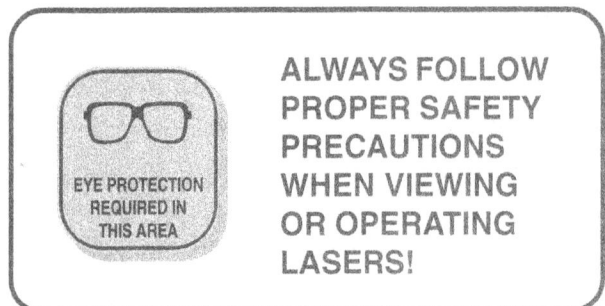

ALWAYS FOLLOW PROPER SAFETY PRECAUTIONS WHEN VIEWING OR OPERATING LASERS!

EYE PROTECTION REQUIRED IN THIS AREA

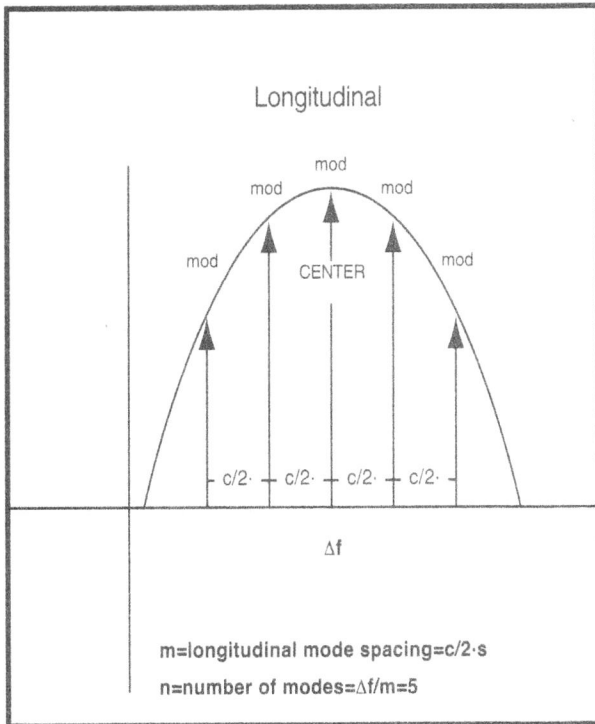

Longitudinal

mod
mod   mod
mod   mod
CENTER
mod   mod

c/2· c/2· c/2· c/2·

Δf

m=longitudinal mode spacing=c/2·s
n=number of modes=Δf/m=5

Diagram I

and defined geometrical paths. The most fundamental path will produce a clear, uninterrupted spot when projected on a target. Other paths, or "modes" will have dark irregularities (called **nodal lines**) separating the spot. The nodal lines can be either vertical or horizontal. (See diagram J.)

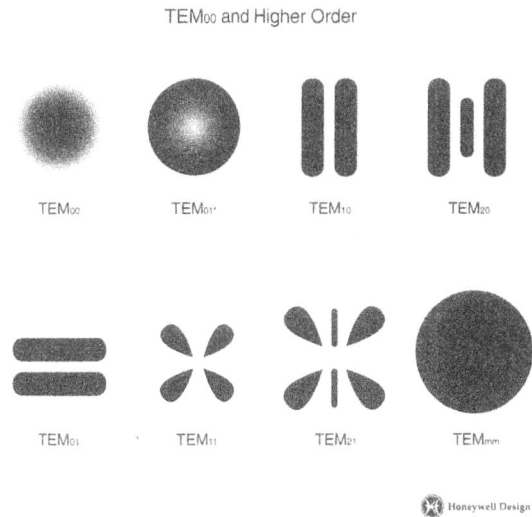

TEM00 and Higher Order

TEM00   TEM01*   TEM10   TEM20

TEM01   TEM11   TEM21   TEMmm

Honeywell Design

Diagram J

The longitudinal mode spacing is determined by the separation of the cavity mirrors (abbreviated S, also called the **cavity length**, it is measured in meters) and can be calculated with the following equation:

$m=c/2lS$

In terms of wavelengths, longitudinal mode spacing (**M**, in nanometers) can be calculated:

$M= mll^2/c = l^2/2lS$

The number of longitudinal modes (n) in the bandwidth can be determined by the equation:

$n = Df/m$

In terms of bandwidth or linewidth, the number of longitudinal modes should be identical. The equation in terms of linewidth is:

$n = (2lSlDl)/l^2$

Because the mode spacings are very close together (generally less than 1/1000 of anm), very small changes in the cavity length can cause the modes to move. In argon ion and other lasers that generate a significant amount of heat, the cavity can expand or contract enough to cause the longitudinal modes to literally jump over each other. This phenomenon is called **mode hopping**. In many laser applications, such as holography, mode hopping can cause undesirable effects.

Another mode that manifests itself inside the cavity is the **transverse electric and magnetic mode**, more commonly called the **TEM mode**. Light will propagate with distinct

These patterns are classified with subscript using the form TEM_{vh} (where "v" designates the number ofnodal lines in the vertical direction and "h" designates the number of nodal lines in the horizontal direction) for example TEM_{10}, TEM_{20}, TEM_{11}. One of the more interesting patterns is the TEM_{01*}, a mode that produces a large circular node in the middle of the beam. Because of its distinct pattern, TEM_{01*} mode is often referred to as the "**donut mode**."

Lasers can be built to produce only the fundamental mode TEM_{00} by reducing the ratio of bore diameter to the diameter of the TEM_{00} beam, usually to less than 3:1. This can be achieved by properly selecting mirror combinations that encourage TEM_{00} transmission.

Generally, long thin bores produce TEM_{00} mode more readily than fatter ones. In holographic applications, TEM_{00} transmission is essential.

TEM∞ Gaussian Energy

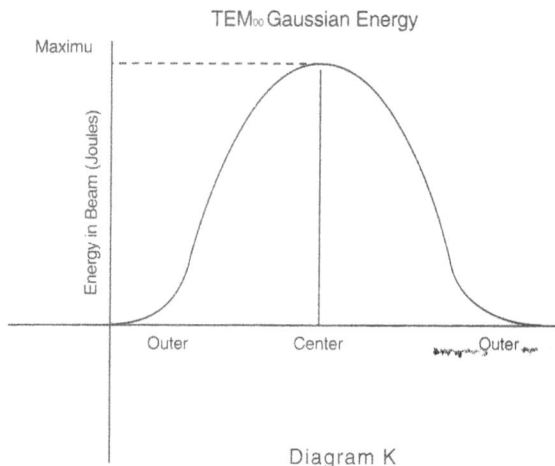

Diagram K

TEM$_{00}$ beams can be focused down to the smallest possible size spot known as the diffraction limit. In theory, the **diffraction limit** of a beam is its wavelength. In real applications, the smallest possible size of a focused spot is slightly higher. In holography, the ability to focus down to a "tight" spot is not an advantage since the beam is expanded.

The beam of a **multimode laser** has a combination of many modes resulting in an uneven energy distribution in its cross section. Multimode beams (commonly written **TEM$_{mm}$**) deliver more energy than, but have none of the advantages of, those modes listed above. A TEM$_{mm}$ beam can also be focused down, but not nearly as tightly as TEM$_{00}$. Although multimode lasers are used in a variety of applications where laser power is the primary concern, they can not be used for holography.

Mechanical instabilities in the optics and tube can cause the beam to wander. A laser's **beam-pointing stability** (also called **angular drift**, measured in microradians) measures how much the laser beam drifts from the beam axis. Beam pointing stabilities of 60 microradians or less are considered good.

Types of CW Lasers Used in Holography

The three primary types of CW lasers used in holography are argon ion, HeCd, and HeNe lasers. Each has distinct advantages that are related to the holographer's needs . Typically, the type of recording material, size of the hologram, and operator budget determines which laser is best suited.

In embossed holography, photoresist is the primarily medium used for recording images. The use of photoresist enables mass-produced holograms. Photoresist chemically etches the holographic image onto a glass plate. The optically-engraved glass plate (called a **master**) is made conductive, and then electroplated-which produces a **shim**. The shim is placed on an embossing machine for mass stamping of embossed holograms

Because photoresist is extremely sensitive to wavelengths between 420-450 nm, HeCd lasers (which lase at 441.6 nm) are most often chosen. HeCd lasers are much more cost-effective to operate than comparable lasers, enabling originators of holograms to keep their operating costs down.

Artistic holography isn't constrained by the necessity to mass produce. This gives the holographer freedom to choose from a variety of emulsions to produce holograms. In such cases, most holographers prefer to use emulsions sensitive to the primary wavelength (514.5 nm) of an argon laser. Argon lasers provide an attractive combination of high power and long coherence length, enabling holograms that are both large and visually striking.

In most forms of holography, the coherence length of the laser determines the size of the hologram. Generally, a 10 cm coherence length laser produces holograms that are 10 cm x 10 cm (4"l x 4"). Seasoned holographers routinely extend the size of their holograms without increasing their coherence length, some being able to shoot a 6" x 6" while using a source with 10 cm coherence length.

Novice holographers typically can't justify the hefty expense of an argon laser. A HeCd laser, though much more attractively priced, still costs over ten thousand dollars. For "weekend shooters" on a budget, the HeNe laser provides a cost-effective alternative to the higher-priced argon and HeCd lasers.   .

Certain laser parameters are essential to all forms of continuous wave holography. The beam must be TEM$_{00}$ mode. A polarized beam (at least 100:1) is necessary in all forms of holography except dot matrix; the low exposure times of dot matrix holography, typically 5 milliseconds per dot, reduce the need for polarization. The laser cannot generate excessive heat or vibration, two mortal enemies of the holographer.

With the exception of dot-matrix holography, high power is very desirable. Less photon power requires longer expo- sure times. Not only do longer exposure times raise the cost of labor (professional holographers do not work for free), but they increase the odds of something going wrong. As every holographer will gladly tell you, things do go wrong.

Argon Ion Lasers

Before the advancement of the helium-cadmium laser, argon ion lasers were the preferred choice for almost all forms CW holography. The argon laser provides a generous amount of power, polarization and coherence-three of the more prized parameters in holography.

Argon ion lasers produce lasing at many wavelengths between 454.5-528.7 nm-usually eight to ten-and can be equipped with a **prism wavelength selector** in the cavity to allow the operator to select a specific wavelength. Of primary interest in holography are the 514.5, 488.0 and 457.9 nm wavelengths. The 514-nm line is the most powerful, followed closely by the 488.0 nm line. Large-frame argon lasers can provide nearly 10 Watts of power at 514.5 nm and over 6W at 488.0 nm. Most argon lasers have a polarized beam with a polarization ratio of 100:1 or greater. The combination of high power and high coherence length make argon ion lasers the best laser for the serious artist who can afford its substantial price.

In embossed holography, the powerful 514.5 and 488.0 run wavelengths of an argon laser have little effect on the photoresist used to record the holograms. Because photoresist is extremely sensitive to wavelengths between 420 and 450-run, only the 457.9 run line can effectively expose it. Unfortunately, the 457.9 run wavelength is relatively weak in comparison to the more powerful 514.5 and 488.0 run lines-less than 20% of the relative power. Large frame ion lasers produce approximately IW of power at 457.9 nm.

Argon ion lasers can be equipped with another intracavity device called an **etalon**. An etalon is a wedge-shaped piece of high quality optical glass which, by means of constructive and destructive interference, can eliminate longitudinal modes from the laser cavity. The etalon acts as a separate laser cavity inside of the main laser cavity. When the beam enters the etalon, it is reflected inside the wedge t of onn its own optical feedback before exiting. By adjusting the angle where the beam enters and the temperature (which changes its index of refraction and cavity length) of the glass, the etalon can choose which longitudinal modes will survive (through constructive interference) and which ones will not (through destructive interference). The beam exits the etalon, plus or minus a few longitudinal modes, to resume amplification in the main cavity.

By reducing the number of longitudinal modes, the etalon reduces the bandwidth. Lowering the bandwidth raises the coherence length. In essence, the etalon allows the holographer to control how much coherence length the laser beam will have. While most lasers refer to their coherence length in centimeters, an argon laser with an etalon can attain coherence lengths ofmany meters. The trade-off, however, is power. An etalon will produce losses of 30% or more, depending on degree of bandwidth reduction.

The active medium in an argon laser is a pure argon gas. Because argon is a good energy absorber, it needs no transfer medium. Argon atoms are excited by passing a high-current density discharge through a ceramic tube. An initial spike of a few thousand volts breaks down the low pressure gas (approximately 1 torr), then the voltage drops to 90-400 volts while the current jumps to 10-70 amps (dc). The discharge current is concentrated in a small-diameter bore in which stimulated emission takes place and must be high enough to ionize the argon gas (hence, the title argon ion laser). An external magnet placed immediately outside the tube produces a magnetic field parallel to the bore axis that helps confine the discharge to the bore.

The conditions inside an ion-laser plasma tube are extremely harsh. Highly charged electrons and ions violently collide with the tube and bore eroding the surfaces, and contaminating the gas. A sizeable amount of UV radiation is generated in the cavity which, over time, tends to damage the Brewster windows and other optical surfaces.

Many times, microscopic particles from the walls of the bore can peel off and fall into the beam path. These pesky flakes can cause minute losses in power called **drop offs.** More advanced materials used in the inner cavity of argon ion laser tubes have reduced the amount of lasers experiencing drop off problems.

The average lifetime of an argon ion laser tube is directly related to how high the operator sets his tube current. Those trying to achieve maximum power will find an average tube lifetime of around 2,400 hours. More prudent operators can expect approximately 3,000 hours per tube.

Although argon lasers are capable ofproducing a significant amount of optical power, the energy efficiency of the argon laser is actually quite poor. Because stimulated emission takes place in energy transitions far above the ground state, much energy is required to pump the electrons up to an excited state.

The inefficient conversion of energy creates a substantial amount of heat. To remove this heat, metal disks are brazen inside the tube to allow it to transfer from the bore to the outside of the tube. A metal duct outside of the tube circulates water which transports excessive heat away from the tube. To keep the temperature of the tube low enough to operate, large-frame ion lasers require a water flow rate greater than three gallons per minute,

Once the water exits the laser head, it must either be recirculated or dumped down the drain. To recirculate water, a heat exchanger is placed in the water flow loop. The exchanger works much like a radiator, extracting heat from water, exiting the head and transferring it to a different medium (usually air or other water). Cooler water exits the heat exchanger and is sent back to the tube.

Although heat exchangers can be relatively effective, their high vibration makes them unappealing in sensitive applications like holography. Most holographers prefer an open-cycle system in which the water flows from the tap to the laser, and then down the drain.

The power supply must also be water cooled, although water circulating in the middle of high-voltage circuitry is not a favorable combination. Condensation and leakage jeopardize the perfonnance and safety of the unit.

Due to the high-current requirements needed to produce stimulated emissions, argon lasers operate from 208VAC, three-phase line voltage. The extreme conditions of the la- ser tube tend to induce nontrivial expansions and contrac- tions in the laser cavity, making argon lasers susceptible to mode hopping and poor beam pointing stability.

Though not perfect, argon ion lasers offer the best combination ofpower and coherence length for artistic holog- raphy. The price of a large-frame ion laser generally starts at $30,000. Re-tubing costs range from $13,000 to $15,000.

Helium Cadmium (HeCd) Lasers The 441.6 nm line of a HeCd laser exposes photoresist used in embossed holography ten times more effectively than the 457.9 nm line of an argon laser. Unfortunately, for many years HeCd laser were only capable of delivering a maximum of 40 mW TEM$_{00,}$ The marginal power resulted in unusually long exposure times for originators of 2D/3D (also called multiple plane), 3D and composite holograms. A 1 W argon laser at 457.9 run, comparable to a 100 mW HeCd laser in its effective exposure rate, could expose photoresist 2 1/2 times faster than the most powerful HeCd laser. Further, a HeCd laser could only provide 10 cm of coherence length, adequate enough for a standard 4" x 4" master. As noted earlier, only the more resourceful holographers could use a HeCd for the more illustrious 6" x 6" image. In comparison, Argon lasers can provide an unlimited amount of coherence length-albeit at the expense of power-by having an etalon installed.

As for originators of dot matrix holograms, in which the exposure time is trivial, HeCd lasers were readily adopted. A HeCd laser costs less to buy and operate, is easier to use, has lower maintenance, and lasts longer than a comparable argon laser. HeCd lasers do not suffer from common argon laser malfunctions such as thermal lensing, color centering, power supply leakage and mode hopping. Further, HeCd lasers operate off of standard 117VAC, need no water to cool the tube and, therefore, no special plumbing is required in the facility.

In the last four years, HeCd lasers have quadrupled their power output. A typical HeCd laser can deliver more than 150 mW TEM_{00} and provide 30 cm coherence length-more than enough for a 6" x 6" hologram. The effective exposure on photoresist for HeCd lasers now meets or exceeds large-frame argon lasers, while saving the average holographer in excess of $800 per month per laser on electricity and water bills. This improvement has created a profound change in the embossed holography marketplace. Today, seven out of every eight lasers bought for commercial embossed holography are HeCds.

The active medium used in a HeCd laser is cadmium. Standard HeCd lasers use naturally reoccurring cadmium that consists of a blend of three isotopes: Cd_{112}, Cd_{114} and Cd_{116}. It is abundant in nature, and therefore, costs pennies per gram. The strmulated emissions of cadmium provide a primary lasing wavelength of 441.6 nm, with a secondary, wavelength of 325.0 nm. One manufacturer has patented a 353.6 nm HeCd laser. With naturally reoccurring cadmium, the spectral bandwidth is 3.0 Gigahertz, which translates into 10 cm coherence length. Standard HeCd lasers can produce up to 120 mW TEM_{00} at 441.6 nm.

By processing naturally reoccurring cadmium in a manner similar to processing nuclear grade uranium, anyone of the three isotopes can be isolated, producing **isotopically enriched cadmium**-more commonly called **single isotope cadmium**. Because the technology and equipment required to produce single isotope cadmium is restricted, the processing of it is extremely expensive. This significantly raises its price. Currently, single isotope cadmium sells for over $1,600 per gram.

An average HeCd laser consumes from five to eight grams of cadmium. Using single isotope cadmium can increase the price of the laser many thousands of dollars. The highest powered HeCd laser with naturally occurring cadmium is priced around $18,000. An equivalent laser with isotopically enriched cadmium sells at $25,000.

Having only one isotope lasing in the cavity produces a third of the spectral bandwidth (1.0 GHz). One third of the bandwidth nets three times the coherence length. In essence, the use of single isotope cadmium raises the coherence length from 10 to 30 cm. As an added benefit, single isotope cadmium also produces more power-about 30% more at 441.6 nm. Single isotope HeCd lasers deliver up to 170 mW, TEM_{00} at 441.6 nm. This translates into 1.7W of equivalent power when compared to the 457.9 nm line of a large frame argon ion laser in effectively exposing photoresist.

Similar to argon ion lasers, the pumping process in HeCd lasers is accomplished by a high density discharge that breaks down helium gas (the transfer medium). The discharge is produced by an anode and a cathode and confined to a long, narrow glass bore. The spike voltage is typically 13,000 volts and after breaking down the gas (ionization), is reduced to approximately 2,000 volts.

After breakdownpfthe helium gas, cadmium placed near the anode in a **cadmium reservoir** is heated to about 2500 C by a heater wrapped around the reservoir. In approximately five minutes, the cadmium starts to vaporize. Through a natural process called catephoresis, the cadmium vapor migrates uniformly from the anode, through the bore, and towards to the cathode. Because **catephoresis** exists, a uniform distribution through the bore is possible. Otherwise, the lasing process would be impossible to control.

It is critical that,the cadmium is properly "trapped" before reaching the cathode. Cadmium ions being deposited on the cathode would drastically alter the electrical properties of the laser, affecting the laser's performance. A cold trap is placed centimeters from the cathode to stop the advancing cadmium. Another cold trap is placed in front of the Brewster window, to eliminate the deposition of cadmium on the window.

The strict regulation ofboth the helium and cadmium vapor pressures are vital to the performance of the tube. The amount of cadmium in the bore can be detected easily by measuring the tube's voltage. A feedback circuit is placed in the head that adjusts the cadmium heater should the tube voltage read too high or too low.

Helium must also be kept at a proper pressure. Helium, being a very small atom, can diffuse out of the tube; the amount is seldom of consequence. During operation, though, the cold trapping process tends to trap the smaller helium atoms under the larger and more massive cadmium atoms. The helium atoms get buried, and the tube pressure eventually lowers. To replenish the lost helium, a refill bottle is placed in the tube with a supply of high pressure helium. When the tube senses low helium pressure, a heater wrapped around the refill bottle is switched on. The elevated temperature raises the kinetic energy in the bottle, creating even higher helium pressure in the refill bottle. The higher temperature also widens the atomic spacing of the container. The combined effect creates an increased diffusion rate of helium from the refill bottle to the tube.

Helium-cadmium lasers that are stored risk subtle migration of helium atoms from refill bottle into the tube. This creates an excess of pressure in the tube, which can make the laser difficult to start. HeCd lasers, when stored, should be started at least once a month and operated 2-4 hours to stabilize the system.

A typical HeCd laser tube typically lasts 4,000 hours. Single isotope cadmium laser tubes, because manufacturers tend to limit the amount of cadmium in the reservoir, last about 3,500 hours.

HeCd lasers have quickly reached a position of dominance in the embossed holography marketplace. The combination of high power, effective exposure rate, low cost of operation, and ease of use make it the laser of choice for embossed holography.

Helium Neon (HeNe) Lasers

The HeNe laser was the first gas laser to be commercially available, brought to market in 1961. Over 30 years later, the HeNe laser is still the most commonly used laser. Supermarkets use HeNe lasers to scan the bar codes on packages for quick and efficient customer check out. High schools and universities find their low price, ease-of-use, good beam pointing stability and long tube lifetimes extremely attractive. HeNe lasers operate from a 117VAC source and are air-cooled.

An average HeNe laser costs a few hundred dollars, making it an affordable tool for those who normally could not afford the expense of an argon ion or HeCd laser. HeNe lasers are low powered, typically delivering between 0.5 and 1 mW, TEM_{00} at 632.8 nm. More expensive models are available, delivering up to 35 mW, TEM_{00} at 632.8 nm. The beam is generally polarized with a coherence length between 20 and 30 cm. An intracavity etalon may be installed for greater coherence, but the corresponding loss of power tends to create extremely long exposure times. The average lifetime of a HeNe laser tube is about 15,000 hours.

The first HeNe laser ever demonstrated emitted an 1,153-nm wavelength, but almost all HeNe J.asers are utilized at the 632.8 nm line. HeNe lasers emitting green, yellow and orange wavelengths are also available, but their low power makes them ineffective in most commercial applications.

The active medium in a HeNe laser is neon. As in HeCd lasers, helium is the transfer medium. The laser tube (see Diagram N) consists of a glass envelope (bulb containing the cathode) with a narrow bore through its center. The bore can be anywhere from 10 to 100 cm in length, depending on how powerful the laser is.

A 10,000 volt (dc) spike breaks downlhe two gasses in a narrow capillary tube. The voltage dro~sto between 1,000 and 2,000V with a current of a few milliamperes. Electrons in the discharge pump both helium and neon-atoms to excited states. The more abundant helium atoms collect most of the energy, then transfer it to lower energy or ground-state neon atoms through a series of inelastic collissions.

This transfer of energy is very efficient for two reasons. first, both the helium and the neon gasses have two higher energy states with comparable energy values. Second, both pairs ofmatching higher energy states are characterized by prolonged delays before allowing their electrons to drop. Energy states that hold their electrons longer (up to many milliseconds) are called **metastable states.**

When the abundant helium atoms in the tube are excited by the discharge current, more of them will have electrons in one of the two metastable states than in other higher energy states. A leaky bucket in a rain storm that holds water longer than a comparably sized peach basket, will more likely have rainwater in it the next day.

A sizeable population of helium atoms with electrons in the metastable state is built up. The excited helium atoms wander in the tube, collide with non-excited (ground state) neon atoms and transfer their energy to them. Since the higher energy levels of the two gases match, the amount of energy transferred to the neon atom is just enough to raise its electron to a metastable state. Soon, most neon atoms have their electrons in metastable state. A population inversion soon develops and lasing begins.

HeNe lasers are being challenged by low-powered laser diodes that emit at the 650 and 670-nm wavelengths. Laser diodes are extremely small, light, use very little current, need extremely low voltages, and are less than one fourth the price of a HeNe laser. Laser diodes have inherent properties that are detrimental to holography, however. Low coherence lengths (less than 2 cm), and wavelength instability make the use of laser diodes in holography unfeasible.

The low-cost HeNe will always be the perfect laser for novice holographers. The combination of affordability, high-reliability and ease-of-use makes this laser perfect for the production of budget holograms.

Editor's Acknowledgment - Special thanks to Michael Fisk for his contribution to this chapter.

8

Lasers - Current Trends

The majority of holographers once used red sensitive emulsions and Helium-Neon lasers. As the industry grew, new recording materials were utilized and commercial-production required more powelful and more versatile CW lasers. Argon Ion, Krypton, HeCd and mixed gas lasers became standard equipment in the studio and in the factory. Diode Pumped Solid State lasers have recently been integrated into the' manufacturing process.

This chapter first examines the "holographer-friendly" features that define the latest generation of ion lasers. It then introduces DPSS laser technology, including its advantages and current limitations. An overview of some of the lasers currently available from four.of the world's major laser manufacturers follows. The chapter concludes by discussing alternative sources for laser equipment.

Ion Lasers Improve; Solid State Diode Pumped Lasers Introduced

Until recently, successfully performing holography in most R&D or production applications usually required the use of an ion laser. The ion laser, in turn, required hands-on operator attention to maintain optimum alignment and achieve single-frequency operation. Fortunately, recent developments in laser technology have eliminated both these requirements. Specifically, ion lasers have undergone significant maturation in terms ofease of use and implementation, resulting in true "hands-free" operation, even for applications as demanding as holography.

The first argon ion laser was made in 1964. Ion lasers opened the door to practical holography and have dominated this application because of their low-noise, high-power output characteristics. Indeed, photopolymers and other materials have been developed with sensitivities that specifically match ion laser wavelengths. As listed in Table 1, argon lasers can produce several watts of naITow-line output at several wavelengths in the UV through green. Krypton lasers have slightly lower output power, but can produce over 2 watts of single-frequency output in the red (647.1 nm) as well as lower powers at other visible wavelengths. Mixed gas (krypton/argon) lasers produce a combination of these output wavelengths.

The basic elements of a single-frequency ion laser head for holography are shown in figure 8.1. A prism is used as a wavelength filter to select the wavelength line of choice (e.g., 488 nm). Each laser output line has a typical width of 0.004-0.01 nm. This means that the coherence length of the laser is less than 50 mm. Unfortunately, successful holography requires a coherence length many times longer than the optical path difference. For most holographic applications, the linewidth of the laser output must be nar-

Output Line (nm) Argon Ion	Small Frame	Large Frame
All lines, multimode	5W	20W
457.9	0.2W	0.8W
488.0	0.9W	4.2W
514.5	1.2W	5.4W
Output Line (nm) Krypton Ion	**Small Frame**	**Large Frame**
All lines, multimode	1.0W	4.6W
413.1	0.15W	1.1W
647.1	0.5W	2.1W

Table 8.1. Typical single frequency outputs for the higher power visible wavelengths of small and large frame ion lasers.

rowed by using an intracavity etalon - a simple optic that serves as a narrow-bandpass filter.

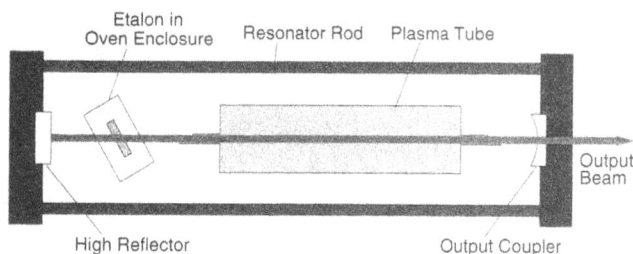

Figure 8. 1. Schematic of a laser head.

Figure 8.2 illustrates how the etalon achieves narrow line operation. Each laser line actually consists of many discrete wavelengths, or longitudinal modes. These modes satisfy the equation, $n\lambda = 2L$ where n is a very large integer, λ is the wavelength, and L is the cavity length, i.e., the distance between the rear mirror and the output coupler. As shown in Figure 8.2, the loss due to the etalon restricts laser oscillation to a single longitudinal mode.

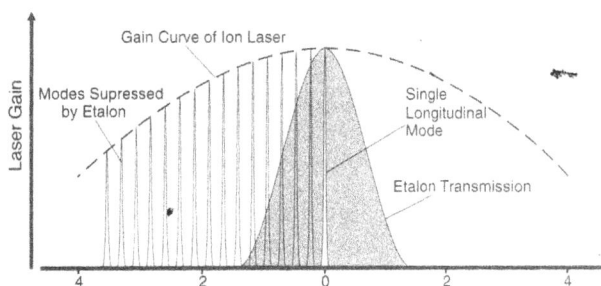

Figure 8.2. Narrow-line operation is achieved by using an etalon to select a single cavity mode.

Ion Lasers Continue to Evolve -- Automated Stabilization

In the past five years, manufacturers have made tremendous progress in the area of active stabilization. Early ion lasers had serious stability problems, particularly when operating narrow-line, i.e., with an etalon.

Waste energy from the lasing process generates many kilowatts of heat. Consequently, minor changes in cooling-water flow or tube current can cause measurable changes in the overall head temperature. For this reason, in high-end products, the cavity mirrors are supported on a low-expansion alloy (SuperInvar) structure. But the distance between the cavity mirrors is typically greaterthan I meter. Thus, even with a low-expansion resonator, small changes in temperature have the potential to change the alignment of, and distance between, the cavity mirrors.

Angular misalignment results in a reduction in output power, poor mode structure, and higher noise. For these reasons, the laser required constant attention and adjustment in order to maintain optimum performance.

In 1989, one laser manufacturer, Coherent, eliminated this alignment problem using an approach known as PowerTrack™ With this system, internal photodiode sensors detect changes in the beam power. Signals from these sensors are interpreted by a microprocessor that directs actuators to make compensating angular adjustments on the mirror mounts. Alignment is thus continuously maintained with no operator intervention - assuring maximum power at a given current, highest transverse mode quality, and lowest optical noise.

Mode Hopping

With single-frequency operation, however, there is still the problem of mbde-hopping: the sudden shift from one longitudinal mode to another. Shifts are accompanied by a drop in output power and fluctuations in coherence length. This phenomenon occurs because temperature affects both the etalon thickness and the laser cavity length, thus changing their wavelength characteristics.

If the head temperature drifts, the cavity mode may drift off the center of the etalon transmission peak. (See figure 8.2.) This causes a drop in laser power. As the drift becomes larger, the next cavity mode will eventually come close to the etalon center and the laser output will suddenly "jump" to this mode.

Because of the associated power and wavelength fluctua- tion, mode-hops during hologram mastering, or HIto H2 transfer, have the potential to completely ruin the hologram. This has long been cited as the leading drawback of ion lasers in holography.

To deal with this problem, a number of stabilization schemes have been employed by laser manufacturers. In 1995, Coherent introduced the v-Track™ system on its Sabre[R] ion lasers. In this system, the etalon is enclosed in a stabilized oven to maintain constant temperature. The cavity mirrors are supported on mounts that permit both angular and translational adjustments. To maintain single-mode operation, the laser's microprocessor senses any temperature-induced mode drift and directly compensates for cavity expansion or contraction by lengthening or shortening the distance between the mirrors. This stabilizes the length of the cavity.

This automated system virtually eliminates the effects of temperature drift in the laser head, ensuring hours, and even days, of mode-hop-free operation. Also, warm-up time is reduced to a few minutes. The benefits are greatly increased throughput for any holography application, as well as enabling longer-exposure holograms, such as rainbow holograms, to be recorded with consistently high quality.

To offer additional operational simplicity, these Sabre lasers can be operated by a hand-held remote module. Besides convenience, the use of a remote control can help reduce unwanted motion in the holography studio.

DPSS Lasers

Like many other areas of photonics, laser technology is inexorably moving towards all-solid-state solutions. For holographers, the area of most current interest is DPSS (diode-pumped solid-state) technology. These lasers have been under development since the mid-1980's and are now delivering suitable perfonllance for a number of holographic applications, most notably for non-destructive testing (NDT) and HI to H2 transfers of masters.

A DPSS laser uses a gain medium ofN d: YAG (neodymium yttrium aluminum garnet) or Nd:YY04 (neodymium yttrium aluminum vanadate). These are efficient lasing materials that produce near-infrared (1,064 nm) output. This near-IR light can be frequency doubled to the green (5'32 nm) by a non-linear crystal such as KDP (potassium dihydrogen phosphate).

Many lasers incorporating Nd:YAG or Nd:YY04 use high-energy flashlamp s to optically pump these materials. Such lasers are used in holography for generating pulsed masters of moving objects. In DPSS lasers, the pump en- ergy is supplied by laser diodes. These are more powerful versions of the same types of laser found in CD players. They efficiently convert electricity to light and do not require water cooling. Also, their output can be tuned to match the absorption profile of the crystal gain medium so that very little of their pump light is wasted. In multiwatt lasers, these laser diodes can be located in a compact power supply and the pump light supplied to the head via fiber optics. This results in an extremely compact, high-power laser head and minimal heat loading, requiring no external cooling.

Until very recently, no DPSS laser offered the appropriate characteristics for the holography marketplace. Some companies concentrated on developing high powers (up to 1 watt by 1995) but with multi-mode output and, hence, poor coherence. Other lasers were designed to capitalize on the potentially long coherence length of Nd:YAG and Nd:YY04. Originally producing 10s of milliwatts ofgreen output, these single-mode lasers had reached the 100s of milliwatts by 1995. In 1996, however, there was a quanturn leap in performance with the development of a laser (Coherent'sVerdi ™) which produces twatts of cw, single-mode output.

As shown in figure 8.3, this laser use's a ring configuration for efficient single-mode operation and overall compactness. The compact cavity results in favorable mode-spacing, making it easier to select a single mode using the internal etalon. The probability of temperature-induced mode-hopping also is greatly reduced. Nonetheless, to ensure mode-hop-free operation, active stabilization is employed. This closed electronic system is completely transparent to the user. In fact, the only control on this laser is the power switch.

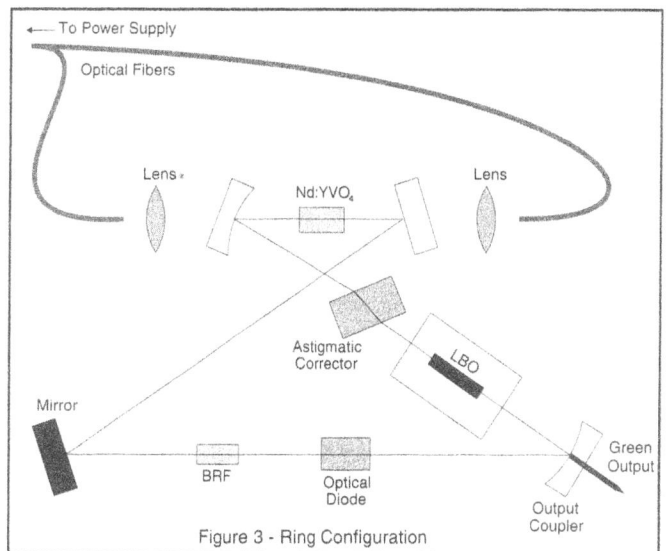

Figure 3 - Ring Configuration

Figure 8.3. The active medium in a DPSS laser is a neodymium-doped crystal that is optically pumped by light from laser diodes.

This laser offers several important benefits for holographers because of its all-solid-state construction. It is rugged, reliable, compact, requires no cooling-water, operates from a standard 110V supply and is highly portable. Furthermore, the spectrallinewidth of the output is much narrower than an ion laser-making for a much longer coherence léngth.

At this time, the only significant limitation of DPSS lasers is that they are restricted to a single visible output wavelength: 532 nm. Ion lasers will therefore continue to be the preferred lasers in applications requiring other visible wavelengths.

Conclusion

To summarize, it is unlikely that ion lasers for holography will ever be as simple and reliable to operate as a light bulb. However, recent progress in automated operation and active cavity stabilization systems has certainly moved ion lasers a long way in the direction ofthat ideal goal. At the same time, a new laser technology that does have the potential for light-bulb simplicity and reliability has become available to holographers. With these tools, expect continued expansion in the applications of holography.

Information contributed by Paul Ginouves.

Below: A picture and caption from Popular Mechanics magazine shows how CW gas laser prices have dropped ill the last 34 years.

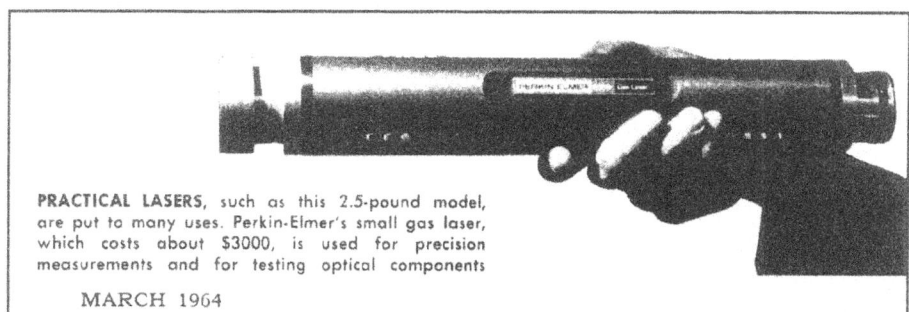

PRACTICAL LASERS, such as this 2.5-pound model, are put to many uses. Perkin-Elmer's small gas laser, which costs about $3000, is used for precision measurements and for testing optical components

MARCH 1964

Current Products from Major Manufacturers

Coherent, Inc. Laser Group

Coherent, Inc., is a major manufacturer of lasers and la- ser systems for scientific, medical, micromachining and entertainment applications. The Laser Group, one of four divisions within the company, designs and manufactures ion, CW, YAG, YLF, ultrafast, CO2, tunable-dye, diode and diode-pumped solid-state lasers. Headquartered in Santa Clara, California, the Laser Group was the foundation upon which Coherent was built and today is the most diverse business group within the company, providing laser products targeted at many different electro-optical applications, including holography. Below is a partial list of their lasers that are useful to holographers.

Verdi ™ DPSS Laser

The Laser Group announced that, as of mid-September, 1996, their Verdi™ diode-pumped, solid-state laser has begun shipping from production. Coherent's Verdi™ is a single-frequency, ring laser design producing a compact and efficient source of CW green (532 nm) light, suitable for exposing photopolymer.

As of fall '96, at least one U.S. based photopolymer hologram manufacturing facility had ordered a Verdi™ and planned to integrate it into its mass replication machine.

The Verdi™ is also particularly well-suited to NDT applications, as its high power and long coherence enable faster testing of larger parts. In addition, because of the laser's portability, there is the potential to test these larger parts in the field instead of hauling them into a laboratory. The laser is capable of transferring (copying) H1 holograms made with high-power pulsed Nd:YAG lasers. Transferring a hologram with the same wavelength used in the original recording greatly simplifies the transfer process, and can increase production yields.

The two-watt Verdi ™ uses the same ring laser cavity as the existing five-watt Verdi , resulting in a single-frequency output with exceptionally low noise. In the two-watt version, a single, fiber-delivered diode bar is used as the pump source, rather than the two diode bars used in the five-watt version. Both the two-watt and five-watt versions require only standard 110/220 volt single-phase power and no external cooling. Numerous other laser systems, based on the Verdi™ platform and utilizing these technologies, are under development.

Water cooled Ion lasers for Hglography

INNOVA 300 Series Ion Lasers

The INNOVA 300 Series Ion Laser System is a full-feature small-frame ion laser, offering high power, low RMS and peak-to-peak noise, power and mode stability, and actively stabilized single-frequency operation. The 300 Series is available in either argon or krypton. Multiline visible power up to 10W, TEM_{00} and multiline UV power up to 1W are available. To enhance productivity and performance, PowerTrack™ - Coherent's actively stabilized optical cavity is a standard on the INNOVA 300. And, for single-frequency applications, ModeTune simplifies etalon optimization and ModeTrack eliminates mode-hops.

Type	Model	Power (W)
Argon	INNOVA 304	4
Argon	INNOVA 305	5
Argon	INNOVA 306	6
Argon	INNOVA 307	7
Argon	INNOVA 308	8
Argon	INNOVA 310	10
Krypton	INNOVA 301	1
Krypton	INNOVA 302	1

The INNOVA Sabre Ion Lasers

The INNOVA Sabre Ion Laser System is a large-frame laser that combines a very stable basic design with performance-enhancing active components. Sabre's Sentry system will automatically acquire lasing, tune to a specific wavelength, and peak for maximum power, TEM_{00} mode, and minimum noise. PowerTrack™ and v-Track™ provide immediate warm-up in multiline, single-line, or single-frequency applications, and ensure stable performance even in changing environments.

Type	Model	Power (W)
Argon	Sabre TSM 10	10
Argon	Sabre TSM 15	15
Argon	Sabre TSM 20	20
Argon	Sabre TSM 25	25
Argon	Sabre DBW 10	10
Argon	Sabre DBW 15	15
Argon	Sabre DBW 20	20
Argon	Sabre DBW 25	25
Krypton	Sabre DBW Kr	4.6

Infinity Nd: Yag

In addition to the CW lasers listed above, the company has recently produced a pulsed Nd:Yag laser (the Infinity Series) capable of >20W output at 532 nm, which could interest researchers doing pulsed holography.

Spectra-Physics Lasers

Spectra-Physics Lasers, based in Mountain View, California, is the world's largest and oldest designer and manufacturer of Helium-Neon lasers. (Their first HeNe was introduced in 1963.) Over the years, numerous silver-halide holograms have been mastered and copied using the company's red emitting lasers. These lasers have proven reliable and extremely durable. Consequently, the company has established a very solid reputation within the holography industry. The company's line has since expanded to include other CW lasers that are suitable for exposing photoresist and photopolymer emulsions, including Water-cooled Ion lasers and DPSS CW lasers. The company also manufactures Nd:Yag Pulsed lasers that can be used for some holographic applications.

HeNe Lasers

Currently the company offers two versions of high power Helium-Neon lasers with a long (approx. 1 m) coherence length: Model 127 (a complete unit) and Model 07B (for OEM applications). Both versions deliver either 25 or 35 mW of TEM_{00} polarized 632.8 nm light, though most working holographers will order the former.

According to company literature, Model 127 features an improved mirror configuration, mounted on adjustable plates, that enhances stability. The resonator design minimizes the effect of temperature changes on output power. Hard-sealed plasma tube Brewster windows make the plasma tube impervious to contamination and provide unlimited shelf life. A large cathode and gas reservoir greatly increase operating lifetime, which typically exceeds 20,000 hours. The plasma tube and the power supply are integrated into a single compact unit. The laser comes equipped with a threaded mounting bezel for various optical accespier. A small portion of the beam is split off outside the laser cavity and is used as a reference beam into the quad cell. An error signal is generated for the piezo-mounted output coupler. With the output mirror moving in the X and Y axes, beam pointing is maintained and wandering eliminated. sories .

Stabilizing Water-cooled Ion Lasers

As new recording materials became available to the holography industry, other types of lasers become necessary for commercial production. However, unlike the simple air cooled HeNe's, the more powerful Argon-Ion and Krypton lasers were water cooled and a bitlmore finicky. Beam stability and output reliability becamlt'even more crucial in manufacturing environments.

As mentioned previously, beam motion, or wander, is caused by changes in the optical cavity alignment of a laser. These changes result primarily from temperature variances as the laser warms up, alterations in the ambient environment or cooling water temperature, and changes in the operating current to the laser. Beam motion obviously hinders hologram production. The engineers at Spectra-Physics designed several methods to ensure reliable single frequency operation, called Beam-Iok®, Z-lok®, and 1- lok®.

The Beam-Lok design is centered around a quadrant cell detector and an actively steered piezo-mounted output cou-

pier. A small portion of the beam is split off outside the laser cavity and is used as a reference beam into the quad cell. An error signal is generated for the piezo-mounted output coupler. With the output mirror moving in the X and Y axes, beam pointing is maintained and wandering eliminated.

Z-Lok was developed to address the common problems that occur in single frequency operation: mode hops. Mode-hops occur when the cavity and etalon lengths change rela- tive to each other, primarily as a result oftemperature varia- tions. This can plague the holographer during critical single frequency exposures. To eliminate mode-hops, Z-Lok uses a temperature stabilized etalon as a reference, locking the laser's cavity to it. The end result is minimal absolute fre- quency drift and mode hop elimination.

J-Lok: In addition to mode hops, short term j itter caused, for example, by cooling water flow can also affect single frequency exposures. Reduction of this jitter is especially important for holography, with exposure times in the microseconds-to-minutes time scale. Using the inherent response capability of the piezo electric crystal employed in all Beam J-Lok reduces the low-frequencyjitter in the 10-500 Hz range.

The BeamLok® Series

Beam-Lok is the trade name of Spectra-Physics Lasers' family of argon, krypton, and white light ion lasers that employ the aforementioned stabilization methods. The Beam-Lok series offers a range of output power options in visible wavelengths ranging from 454.5 to 514.5 nm and TEM_{00} operation. The small frame 2060/65 systems provide up to 10W multi-line visible output. The large frame 2080/85 systems are configured for 12 to 30W multi-line argon, up to 5W multi-line krypton, and 14W multi-line white light outputs.

Diode Pumped Solid State Lasers

Spectra Physics also makes a line of solid state lasers (Millennia) that can be used to expose green sensitive photopolymer emulsions. These lasers could be very effective for contact copying H1s to H2s, though the coherence length probably makes mastering impractical. These lasers are air cooled, which makes them inherently more stable and more convenient to operate than their water cooled counterparts . The solid-state technology offers higher efficiency, lower utility requirements and operating costs, smaller size and weight, and longer lifetime. In terms of performance, the diode-pumped source provides superior spatial mode quality and greatly reduced amplitude noise, which directly benefits many holographic applications. As a result of the success of their five watt solid state laser, the Millennia technology platform has been extended to include two watt (Millennia II) and ten watt (Millennia X) versions.

Millenia Specs Power   2W, 5W, 10 W @ 532 nm

                     Mode TEM_{00}

                     Polarization   >100: 1 vertical

Melles Griot

Melles Griot, Inc., a worldwide supplier of a wide range of photonic products including lasers, acquired the business of Omnichrome Corporation this past year. Omnichrome is a leading supplier of helium cadmium and ion lasers to scientific and original equipment manufacture (OEM) markets. The Melles Griot product line now includes a full spectrum of helium cadmium, helium neon, argon Ion, and krypton argon lasers with wavelengths from 325 nm to 3.39 mm and power output to 500 mW. Many are appropriate for holographic applications.

HeNe Lasers

Of special interest to display holographers is a new line of economical 35 mW + helium neon lasers featuring the 05 LHP 928 linearly polarized and 05 LHR 928 randomly polarized models. These lasers use the Melles Griot hard sealed internal cavity mirror construction for long lifetime and long term mirror alignment. As with all Melles Griot lasers, beam delivery systems and other accessories are available.

Here are some frequently asked questions regarding HeNe lasers:

How do you determine the coherence length of a Helium Neon laser?

M.G. - Coherence length is defined as the length over which energy in two separate waves remains constant. With respect to the laser, it is the greatest distance between two arms of an interfer"ometric system for which sufficient interferometric effects can be obtained: $Lc = c / D n L$. Coherence length will vary from laser to laser as a function of the Doppler broadened gain width; however, for a HeNe, 20-30 cm is typical.

What is the Doppler broadened gain width of a Helium Neon laser?

M.G. - Helium Neon lasers can range from 800 MHz to 1600 MHz full-width-at-half-maximum (FWHM) depending on the design. The typical Red HeNe is 1400 MHz. The width of a single mode located under the gain curve is approximately I MHz.

Can a Helium Neon laser be re-gassed?

M.G. - While re-gassing can provide some extension of the output performance in some gas lasers like the CO" Argon and the higher powered side arm HeNe's (which have external optics), it is not recommended or provided for smaller internal mirror coaxial tubes. Typical end-of-life failure for a HeNe is cathode sputtering. This occurs when the protective oxide layer on the cathode is expended through continuous bombardment by the laser,discharge. There is no cost effective way of regenerating this layer. When the oxide layer is expended, the discharge itself vaporizes the "raw" aluminum and deposits this material, in its vapor state, on other surfaces such as the optics and bore.

Other Suitable CW Lasers

In addition to their Helium Neon lasers, the company carnes a range of other CW lasers that are suitable for commercial display holographic applications, including Helium Cadmium (200 mW / 442 nm), Argon Ion (100 mW / 457-514 nm and 300 mW / 454-514 nm) and Krypton Argon Ion (up to 150 mW / 467-752 nm) lasers. Contact Melles Griot for a full product catalog.

Diode Laser for Interferometry

Of special interest to industrial holographers is a new diode laser with long coherence length and stable power output. For instance, model 56 IMS 663 is a completely selfcontained, temperature-stabilized diode laser system with long coherence length and stable output, especially designed for holographic interferometry. It delivers 18 MHz direct modulation capability and high-power output in a small footprint, and requires only 5VDC to operate. A front panel C-mount is included for beam expanders and spatial filters.

Features:

- 685 ± 10 run
- >3 m coherence length
- < 100 MHz linewidth
- Power stability < 0.05% over 60 minutes
- >25 mW delivered output power
- dl /dT is 0.6 GHz/°< C
- 8 MHz (-3dB) modulation bandwidth

Fiber Optic Coupler

In addition, the company is selling a new laser-to-fiber coupler that may interest holographers and lighting designers. Fiber optic technology has intrigued holographers for years - these strands of glass could theoretically pipe laser light around the holography table and do away with more cumbersome optical components. Or, the thin fibers could be integrated into a stylish and effective laser illumination system for hologram gallery installations. However, one of the problems has been t of ind a way to direct the laser light into a small diameter fiber in an efficient manner, while still preserving the beam's essential properties.

The new Melles Griot 09 LFM 001 laser-to-fiber coupler provides a simple and convenient method of launching laser output into single- or multimode fibers, At 633 nm, optical coupling efficiency can exceed 95% into multimode fibers, and 75% into single-mode fibers (10 mm core). Differential X-Y adjustments allow precise adjustment for optimum coupling. A three-element focusing lens with an 8.0 mm focal length with an FC connector is supplied, Modular construction allows easy output connector and lens changes. The coupler mounts directly to lasers with a 36mm mounting hole circle.

LiCONiX

LiCONiX is a leading supplier of UV and blue lasers, based on Helium Cadmium (HeCd) and Diode Pumped Solid State (DPSS) technologies. It is the only manufacturer of HeCd lasers with all 325 nm, 442 nm and 354 nm wavelength options. (The 354 nm line was discovered and patented by LiCONiX.) The company claims that holographers mastering embossed holograms will greatly benefit from using their 442 nm output HeCd lasers rather than using comparable argon ion lasers (458 nm) due to the fact that standard photoresist emulsions are ten times more sensitive to the former wavelengths.

Also, according to company literature, the bandwidth of HeCd lasers using natural Cd vapor is approximately 3 GHz, corresponding to a coherence length of 10 cm. Using isotopic enriched Cd, the bandwidth is reduced to 1 GHz, corresponding to a 30 cm coherence length. Both the single wavelength output and the relatively long coherence length make the HeCd laser a practical choice for embossed holography mastering (the company's "NX" models employ isotopically enriched cadmium for enhanced coherence length and higher power).

HeCd lasers range in power from a few mW to almost a quarter of a watt. As with all CW lasers, the output power is determined by the gain length - the size of the laser. Li-CONiX offers HeCd lasers in three sizes: small frame 61 cm (24 inch), medium frame 102 cm (40 inch), and large frame 140 cm (55 inch). The company 's medium and large frame Embosser Series was designed to give holographers making embossed holograms and diffraction gratings the high power they require.

There is, as always, a trade-off between power and mode. LiCONiX offers all frame sizes in TEM_{00} and TEM_{mm} configurations. For holographers, the company states that their TEM_{00} mode lasers are "true TEM_{00}", rather than < 1 the less exact "visual Gaussian." The M squared value is typically less than 1.10. For more information on laser beam characterization, contact a company representative.

Recently LiCONiX released a new CAD Line series of laser to comply with current and projected CE mark regulations for sale into the European community. Their performance is identical to the older lasers listed below, but European model numbers may be different than those listed.

Large-Frame HeCd Lasers

Model:	Embosser II Series
	325 nm, 354 nm, 442 nm
Application:	Embossed Holography
Features:	Worlds highes power HeCd laser.
	Available in two different coherence lengths: 10 cm and 30 cm.
	Expected Life: >4,000 Hrs
	Polariztion, plane vertical >55:1.

Large-Frame HeCd Lasers (all TEM00 Mode)

Model	A	Power	Dia.	Coh.Len
Emboss II	*(nm)*	*(mW)*	*(mm)*	*(cm)*
3620N	325	20	1.2	10
3630NX	325	30	1.2	30
46120N	442	120	1.3	10
46150NX	442	150	1.3	30
46170NX	442	170	1.3	30

Mid-Frame HeCd Lasers

Model:	Embosser 1 & Series 200
	325nm, 354nm, 442nm
Application:	Embossed Holography
Features:	Convention cooling and carbon-fiber composite resonator rods provide outstanding stability.
	Excellent mode quality.
	Low Noise, 200 series incorporates LiCONiX "Noise Lock" feature.
	Expected Life >4,000 Hrs.
	Polarization, plane vertical >500:1

Mid-Frame HeCd Lasers (all TEM00 Mode)

Model	A	Power	Dia.	Coh.Len
Emboss I	*(nm)*	*(mW)*	*(mm)*	*(cm)*
3210N	325	10	1.0	10
3216N	325	16	1.0	10
3220NX	325	20	1.0	30
7205	354	5	1.0	10
4270N	442	70	1.2	10
4290NX	442	90	1.2	30
Series 200				
3207N	325	7	1.0	10
321 IN	325	11	1.0	10
3215NX	325	15	1.0	30
7203N	354	3	1.0	10
4230N	442	30	1.2	10
4240N	442	40	1.2	10
4250N	442	50	1.2	10
4260NX	442	60	1.2	30

The company also produces a range of optical equipment for the holography studio, including a power meter designed for low-power lasers.

Alternative Sources for Lasers Suitable for Holography

For those that cannot afford to purchase a new laser from the original equipment manufacturer (or an authorized distributor), there are alternative sources for lasers suitable for holography. Specifically, there are companies that specialize in selling never-been-used "surplus" equipment and others that resell used and/or refurbished hardware, at prices far below retail. Besides offering. low prices, these companies often have access obsolete lasers and related equipment which holographers still find desirable.

It is also possible, but often less practical, for knowledgeable buyers t of ind decent deals on lasers from various other sources - such as optical laboratories, universities, manufacturing plants and corporate auctions. It is worthwhile to hunt for bargains at these places if you have the time and the expertise. Unfortunately, these sources do not provide the technical support, selection and service needed by most professional holographers and hobbyists.

For these reasons, we recommend that you contact those companies whose primary business is the resale and/or repair of electro-optical equipment (see the Business Directory in this book for a listing of such companies). These companies have developed a close working relationship with the laser industry and thus can provide a high level of customer service.

Since many ofthese companies are staffed by experienced laser aficionados, they are often able and willing to answer technical questions before and after the sale. Salespeople are typically encouraged to spend some extra time with novice shoppers in order to ascertain their specific needs and steer them to the right equipment (as in most industries, the factory's sales force is typically geared to service the larger corporate accounts and don't have the resources to answer a lot of basic questions from small businesses and hobbyists interested in purchasing single units). In addition, most of the companies we know about in this business have one or two people on staff who are especially familiar with the lasers required by holographers.

Another good reason to shop with resale businesses is product availability and selection. Most of these companies carry a wide selection of lasers from different manufacturers in various price and power ranges, rather than only a single product line. Prices usually range from a few hun- dred dollars for low powered units to thousands of dollars for more industrial gear. Although most holographers are hunting for HeNe and Argon lasers, it is also possible t of ind an assortment ofmore esoteric hardware, such as Heed, Nd:Yag and ruby lasers.

Most notably, these companies often stock models which the manufacturer has discontinued, even though they have proven extremely useful to holographers in the past. These companies also sell components (laser heads, power supplies, optics, etc.) for customers interested in assembling their own units and spare parts that might otherwise be unobtainable. Several US companies mentioned that they routinely customize equipment to make it suitable for countries with different electrical systems.

Finally, it is important to consider warranty protection. Gas lasers do have a finite life expectancy and are rather aelicate pieces of equipment. Reputable companies should provide some sort of guarantee, especially on used equipment (some buyers might prefer buying refurbished lasers, as the life expectatlcy can be more easily ascertained). Every company we surveyed offered a warranty, ranging from 30 days to a full factory equivalent. Obviously, it benefits the customer to have one, especially if the laser is being shipped.

Where does this surplus and used laser equipment come from? One company we interviewed specializes in purchasing large quantities ofsurplus components directly from the OEM and resets both pre-assembled packages or individual parts. This "factory fresh" equipment is typically obtained from excess inventories of discontinued models, spare parts and production overruns. Another company we asked acquires large lots of used equipment from cor- porate users who are upgrading their equipment or switch- ing technologies. For example, many inexpensive HeNe gas lasers became available when supermarket chains purchased new bar-code scanners built around solid state diode lasers. Other businesses purchase older units from a variety of sources and refurbish them t of actory specs. All these companies are potential sources of good equipment at discounted prices.

Since the primary reason for buying surplus, used, or repaired lasers is price, we surveyed several businesses t of ind out how much a customer could reasonably expect to save. SteveGarret,ownerofMidwestLaserProducts,re- plies, "The amount of money saved depends on the cost of the laser; however, one can often save (an average of) 50% off the price of new equipment." At the time of this survey, his company was selling a seven mW HeNe laser and power supply for one-third the cost of a similar new unit, and a new Argon for 30% below the factory price.

Martin Hasa of MWK Industries concurs: "The clear advantage of purchasing a surplus or used laser is cost. A new HeNe laser purchased directly from a manufacturer or one of their distributors will usually cost about three to six times more than a similar unit purchased from us. "For example, in a recent catalog we listed a surplus ten mW HeNe and power supply for one-third the cost of a similar system offered by a major distributor." Substantial savings are available for refurbished equipment, too. Hasa concludes, "Though the price of new laser equipment can be far beyond the means of the average individual, used (and surplus) equipment makes holography an affordable pursuit for people, companies and institutions on a tight budget."

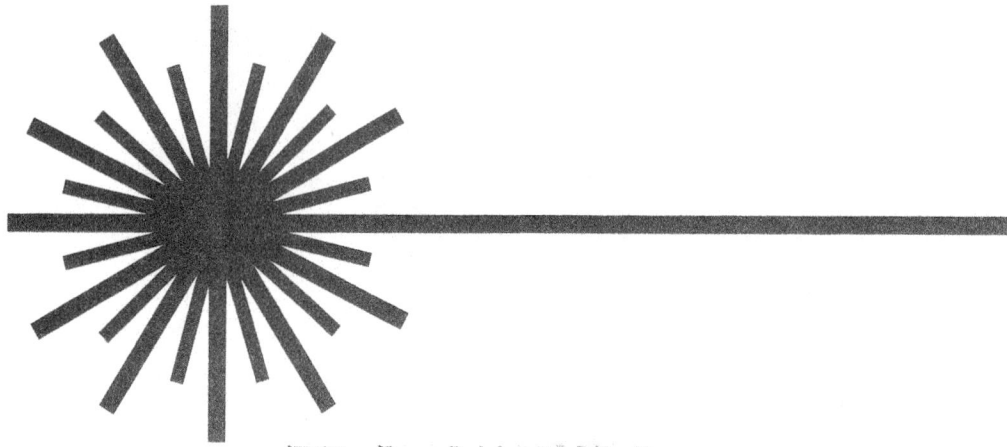

PULSED RUBY LASER HOLOGRAPHY

FOR SCIENCE & INDUSTRY

LASER
SALES & SERVICE

- 1, 3, & 10 joule output at 694nm
- Single, double, triple, & four pulse Q switched
- Coherence length greater than 1 meter
- Unequaled reliability

Tel. (610) 982 0226
email: lasertek@ptd.net

LASER TECHNICAL SERVICES
1396 River Road
PO Box 248
Upper Black Eddy, PA 18972

Laser Technical Services is the sole authorized agent for American sales and service of
this equipment now manufactured by **INNOLAS Ltd**, Rugby, England

Pulsed Lasers for Holography

by David Ratcliffe (GEOLA)

Pulsed lasers offer one very large advantage over conventional CW systems for display holography - their complete indifference to vibration and motion. In no uncertain terms these lasers free the holographer from his normal rulebook. With the nanosecond exposure times characteristic of the Q-switched pulsed laser; human and animal portraits,unstable models, falling leaves,waterjets, smoke, speeding bullets and even fast impacts become possible to record. All definitely out of the question when using CW lasers - but possible with pulsed ones.

of course, until now there have been several catches. First, there have been few available pulsed laser systems designed especially for holography - and this is pretty important unless you are a trained laser engineer and you don't mind redesigning a laser intended for another application. Second, the holography laser systems available until now have operated exclusively in the very deep red; a wavelength characteristic of the Ruby crystal and which can at best be described as "inconvenient" for display holography.

A new Ruby laser with enough energy to be useful usually comes with a price tag approaching six digits. Second-hand systems admittedly do sell for less. But put all this together with the fact that no one has ever really offered a professional commercial camera system based on an available laser (and therefore the nontrivial task of designin& and putting together the optical scheme outside the laser has always been the holographer's job) then one can perhaps understand why pulsed lasers have remained until now somewhat untouchable dreams to all but a few.

T of urther exacerbate this situation, Agfa has pulled out of the (red) holographic emulsions market in 1997 leaving no clear producer for film or plates suitable for Ruby laser use. (Ed Note - Slavich Co. is reportetfly making such film.)

This rather depressing situation is hbwever definitely not the end of the story. In a somewhat serendipitous fashion, there have recently appeared several manufacturers of holography laser systems whose products are based on the more advanced Neodymium technology and who have done what no Ruby laser manufacturer ever did: supply entire camera systems for mastering and transfer copying.

Ruby vs. Neodymium

Notwithstanding the rather crucial issue ofemulsion avail- ability for Ruby lasers that has recently arisen, there are several important reasons why today 's holography laser manufacturers have opted for solutions based on Neody- mium technology rather than Ruby.

Ironically, the Ruby laser has become to be regarded amongst the holographic community as the only seriously viable laser for pulsed holography. However, this has happened mostly for the reasons of availability and appears not to be based on a sound appraisal of the relative merits of the various different possible systems.

We must firstly observe that Ruby is actually a three-level quantum system. Lasing in Neodymium is, by contrast, governed by a four-level scheme, making the Neodymium laser inherently more efficient than Ruby. In practice, this difference becomes visually evident when you compare typical Neodymium Glass laser power supply units with those of Ruby lasers . The Neodymium Glass units are usually small single phase modules that can be plugged into any household supply. Ruby units are large industrial type racks requiring hefty 3-phase outlets. And because Ruby lasing is fundamentally less efficient than Neodymium, optical pumping in a Ruby laser must be much higher. This leads to complications that do not arise in a less pumped system.

Perhaps even more important than these size and power consumptionjssues is the clear fact that the green frequency-doubled Neodymium light is simply far more appropriate as a lighting source for display holography than the far-red light of Ruby. For portraiture applications, it has been known for many years that Ruby light penetrates the outer skin layer and is reflected only by the relatively uniform subcutaneous layer. This is what is responsible for those waxlike images you see in Ruby portraits devoid of any surface detail. Neodymium, ofcourse, doesn't do this. Light at around 530 run behaves as we are used to in our normal everyday experience. In addition, our eyes are much more

sensitive to this part of the spectrum.

Finally, every CW holographer knows how useful spatial filtering is when you are making transfer copies and even when mastering. It is extremely difficult and highly unrecommended to use spatial filtering techniques with a Ruby laser], as back reflection from an accidental optically-induced spark may do serious damage to the laser. With Neodymium lasers the frequency-conversion crystal that converts the infrared lasing light to the visible green acts as a kind ofvery efficient optical protector, effectively preventing back-reflected light from re-entering the laser and causing damage. This difference means that Neodymium laser beams can be cleaned up a lot easier than Ruby beams. Hence, systems based on Neodymium can do most things a CW laser can, including the production of transfer copies.

Choosing a Pulsed Laser

Neodymium lasers come in various shapes and forms. Those most useful for holography applications are based on the Neodymium doped Yttrium Aluminium Garnet Crystal (Nd:YAG) or the Neodymium doped Yttrium Lithium Fluoride crystal (Nd:YLF). Both crystals are optically pumped by flashlamp (as in the Ruby laser) and emit in the infrared. YAG lases at 1,064 nm and YLF a little lower at 1,053 nm2 A non-linear frequency-doubling crystal such as KDP, KTP or DKDP is used to convert the emission into a green output at 532 nm (YAG) or 526.5 nm (YLF). Essential features for any holography laser are a linearly polarized TEM00 transverse mode with a high quality quasi-gaussian beam profile and true single longitudinal mode operation.

YAG Lasers

Most of the large laser manufacturers such as Continuum, Coherent, Spectra, Spectron and Quantel make standard model Neodymium YAG lasers giving energies in the green of up to around 1 to 1.51. Such lasers use a Neodymium YAG oscillator followed by one or more YAG amplifiers. Depending on who you talk to and when, YAG lasers are, or are not useful for display holography. The true answer is that such lasers can be used for display holography with some effort but are not designed for this. Therefore, they usually represent an inappropriate choice for such an application. An exception would be for the shooting holo graphic movies which entail rapid sequential exposures.

All standard YAG lasers achieve "Single-Longitudinal Mode" operation - which is the mode you need to ensure long coherence - by the use of a complicated and costly technique known as injection seeding. This requires continual firing of the oscillator and necessitates pulse shuttering for standard holography use. Injection seeded systems are moreover difficult to realign and require the use of expensive scopes with nanosecond resolution. In addition, YAG lasers are usually designed to pump doubling or tripling crystals whose conversion efficiency is critically dependent on the beam profile. Thus, it is unusual t of ind such a laser (having the appropriate amount of energy per pulse) with the quasi-gaussian beam profile that one re- quires for display holography. Of course, one can filter a non-optimal profile - but then your useful energy decreases. And lets face it, even without filtering, one joule (1) of energy for an approximate figure of $150,000 (including laser, injection seeding system and scope) is just too much!

Hybrid Neodymium Glass Lasers

Hybrid Neodymium Glass lasers are undoubtedly the best solution for display holography available today. Those that may be considered for this application are lasers in which the oscillator is either Neodymium YAG or YLF and which employ one or more glass amplifiers. Usually a ring cavity oscillator is used with passive Q-switching. Single longitudinal mode is thus produced without the need for injection seeding. The advantage of a glass amplifier is that it is effectively more efficient than a YLF or YAG crystal and additionally, the glass material can be produced cheaper and easier in the large sizes and high optical standards required for amplification. Practically this allows one to

make a smaller and cheaper laser with higher output energy. The disadvantage, which in fact is not a disadvantage for holography at all, is that this type of laser can only be typically fired once every few minutes. This is due to the lower thermal conductivity of glass.

One practical solution for the design of a Nd:glass holog- raphy laser is to a pair a Neodymium YAG oscillator with Silicate glass amplifiers. Such a system has been successfully constructed by Ron Olson of Laser Reflections and has been reported on in the Sixth Edition of this publication. Ron has been producing and continues to produce many high quality holograms using this laser. He uses his system.b9th for mastering and copying.

Another soluuon to the problem is to pair Neodymium YLF with phosphate glass amplifiers. Phosphate glass has significantly more gain than silicate glass and this is a distinct advantage. In addition, YLF, with lower gain than YAG, allows more energy to be stored in the oscillator without the risk of superiuminescance, an effect that can be a problem when designing a YAG/Silicate system. YLF has weaker thermal lensing than YAG and hence, even though it has higher thermal conductivity, YLF can tolerate more optical pumping without propagation modification. This is important when you want to design a laser that has a high thermal stress low-energy alignment mode, in addition to the normal high-energy low frequency mode.

Other Lasers Appropriate for Holography

There are currently no serious candidates for monochromatic display holography applications that can compete with Neodymium glass lasers.

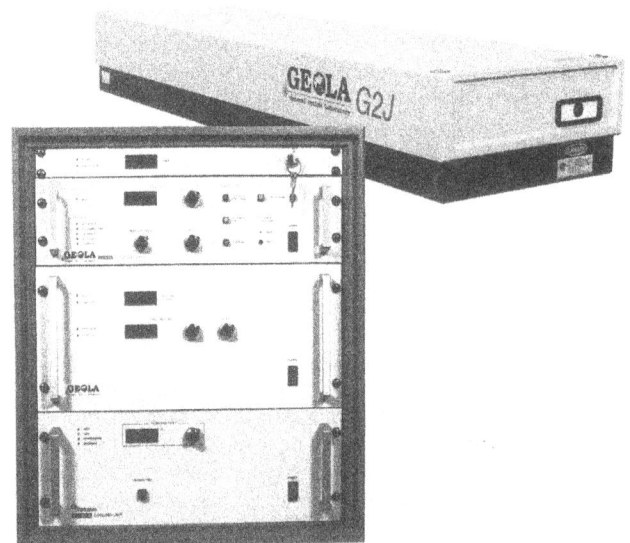

Figure 8.4. GEOLA's compact G2J Holography Laser, based on modern Neodymium technology.

Commercial Pulsed Lasers

GXJ Laser Systems

GEOLA (short for General Optics Laboratory) is a high tech European company located in Vilnius, Lithuania. In addition to producing custom lasers for the science community, this company has recently started manufacturing a family of advanced hybrid Nd:YLFI Nd:Phosphate glass lasers designed specifically for display holography. All GEOLA's "GxJ" lasers are based on a high-stability single transverse and longitudinal mode ring-cavity master-oscillator and are commercially available as models G11, G2J, G5J and G8J with respective output energies of 1, 2, 5 and 8J at the second harmonic wavelength of 526.5 nm. For higher energy applications, GEOLA offers multi-channel holography laser systems such as the GM32J model that produces a total energy of 34 Jin four object channels (32J), plus one reference channel (2J).

Inside the GXJ Lasers

All GxJ lasers offer a fast repetition mode that can be used for the alignment of external optical elements. This mode energizes only the oscillator and produces a lower- energy green beam that has identical propagation charac- teristics to the main high-energy beam. Such a feature seems to be a significant improvement over the more common technique of using an additional CW alignment laser that inevitably is of a different wavelength, and that has different beam parameters compared with the main laser emission. The G11 and G2J lasers employ a single YLF ring oscillator incorporating passive Q-switching by Cr+4:GSGG and a single double pass phosphate glass amplifier with SBS mirror.

There are several features worth n~ng about the GEO-LA oscillator. The choice ofCr+4 :GSGG over such better known materials as LiF not only ensures reliable Q-switching, but also guarantees single mode operation without the use of an etalon. The use of very high purity YLF produces all almost unheard of oscillator output energy - more characteristic in fact, of a YAG laser with one amplifier than a simple oscillator. This makes the design of the rest of the laser much easier and less critical when it comes to setting the transverse beam profile. Since the oscillator is what actually determines the coherence length and basic mode of the final beam, one can appreciate just how vitally importantjt is that the oscillator does not respond in a sensitive fashion to external conditions. Besides using high-sta- bility precision component holders for all optics, GEOLA has opted for a three-mirror vertically mounted cavity de- sign that is very insensitive to the usual bending moments felt by the optical base.

All GxJ lasers use the standard technique known as "diffraction cleaning" (optical elements M4, M5, T2, PR, T3) to improve the spatial distribution of the oscillator output before driving the Nd:phosphate glass amplifier. GEOLA uses a two-pass phosphate-glass amplifier incorporating phase-conjugation by Stimulated Brillouin Scattering (SBS) in order to guarantee high quality spatial distributions. The use of a Brillouin cell is very important for two reasons. First, SBS allows the formation of a diffraction-limited beam by compensation of the aberrations and distortions in the wavefront which are produced by the hot Nd:phosphate glass rod. This assures identical beam divergence and propagation direction in both the high repetition low-energy alignment mode and in the usual high-energy low repetition mode. The second reason is the greater energy extraction possible with a double-pass scheme without self-excitation. Here the Brillouip mirror serves as a selective reflector which reflects only a coherent signal and not the noise from any amplified spontaneous emission.

An important general point to make is that GEOLA has paid special attention to the mechanical stability of their laser heads.' First, they have used a floating 3-point suspension system for the honeycomb superinvarlstainless optical base. In addition, they have designed all optical mounts to the exact precision and function required. Finally, an optical scheme has been designed that cancels out temperature induced expansion - making the GxJ highly stable devices.

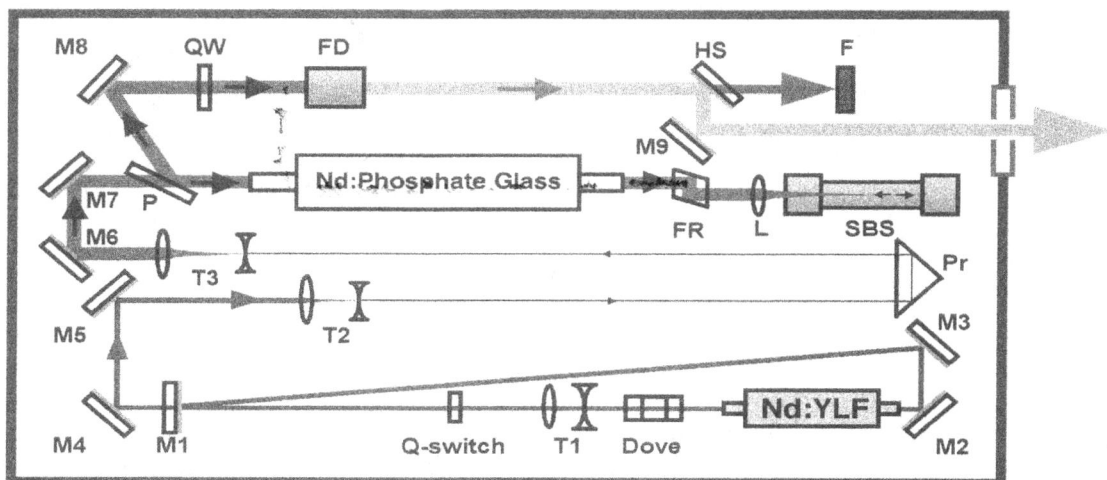

Figure 8.5. Optical Scheme of the G1J and G2J lasers. Notable features include a simple high energy ring oscillator, double-pass glass amplifier with SBS mirror and frequency doubling by DKDP.

Since YLF lases in the infrared at 1,053 nm, an harmonic generator must be used to double the frequency of the emission output. For this, the GxJ use a DKDP crystal sealed in a temperature-controlled dry cell as the harmonic generator. A precision engineered angle tuning mechanism (based on a simple mechanical design) ensures excellent long-term stability of the harmonic output. The technique permits an energy conversion efficiency to 526.5 nm of up to 60%. Harmonic separation is achieved by pairs of dichroic mirrors, giving > 99.7 % separation.

As I have mentioned above, two of the most important parameters for a holography laser are beam distribution and coherence. The GxJ lasers have been constructed to have output pulse durations of no more than 25 ns, giving coherence lengths in excess of 3 m. The spatial profiles have been optimized for holography and all lasers may hence be used for both mastering and copying.

The main differences between the G1J and G2J lasers are a larger diameter phosphate glass rod in the G2J amplifier ($d = 12\ mm$ instead of $10\ mm$ for the G1J) and special higher damage-threshold dielectric coatings on the optical elements. The higher energy models, G5J and G8J additionally utilize a final-stage large diameter amplifier. This amplifier is single pass and uses focal plane translation incorporating vacuum spatial filtering. Vacuum filtering assures effective decoupling between amplifiers, thus preventing unwanted lasing, while also improving beam quality.[3]

Power Supplies and Control Electronics

The main features of the GxJ control electronics are: remote operatiop by wireless push button; an LCD shot counter; digital countdown to next pulse; digital programming of the amplifier and oscillator voltages; push-button selection of "Aligrunent" and "High-Energy" modes; digital laser-coolant temperature stabilization and readout; and a variety of interlocks against over-voltage, flash-tube activation prior to rod cooling and accidental shots. The wireless push button has proved to be a very successful feature in holography applications where control of the laser pulse from awkward positions is often essential when working alone. All laser models are produced with a variety of parameters that are controlled by 10 turn potentiometers and displayed by LED digital readouts. These include the flashlamp voltage, pulse repetition rate and triggering delay (internal and external triggering is provided). Panel indicators signal interlock input and various status indications.

Multi-Channel High Energy Laser Systems

There are some unique applications in large format display holography that require even more energy than an 8 joule laser can give. However, in nearly all cases most of the energy is required for object illumination, and therefore the spatial distribution is relatively unimportant for the majority of the laser light. GEOLA's GMxJ (multi-chan- nel) series of custom ultrahigh energy holography lasers have been designed to take advantage of this fact by pro- ducing separate beams for the reference and objects. The reference beam in all GMxJ lasers is adjustable up to 2 J of energy in the green and has near perfect beam parameters.

One or more object beams produce up to 8 joules each. Since the technology used to produce these lasers is scalable, systems with any number of object beams may be produced. The GMxJ series of lasers appear ideal for the creation of ultralarge format rainbow, master-reflection or transmission holograms.

The Greenstar Lasers

In addition to GEOLA, there is another interesting source for holography lasers. For many years the S.l. Vavilov State Optical Institute in St.Petersbough, Russia has been producing holography lasers based on the Neodymium technology, for mostly Russian clients. Recently, the institute has spawned a commercial company called Holor that markets specifically this type of laser. The S.l. Vavilov State Optical Institute also makes entire holography camera systems based aroudd the Greenstar series of lasers.

Holor's Greenstar lasers use similar optical designs to GEOLA. Two models are offered, the GS-2 and the GS-4 which respectively produce 2 and 4 joules of output energy in the green. The lasers use a ring oscillator based on the YLF material and passive Q-switching. Amplifiers are

Parameters	GS-2	GS-4
Wavelength (nm)	526.5	526.5
Duration (ns)	30	30
Energy (J)	2	4
Pulse/Min.	1	1
Coherence (m)	>3	>3
Beam Dia. (mm)	10	10

based on phosphate glass as in the GEOLA systems. Table 1 lists a summary of essential parameters.

Table 8.2. Main technical parameters of the Holor Greenstar lasers.

Holography Camera Systems

The S.l. Vavilov State Optical Institute in St. Petersbough were really the first people to offer a compact, integrated holography camera system based on a Neodymium laser. Their system, now named GREEF and based on the Greenstar lasers, is a compact unit containing a laser and all optics and accessories that are required for shooting a 30 x 40 cm master hologram suitable for the production of a reflection copy by CW transfer. Due to its size and construction, the camera can be transported easily from location to location.

GREEF is a concept that is vitally important to the holographer, as the unit is complete. You no longer need to design your own external optical scheme because it is done for you . All the display holographer needs to do is to adjust the lighting and press the button to take the portrait! If one imagines where photography would be today if photographers had to design and make their own lens and shuttering systems, one can start to understand the importance of the availability of such integrated holography cameras. Particularly when one also understands that pulsed laser light is a lot harder and more dangerous to play with than the light from your usual HeNe or Argon laser. Such holographic mastering cameras are now also produced by GEOLA.

Truly Integrated Master/Copy Systems

GEOLA has recently taken the holography camera idea one stage further. That is, to produce an integrated compact machine that does both the mastering and the transfer copying, to either reflection or rainbow format.

One of the big problems in the past has been how to make the copies. If you used a pulsed laser, then you ended up with a transmission master that had to be transferred somehow to produce the white-light viewable reflection hologram. To do this, one usually used a Krypton pulsed laser on a large conventional optical table. Usually the pulsed laser that was used was a Ruby, which really ruled out HeNe copies.

Even with Krypton, color control was a problem and the transfer process would take far longer than master- ing. When you realize that the transfer setup could cost nearly as much as the pulsed laser, you start to understand why it might be a good idea to use the same pulsed laser to do the copy. In addition, using the same frequency of laser light for the copy and the master is, of course, ideal. As I have pointed out, such a system is a practical impossibility with Ruby lasers and has been the natural outcome of the newer Neodymium technology.

The GP2J System

The GP2J is GEOLA's first Integrated Master/Copy Camera system. The first commercial units should be

available for sale as of publication and a model is currently available for demonstration by appointment at GEOLA's he office in Vilnius. The GP2J is based on a G2J laser and comprises a central unit containing laser, optics and the master plate holder, an integrated power/control unit, several mirror's and a small transfer rig. The entire machine fits into room 5 m x 5 m with a low to medium height ceiling. This space leaves ample room for scene construction and for walL ing between the various components.

The GP2J is highly automated. No manual adjustments are required either to the laser or any of the high energy optics during mastering, copying or in switching between the two modes. Sophisticated electronic controls activate servo-motors in all crucial components, meaning that practically all the holographer has to do is load plates, press buttons and look at digital displays. The GP2J is capable of producing up t of ifty 30 x 40 cm or 40 x 60 cm reflection hologram per day, with an operating staff of two technicians.

During 1998/99 GEOLA plans to introduce a range a master/copy systems covering the needs of both conventional holographers and a new breed of client that the company foresees. These newcomers to the market will start to use holography without actually understanding how it works: much like how the guy in the drugstore who processes photos with an automated 1-hour machine, hasn't got a clue what actually happens inside this device.

Color Systems

Both GEOLA and Holor are currently working on color lasers and color camera systems. Neither of the companies currently have a finished commercial product, so don't hold your breath too long. However, it will come. GEOLA has been concentrating on the concept of Raman scattering of doubled Nd radiation (RSDN). The company's published work currently appear to show results significantly ahead of any other group regarding hydrogen conversion efficiencies. GEOLA is now turning to studies of Deuterium Raman conversion in the hope to realize a two color Green/Red laser within the next 12-24 months. This laser would be used in GEOLA's planned next stage 2-color master/copy cameras, which would look and work almost exactly as the GP21 - except for the fact that the final reflection hologram would have two colors. For most portrait applications, the lack of the final blue color is of a far lesser importance than the inclusion of the red.

An important point to note here is that color laser cameras are not much use unless you have the proper holographic emulsions. I have already mentioned that there is a current prob- lem with the availability of Ruby emulsions. Concerning this, GEOLA has been collaborating closely with the Russian emulsion producers Slavich, in the development of both green and near-red sensitive materials. Excellent emulsions now exist for both Neodym-

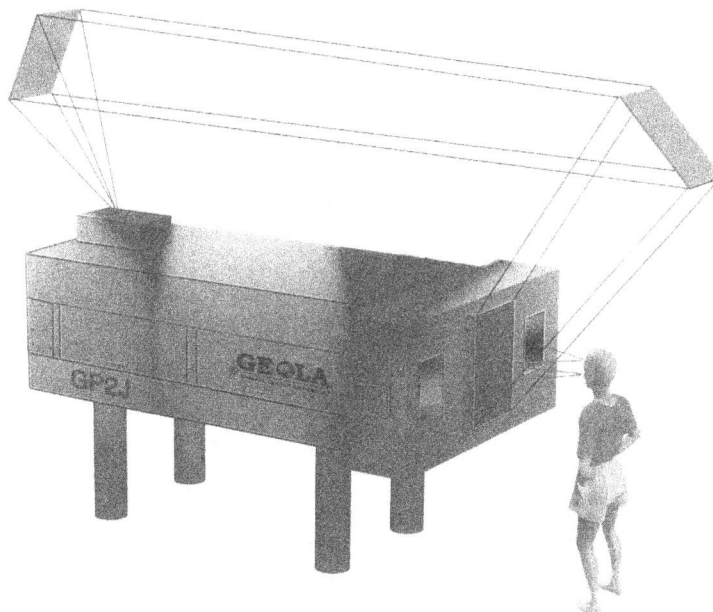

Figure 8.6. Concept drawing of the GEOLA GP-2J Integrated Master/Copy Camera.

ium green (526.5 nm) and for Hydrogen RSDN at around 620 run, both in the form of glass and film.

Holor is hoping to produce a three-color laser by a rather different mechanism to GEOLA's, with output energies of 0.4 J at 440 nrn, 1joule at 526 nm and I joule at 660 nrn. Their estimated time to market is 12 months.

Conclusion

Pulsed laser holography is really in the process of a quantum jump. Not because fundamental new discoveries have occurred in laser physics or optics; but because a few well known concepts have finally been carefully and systematically applied to what must be regarded as a pretty simple problem for laser physics - display holography.

The Ruby laser has been largely responsible for the rather pessimistic situation up until now. But introduce Neodymium technology (which by the way has been around for ages in Russia in the form that is needed for holography) and things get easier! Cost and size come down; beam quality goes up. The color is naturally optimum. Alignment is easy. And spatial filtering lets you really do anything you like.

The GxJ series of lasers are the first real commercial pulsed holography lasers to become available on the market and others will follow. These type of lasers are highly competitive with CW Neodymium or Argon lasersJiom both a dollar point of view and in what the laser can offer. They will also make portraits.

With reliable Neodymium Glass lasers finally here, it has been an obvious step to make commercial integrated systems for people who don't like playing with laser beams, and to make a single automated device for mastering and copying. With Neodymium, things can be made small. And with computer automation, thermal color control and automated processing; perhaps the day when you will be able to walk down the street to your local "holo-mat" and get your 1 hour 3-D portrait for fifty bucks, has gotten just that bit closer.

Finally, for those who are skeptical about holographic emulsion availability, I should stress that Neodymium lasers allow you to continue using available Agfa products. After all, Agfa has never indicated that it would stop producing photoplates for the lucrative microelectronics market, and of course these plates are just what the new lasers need. As for film substrates, the Russian producer Slavich is now producing an equivalent product to the Agfa 8E56; which I might add is significantly cheaper.

[1] In fact it is technically feasible to modify a Ruby laser such that the output beam transits a Faraday isolator. However this modification is expensive and brings other disadvantages.

[2] It should be noted that YLF actually has a more powerful line at 1,047 nm but this line does not match well with an emission line of glass.

[3] In fact it is more accurate to say that the beam quality is improved by the actual process of image translation which is of course integral to the vacuum filtering.

A Custom-built Nd:Yag Pulsed Laser System

by Ron Olson (Laser Reflections)

We use a passively Q-switched Nd:YAG oscillator operating in the TEM_{00} mode. Two etalons and a passive Q-switch (BDN dye) define a single longitudinal mode. We have a 500 MHz scope and a fast photo-diode to look at the temporal profile, but seldom use it as a diagnostic tool.

A preamplifier of Nd:YAG provides -100 mJ at 1Hz as an input to the Nd:Silicate Glass amplifier, which we use in a double-pass configuration to supply us with just over 2J at 1,064 nm. A 40 mm long KD*P doubling crystal is - 50% efficient and we work routinely with 11 at 532 nm. The frequency-doubling without the final amplifier stage in operation is extremely inefficient - allowing us to work at a manageable level (1mJ) for image composition - we do not have an alignment laser within the system.

We take great care to underfill our final amplifier, assuring a beam spatial profile which is "very nearly Gaussian". Much of the criticism regarding Nd:YAG lasers concerned beam spatial profiles - but the dark days of Nd:YAG crystal growing are long past - to the point that except for fanatics like myself - commercial Q-switched Nd:Y AG lasers as they leave the production floor would be adequate for most pictorial holography.

We split -1 0% of the energy for the reference beam (variable via a half-wave plate and a dielectric polarizer) which is delivered to the plate by a large float glass mirror at Brewster's angle. The remaining 90% of the energy goes to the two ground-glass diffusers which assure that the illumination beams are eye-safe.

We work under red safelights and to maximize the studio session, we often put aside half of the transmission masters for processing the day following a shoot - at this point we have that much confidence in our techpique. In the last two years we have produced more than 1Qb large format transmission masters and more than 300 reflection copies - all on glass.

Safety Concerns

When asked if a holographic portrait system such as ours is truly franchisable -- I caution people immediately. I have almost 20 years experience with high-energy, high-power lasers, and what is natural and instinctive for me is not necessarily the case for people not familiar with 1Joule/l00 Megawatt lasers. The laser company in which I am a founding partner produces more custom-designed high power lasers than any in the U.S. and it cannot be overstated that such lasers should be operated only by trained personnel who are thoroughly familiar with laser optics, laser power supplies and laser safety issues.

We would be happy to design a custom Q-switched Neodymium based laser for anyone with the requisite $150,000 - but a technician trained in the operation and maintenance of a Class IV laser is a requirement at the time of installation and training. No laser system in this power class is truly hands-off. Our experience, with some 500 laser systems shipped worldwide, is that unskilled operators who attempt to learn their craft on-the-job jeopardize the laser system, their own safety, and the safety of others in the studio/lab. In addition, it is extremely unlikely that anyone would be issued studio liability insurance such as we carry without a certifiable history in Class IV Laser System operation.

Images for the Future

Compared to all other 3-D techniques, quality display holography elicits much stronger responses with considerably longer retention. The market for fully dimensional live-subject imaging in art, P.O.P advertising and giftware is growing, feeding on the ever-increasing demands of the been-there / done-that generation of consmers that, in many respects, is miles ahead of the establishment advertising and retailing firms supposedly leading the way.

In an auspicious development, several high profile fine artists and portrait photographers have recently approached us after having seen our latest work. They immediately recognized the potential of our medium. One plans to use our holograms as a working tool for his human figures, citing the incredible detail captured by the pulsed holographic technique. Another wishes to utilize our technology for his portraiture commissions. The success of our newest pulsed hologram decorative products bode well for our studio's continued good fortune and for the fortunes of others (such as our friends at GEOLA) who have made the necessary investment in advanced laser hardware.

Conclusion

No longer thwarted by stationary (or even real-world) subject matter, the main difficulty facing modern holography is overcoming its primitive past and the brutality of a seemingly endless stream of lifeless juvenile images. This association, long reinforced, between holograms and junkware is more of an obstacle to holography's success than any perceived technical shortcomings - real or imag- ined. The days when you can offer a product of dubious merit "because it's a hologram" are quickly slipping away. In their stead, will come a time when holographers - like their 2-D photographic counterparts - will be judged by their images and creativity rather than by their technology and price points.

9

Sales & Distribution of Stock Holograms

This chapter discusses the commercial development of artistic holography, especially the sale and distribution of stock holograms and related products by the giftware industry. Tips on opening your own hologram store are included.

Some Text is MIssing 2nd to last paragraph

The Commercialization of Artistic Holography

In its early stages holography remll'ined unseen by the general public. Only scientists and researchers had access to the lasers and other specialized equipment that were needed to create and view a hologram. When methods were developed in the late 1960s that enabled a hologram to be"- created and viewed in more practical ways, holography slowly left the laboratory and began a journey that has resulted in a multi-million-dollar, worldwide industry. A great portion of this industry deals with "artistic" holography, (i.e. three-dimensional images ofthings) which is also commonly referred to as "pictorial" or "display" holography.

During the 1970s and early '80s holo,rams were made by individual holographic "artists" on a one-by-one basis. The process was labor-intesive and time consuming. Production techniques were developed through trial and error. Raw materials such as film emulsions were scarce, equipment was often homemade, and production quality often inconsistent. Unfortunately, the individuals and small companies that were capable of making high-quality holograms generally did not have the money or marketing expertise needed to get their work into widespread distribution, so holograms were still out of view of the public-at-large.

The handful of galleries and stores that did show holograms proved that the public was fascinated by this emerging medium. Although most holograms were treated as futuristic artworks or novelty items and were relatively expensive - the public constantly asked for more affordable ones. Enterprising gallery owners, retailers and holographers recognized this demand for holographic merchandise, and a small industry slowly evolved. Holographers and their entrepreneurial partners began to create products rather than artworks; well connected retailers began to distribute holograms to other retailers; and hologram aficionados became customers. Everyone recognized the potential of this new industry, yet manufacturing and display/lighting problems still needed to be overcome before holograms and related products could enter the mainstream marketplace. Limited production runs kept prices high.

Over the next decade technological advances enabled holograms to be mass-produced in a variety of ways. This made it feasible for artists, technicians and businessmen to join together to create facilities dedicated solely to producing large runs of affordable, high-quality holograms. These holograms were intended for a variety of commercial aapplications including security, packaging and advertising - as well as products for the giftware industry.

Art holographers copied their most popular images ont of ilm (which was less expensive and easier to handle than holograms produced on sheets of glass) and began to use assembly-line production methods in their labs. Whole catalogues of images soon became available, intended for sale as waII decor. Retail price points dropped considerably. Other holographers perfected methods of mass producing very bright dichromate holograms for use as jewelry. Still others concentrated on developing high-speed automated replication technologies capable of embossing holograms on very inexpensive foils and plastics - perfect for use on toys, optical novelties and paper products.

Once reliable supply lines were established, it became feasible for other companies to package and market these holograms in a variety of ways and integrate them into the normal chain of giftware distribution. Businesses in the United States and England quickly grew into major distributors. Film holograms were matted and/or framed and marketed as high-tech art. Holographic fashion accessories (including watchfaces, pendants and earrings) were developed. Executive gifts and desktop accessories were created. Rolls and rolls of kids' stickers were produced. New toys were invented. Holograms started to appear at national gift shows, in giftware catalogs, and in the media.

Savvy retailers soon realized that holograms and related products were very popular with the buying public, and if displayed correctly, could prove quite profitable. A good display of holograms drew a crowd, generated customer excitement and more importantly, generated dollars! (Most giftware items sold in stores are priced at twice their cost.) Holographic merchandise spread from science museum gift shops to mainstream outlets, and even included a number of specialty stores set up to sell only hologram products.

Increased visibility created greater public awareness of the product and demand for new and better holograms. Artists added color and motion to their images. Manufacturers automated further and invented materials espetrittlly suited for holographic applications. Holograms became brighter and easier to see under typical viewing conditions. Distributors created new product lines by integrating holograms into existing merchandise. Packaging was brought up to commercral standards. Wholesalers adopted more sophisticated marketing techniques, while retailers offered a wider selection of goods.

Today, a variety of holograms are manufactured around the world for the giftware industry, with the highest concentration of factories located in North America and Europe. English holographers have traditionally dominated the silver-halide film replication business. American holographers are actively developing photopolymer replication factories. The production of dichromate holograms seems to have slowed, while facilities capable of producing embossed holograms have multiplied significantly, especially throughout Asia. Surprisingly, very few holograms are exported from Japan.

The number of distributors and wholesalers dealing exclusively in holograms has dwindled as the market has diversified. To stay profitable several major distributors have developed their own custom images in order to target specific consumer groups. One major US distributor has developed a very successful product line based on the ever popular hologram eyeglasses with stock and custom photopolymer images as lenses. Another has developed close working relationships with the product development departments at several major retail chains, thereby ensuring longterm sales. The sale of "licensed" holographic images featuring popular sports figures, cartoon characters and movie scenes has grown steadily, while the sale of more mundane images has stagnated.

There are fewer holography specialty shops in business now than a few years ago. Those that continue to do well have increased the variety of goods that they carry and often include related optical novelties in their product mix. Sales of holographic artwork are practically nonexistent. However, more stores than ever before are carrying some sort of hologram-related product. It is not uncommon t of ind an inexpensive hologram item at the corner store.

The Chain of Distribution

Let's examine the chain of distribution as it typically exists for a holographic product.

The Copyright Holder The distribution process starts with the copyright holder. Any unique worl of art, including a painting, photograph, or computer-generated graphic, can be protected from unauthorized duplication (in most countries) by registering the image.in the appropriate manner. In the case of holography, the original work of art is either a model, a graphic, or a computer program designed to generate a holographic image. Whoever creates the unique work of art that later becomes a hologram is considered the copyright holder of the image. It is also possible to copyright the finished hologram itself as a unique work of art, provided that none of the components that appear in the holographic image belong to another party. There are, however, statutory limits stating that after number of years, a piece of art can become public domain and may be used freely.

Every hologram, if properly copyrighted, has only one owner, the copyright holder. The copyright holder therefore controls all subsequent distribution and is positioned at the top of the distribution chain. The copyright holder can be an individual or a group of people such as a business. Most commonly, a business commissions an artist to make a model or to design graphics and the artist turns over all copyright privileges to the business as part of the arrangement. This is legally known as "work for hire" and each party's responsibilities and rights must be documented to avoid problems concerning ownership. Holograms that are not copyrighted can be copied by whomever owns the "master" hologram.

The Manufacturer

Different companies specialize in manufacturing specific types of holograms. The manufacturing process generally involves three processes - mastering (creating the original hologram), reproduction (producing some quantity of copies), and finishing (lamination, cutting, sorting, etc.). Some companies do everything - others subcontract out some part of the job. Often a company that manufacturers holograms also owns copyrights in order to have a selection of stock images to offer their customers. A few marnifacturers bypass the normal chain of distJibution and sell directly to retailers.

The copyright holder needs to know the exact cost of each unit produced, since the manufacturer's charge will obviously influence the final price billed to the end user. T of igure the unit cost, one would take the total bill from the manufacturer (including any additional shipping and handling charges) and divide it by the number of usable copies

MARKETING
WITH

POLAROID MIRAGE HOLOGRAMS

O V I S I O N

It's a major challenge to make a new product or brochure - or any marketing message - stand up and get noticed. Utilizing our unique Polaroid Mirage holograms, Polaroids expert staff will work with you to achieve this goal.

What Are Mirage Holograms? You've seen them on everything from phone cards and video games to sports trading cards and action figure stickers. Mirage holograms are images of almost magical depth, clarity and realism. Starting from ordinary photos or drawings, they're made by using 3D modeling and laser technology to create dynamic dimension. And they're available only from Polaroid.

Why Mirage Holograms Offer A Visible Difference.

Conventional embossed holography provides only a surface image, limiting three dimensional visibility to a narrow range of lighting conditions. So Polaroid took an unconventional approach to solving the problem. We developed an exclusive Krypton laser process that allows light to reflect through several layers of film. This creates an unprecedented sense of depth that's easily visible under a wide range of lighting conditions. While stock holograms are available

from Polaroid, having one customized is highly cost effective. Prices include creation, production and finishing. Completed holograms come in sheets, rolls or cookie-cut for machine or hand application. And they're convenient, requiring no special printing runs. They're simply applied after printing like any pressure sensitive label. You can expect fast 10 to 12 week delivery too.

For more information call
800-237-5519 Fax: (781) 386-8671

◆Polaroid
Holographic Products

POLAROID MIRAGE HOLOGRAMS

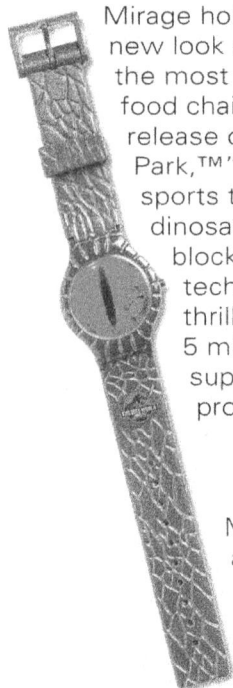

Mirage holograms and Burger King took a new look at wrist watches to create one of the most successful promotions in the fast food chain's history. Tying in with the release of "The Lost World: Jurassic Park,™" a special watch was created. It sports the slowly dilating eye of a raptor dinosaur that appeared in the earlier blockbuster, "Jurassic Park." Mirage technology captured the chills and thrills of the original film clip so well, 5 million watches were sold with supplies running out before the promotion was up.

A new series of children's stickers featuring true 3-D motion will be added to American Greeting's 1998 hologram sticker line. The quality of these Mirage holograms is unmatched in the business, and they represent a very special value because they can be removed. They're also non-toxic, making them a safe bet for collectable fun among kids of all ages.

Mirage hologram sunglasses are another way to put a bright new face on promotions and premiums. They can be customized with any image, name or logo for a 3-D look that's bound to turn heads.

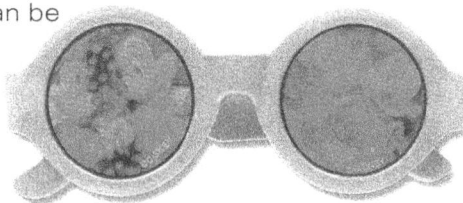

Polaroid Is With You Every Step Of The Way. Our team supports you through the creation, design, development and mass production of your hologram. Since we control the entire process, you're assured of the highest level of quality and security. Contact us to see what we can develop for you.

Mirage holograms are a highly effective tool for product authentication and trademark protection. Super Bowl ticket counterfeiters discovered their efforts to duplicate tickets were "not even close to the actual ticket and were easy to spot" according to the NFL Director of Security. Thanks to the unique physical design of Mirage holograms, they're virtually impossible to copy.

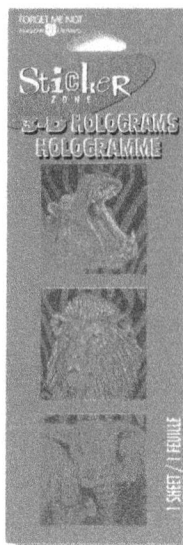

Working with SmarTalk Teleservices, Inc., Polaroid added holograms to phonecards making a popular collectable even more appealing. In 1997, Santa Claus was the debut card offered to major retailers and advertisers. Holographic phone cards make great gifts and provide a unique visual edge in an already booming category.

For more information call
800-237-5519 Fax: (781) 386-8671
email: holography.info@polaroid.com

◆Polaroid

The Brightest Idea in the World for LCDs

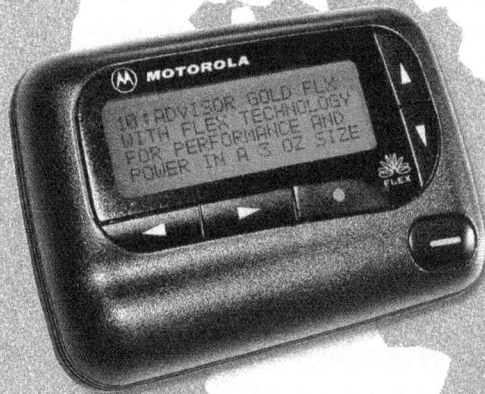

Polaroid High Gain H-Film Reflectors are up to 3 times brighter than any other reflector. That's because Polaroid technology works efficiently to optimize the viewing zone of liquid crystal displays. The corner-to-corner brightness is exceptionally uniform too, which is ideal for both large and small area devices including PDAs, palmtops, watches, marine instruments, GPS navigators and more. While ordinary LCD products can waste as much as 50% of their energy on the illumination of displays, Polaroid reflectors require no power at all. Talk to Polaroid about how we can help you achieve more brilliant designs with Polaroid holographic reflectors adapted for your own products. Examples of applications as well as telephone numbers and addresses are listed on the back of this page.

Polaroid
High Gain H-Film

Polaroid holographic reflectors are making millions of watch faces brighter. Timex, Casio, Citizen, and Seiko have been shipping watches with our film for over a year now with fantastic consumer response. They're a big improvement over ordinary LCD watch displays under daylight conditions. And Timex is also using **Polaroid's** holographic technology to enhance a night time backlite display called "All Day Indiglo."

The POIS Personal Navigation System is a natural for **Polaroid** technology. Qualities like 3 times the display brightness of ordinary materials, glare-free viewing, and corner-to-corner brightness and uniformity are important for any screen that utilizes many layers of information. But they're even more critical when someone is viewing that information from behind the wheel.

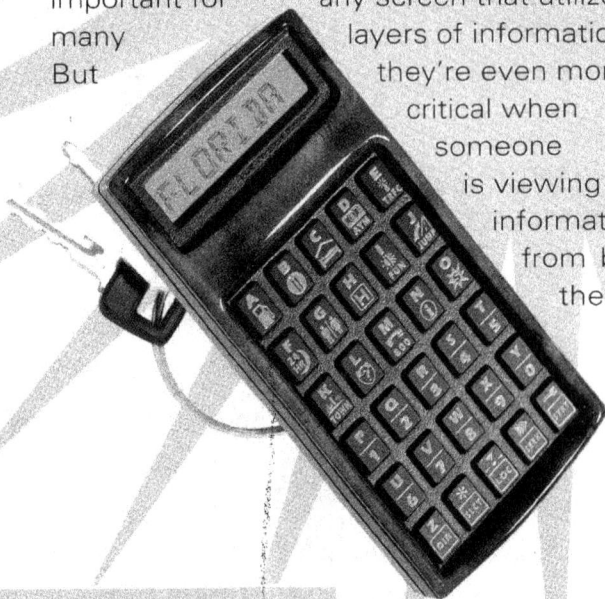

Polaroid's New White High Gain H-Film is now here! Don't forget to ask about our latest high quality holographic film for use in monochrome and color LCDs.

◆**Polaroid**
High Gain H-Film

Contact Polaroid today for more information and assistance:

Polaroid Corporation
Holographic Products Division
Cambridge, Mass. 02139
Phone: 781-386-3354
FAX: 781-386-8671
email: holography.info@polaroid.com

Nippon Polaroid K.K. - Japan
Mori Building No. 30
Toranomon 3-2-2
Minato-Ku, Tokyo 105
Japan
Phone: 81-33438-8883
FAX: 81-35473-8637

Polaroid Far East Limited - Korea
Room 2001, 20/F
Korea World Trade Center
159 Samsung-dong, Kangnam-ku
Seoul 135-729, Korea
Phone: 82-2-551-8633
FAX: 82-2-551-6763

Polaroid Far East Limited - Taiwan
3/F, 82, Kuang Fu N. Rd.
Taipei, Taiwan, R.O.C.
Phone: 886-2-578-2216
FAX: 886-2-579-2771

◆**Polaroid**
Holographic Products

actually delivered. As in most manufacturing businesses, prices decrease as quantity increases. In order t of igure the suggested retail price of their product, it is very common for a copyright holder to multiply the manufacturer's unit cost five times. For instance, a product that costs the copyright holder $4.00 will be resold to a distributor for $6.00, and will end up selling in a store for $20.00.

I f you are having holograms made to your specifications, choose. a company that produces holograms appropriate to your final application. Be aware that manufacturers have not yet standardized their pricing - some itemize production processes, others quote a finished price. Some quote by the square inch, others according to a sliding scale based on quantities ordered.

The Distributor

The distributor is a business that specializes in buying large quantities of product from a copyright holder and distributing it to other businesses that cannot afford to, or are not interested in, stocking inventory. Distributors are often contractually obligated to order large amounts of merchandise, carry an entire line of their supplier's products, and not sell competing products. This alleviates many problems for the copyright holder and the manufacturer who are not usually set up to market their own products to numerous customers.

In return, the distributor commonly receives the sole rights to sell the product in a particular geographic region or to a particular group of customers and pays less than any other customer down the line. Distributors commonly pay 70% below the suggested retail price, resell these products for 60% below suggested retail, and depend upon a large volume of sales to make their profit. For example, if they pay $6.00 for an item from a copyright holderrthey would resell it to a wholesaler for $8.00.

Many distributors also repackage goods under their own names, deal with import/export procedures and constantly work to expand the marketplace. A popular product can make a distributor a lot of money, due to the fact that potential customers have no alternative supplier. The distributor sells mostly to wholesalers.

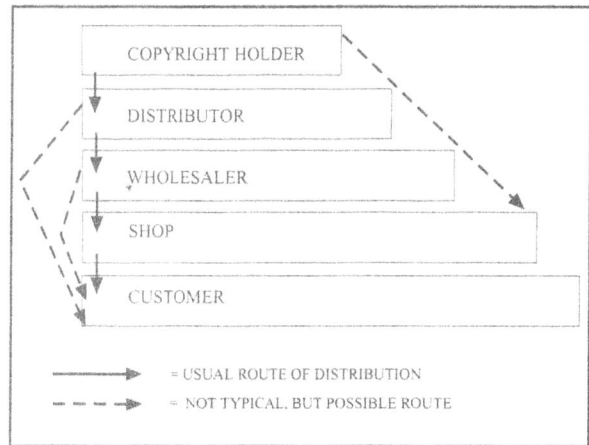

Chain of Holography Distribution

The Wholesaler

The wholesaler connects distributors to retailers. The essential function of a wholesaler is to get the product into shops. Most wholesalers use a combination of in-house salespersons or independent sales reps working on commission to persuade buyers to try a product. Many rely on telemarketing departments, catalogue mailings and trade shows to establish new accounts and service existing ones. A good wholesaler will teach a shop ow,ner how to best merchandise a product, provide point-of-purchase materials, restock displays, update product selection, and generally keep the customer happy.

A wholesaler normally carries many different product lines, which is a convenience to retailers who want to consolidate the number of their suppliers. Also, wholesalers stock far less merchandise than a distributor, which allows them to react quickly to changes in the marketplace.

Wholesalers do not generally have exclusive rights to a product. They often extend payment terms to their customers (net 30 days is common) after a probationary period or credit check. For their efforts, wholesalers receive special pricing that allows them to make a profit when they resell the goods to retailers. Wholesalers typically work off 25% profit margins; if they buy an item for $8.00, they will resell it for $10.00. Some distributors offer additional discounts to wholesalers for larger orders.

The Retailer

Retailers are the point of contact with the public, the place where merchandise is displayed and purchased by the customer. Holograms and related products have been sold in a variety of settings, ranging from temporary tabletop setups (flea markets, trade show booths) to art galleries and department stores. Many entrepreneurial businessmen start with a small cart or kiosk in a busy mall during holiday season and graduate to a bigger store that runs year round. Several single-store operations have expanded to multi-store chains.

Wholesalers have placed merchandise in obvious locales like museum gift shops, technology stores, and poster

Typical costs, margins, and profits for hologram with retail price of $20

Business	Buys For	Resells For	Discount Off Retail	Markup (% of Cost)
Copyright Holder	$4	$6	80%	50%
Distributor	$6	$8	70%	33%
Wholesaler	$8	$10	60%	25%
Retailer	$10	**$20**	50%	100%
Customer	**$20**	N/A	N/A	N/A

shops; less obvious locations include airport shops, nature stores and stationery stores. Other dealers have targeted specific interest groups such as hobbyists and collectors, and sell licensed products to comic book shops, trading card stores, and the like. Larger suppliers have cut deals with amusement parks, national chains, and event promotion agencies. Holograms have even been sold by mail order using catalogs and classified ads, even though a written description or photograph does not adequately capture the wonder of a three dimensional image.

All successful hologram retail businesses have several things in common - they are located in high-traffic (pedestrian) areas in places where people go to buy interesting items (often tourist destinations), they display the merchandise correctly and with flair, and they offer a high level of friendly customer service. Although the retail price suggested by the copyright holder is only a guideline, most retailers double their costs to establish the final price a customer sees in the store. Therefore, an item that costs them $10.00 will end up on the shelf for $20.00.

Categories of Merchandise

Most successful hologram stores sell a selection of holographic merchandise, including pictures (stock wall decor images and limited-edition fine art), jewelry (and related fashion accessories), executive gifts (desktop accessories), toys and optical novelties. There are several ways to categorize this merchandise - by price, by size, by manufacturer, and so on. For now, we'll discuss selection by "format" (which refers to the type of material the hologram is produced on).

Silver-Halide Glass Plate

Traditionany, most holographers have produced their holograms on sheets of glass (glass plates) coated with a high resolution light-sensitive emulsion called silver halide. This is similar, but not identical to, the emulsions used in conventional photographic films.

These glass plates are rather costly, but can be used to make the highest quality holograms - mostly because the glass plates are quite rigid and will not move during the exposure period (movement will ruin the hologram). Working with these glass plates is quite time-consuming, as each plate has to be handled with great care during each production step.

Due to the time and the cost involved in making holograms on silver-halide glass plates, they are mainly used for limited-edition holographic wall art, archival images or custom work. A finished 8" x 10" glass plate hologram typically retails at $300.00 or more for an open edition work.

It is possible to produce striking "deep image" holograms on silver-halide glass plates - that is, images which display considerable projection in front of the hologram's surface, and/or appear to be a considerable distance behind the hologram's surface. To achieve such dramatic effects, glass plate holograms require the best possible illurnination, which usually takes some foresight.

Glass plate holograms are extremely fragile and should always be securely wrapped and handled with care. When the hologram is framed, the actual emulsion is facing ward the back and is protected. The surface which we can see (and touch) is ordinary glass and can be cleaned by gentle polishing with a soft cloth.

Silver-Halide Film

Since glass plates were too expensive and impractical to mass produce, holographers developed methods to produce holograms on silver-halide holographic film, which is cheaper and easier to handle. It is thin and flexible, similar to the film used in an ordinary camera. After exposure and processing, each piece of film is usually sandwiched in a cardboard matte, ready to package and/or frame as wall decor.

An 8" x 10" silver-halide film hologram typically retailsfor $150.00 (less than half the cost of a comparable hologram produced on glass), making this format much more suitable for the giftware market. The best-selling size has traditionally been a 4" x 5" film mounted in an 8" x 10" matte due to the retail price (which has hovered around $35.00). Other standard sizes include a 2" x 2.5" film mounted in a 5" x 7" matte (which typically sells for $15.00 - $20 .00), and a 5" x 8" film mounted in an 8" x 10" matte (which typically sells for $70.00).

Silver-halide film holograms do not display quite the resolution, projection and depth of their glass plate counterparts. However, under proper illumination they are quite striking.

The emulsion of a film hologram is usually covered by a cardboard matte; however, the front surface can scratch easily. Fingerprints and dust can be cleaned by gently polishing with a soft cloth. Framing these pieces behind clear glass is recommended. Like all photographic films, they should be kept away from excessive heat and moisture.

Photopolymer Film

Several companies (notably Polaroid and DuPont) have developed a plastic film material that is especially suited for reproducing holograms. These photopolymers are extremely bright and durable, which makes them appropriate for a wide variety of commercial applications. The automated production process used to reproduce these holograms cuts costs considerably. An 8" x 10" matted photopolymer film hologram can retail for well under $100.00.

Image depth is a bit shallower and projection distance is a bit less than in silver-halide films. However, photopolymer holograms are a magnitude brighter and generally have a much wider viewing angle. Although this allows for more latitude when using a less-than-ideallight source, a single overhead light is still recommended for illuminating wall decor. The material works exceedingly well for items that will be illurninated by normal outdoor lighting, s.uch as jewelry, bookmarks and postcards. The plastic surface of the photopolymer films can be cieaned with a soft cloth.

Dichromate Gelatin (DCG)

Another photosensitive material used to make holograms is "dichromate gelatin." Unlike silver halide glass plates or rolls of film, this material is actually a chernicaVgelatin mix that is coated onto a piece of glass. After exposure, a cover glass is securely attached to the base plate, creating a permanent seal- since moisture or excessive heat can dissolve the dichromate gel and the hologram will eventually disappear. Sometimes plastic is used instead of glass, lowering production costs.

Dichromate holograms are very bright and can be viewed under less-than-ideal lighting conditions, e.g., outdoors. Therefore, they are most frequently used as jewelry items anq fashion accessories. The glass can be cut into any shape, but it is often cut into small discs. These hologram discs have traditionally been used in watchfaces, broaches, key-rings and belt buckles. For years, the $20.00 dichro- mate glass pendant has been a staple for retailers. An 8" x 10" framed dichromate plate usually retails for over $100.00, but they are seldom made anymore.

Embossed Foils

These holograms are reproduced by a process that is not optical and does not use costly photosensitive materials. In this case, the holographic image is transferred onto a mechanical stamping die, which is then used to emboss the image on rolls of very thin plastic films or metal foils. These holograms can be reproduced for cents per square inch, making it the least expensive way to copy an image. These foils are commonly hot-stamped on paper or plastic, or backed with an adhesive layer. The material is very durable.

The embossing process creates very shallow images that can be viewed under poor illumination. They are usually silver-backed and reflect a rainbow of colors. Some of the best-selling holographic products are animated 3-D embossed images which display fluid motion and realistic colors. These commonly retail for $15.00 - $20.00, matted and packaged, ready t of rame. Most of the embossed holo- grams on the market are called 2-D/3-D. They display multi-level graphics and are usually used as inexpensive stickers, or part of keychains, pins, and magnets.

Often a repeating prismatic pattern 'is embossed onto the plastic or foil , which creates a ever-changing rainbow effect. This colorful material is used in toys and optical novelties .

Opening a Shop - Basic Guidelines

1. Choose the best location you can afford. Tourist destinations in "festival" type shopping districts located near dining and drinking establishmentslraditionally do very well. High-traffic locations ensure a iiteady stream of curious shoppers. Malls do very well around the holidays, but can be slow during the off-season. Do extensive market research.

2. Shop for a reliable supplier that can stock you in a timely manner. A good supplier will be able to replace the merchandise you have sold quickly, thereby reducing the number of units you need to backstock. A good supplier will also assist you with displays , inventory selection, and point-of-purchase materials.

3. Stock a wide selection of merchandise in various price ranges. Many hologram stores cover their overhead with the sales of inexpensive items in the $1.00 - $10.00 range. Most retail sales average $5.00 - $25.00. Wall art typically sells well at the $35.00 price point, as do $65.00 watches. Sales of more expensive pieces can be icing on the cake.

4. Plan your displays carefully. Due to lighting considerations and limited viewing angles, holograms need to be merchandised more carefully than many other products. Installing adjustable track lighting is more practical than using stationary fixtures. Halogen spot lights (50-100 watts) work very well. Avoid fluorescent and recessed incandescent lights.

5. Create an entertaining atmosphere. No one needs to buy a hologram. A friendly, informative staff will boost sales dramaticall y.

Start-up Costs

The amount of capital you need depends on how large an operation you want to have. On the low end, you can stock a "cart" operation, as opposed to an actual shop, for as little as $5,000. Buying starting stock for a small shop (several hundred square feet) will cost several times that. A larger space of 800-1,000 square feet can easily hold $50.000 worth of goods. Plan on spending $10,000- $15,000 for high-quality lighting and displays for a store of this size.

Based on a "double markup for inventory," $200,000 worth of annual gross sales should support an owner/ manager making $30,000, a small staff (dividing another $30,0(0), rent! overhead in a high-traffic location ($25,000), and first year build-out ($15,000).

CHROMAGEM INC.

INDEPENDENT HOLOGRAPHIC MASTERING STUDIO

SPECIALIZING IN HIGH END DESIGN AND MASTERING

All types of holograms: embossed, photopolymer, dichromates.
3D, 2D-3D, stereograms (from your computer files or video),
Dot matrix, large format rainbows (hoe's).
In house computer design, and 3d model-making.
Over 175 tileable dot matrix stock images and patterns
geometric, christmas and all occasion).

Chromagem has been in existence since 1982. Our holographers cumulatively have over 45 years of lab and design experience. Our 8,000sq. ft. facility has 4 shooting labs and attendant equipment (high powered lasers, electroforming, high end computer systems, gang up capabilities, etc.).
In addition to our mastering skills, Chromagem has considerable experience guiding high end clients towards an effective holographic project.

SOME SIGNFICANT JOBS:

•PepsiCo. Xmas 24 pack carton largest holographic material run at the time.*
•Hershey Chocolate Co., designed 9 custom dot matrix holograms,
for Lost World, Jurassic Park nationwide promotion (flexible packaging).
•Smith Kline Beecham's Tagamet, full page ad, 3d image and rainbow surround
with registered ink over.*
•7UP wrap around label for 16 oz. bottles.
•Miller Brewing Co. over 60 million labels.
•Merck Pharmaceutical, complex two channel 3D hologram.
•Cambell Soup six multi-color holograms from 3D models and 2d background
for SpagettiOs label collection.*
•DuPont's DryPel hangtag
•German Post Office, Isaac Newton commemorative stamp.
•six 2-channel 3d images for Kellogg's, cereal box insert, 15,000,000 holograms
We have also made holograms used by: Upper Deck, HBO, Holiday Inn, Reebok, McDonald's,
Hasbro Toys, America Online, Packaging Magazine to name a few.

*Award Winning

Phone/Fax: (330) 793-3515   e mail: chromagem @ aol.com
573 S. Schenley Ave. Younsgtown, OH 44509

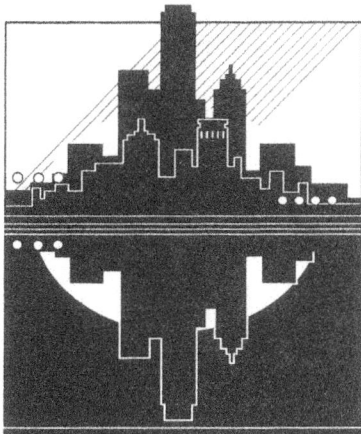

10

Business Listings

In this chapter we list businesses and people involved in the field of holography. There are five lists:

1. *A complete international business directory, with businesses listed in alphabetical order.*

2. *A list of business names, sorted by country.*

3. *A list of individuals, listed alphabetically.*

4. *A list of Internet World Wide Web addresses.*

5. *A list of us. businesses, sorted by Zip code.*

Please assume that all website addf-esses begin with "http://www." Also, please note that several countries are in the process o f changing telephone prefixes. We would appreciate if you notify us of any changes, omissions or mistakes by fax (510-841-2695) or by email (staff@ross-books.com)

International Business Directory (A - Z)

21st Century Finishing Inc,
215 Pennsylvania Avenue
City: Paterson
State/Province: NJ
Postal Code: 07503
Country: United States of America
Voice Phone: (1) 973 279 2100
Fax Phone: (1) 973 279 5659
Contact #1: Anthony Olmo
Business Description: Multi-faceted converting specialists. 11 years experience working with trade and corporate clients. Full range of web ' or sheet finishing services offered including: cutting, hot-stamping, labels, lamination, overprinting, etc. Capable of applying holograms to a variety of substrates.
****

3 Deep Hologram Company
609 California St.
City: Huntington Beach
State/Province: CA
Postal Code: 92648
Country: United States of America
Voice Phone: (1) 714 969 5354
Fax Phone: (1) 714 969 5354
email: acheimets@aol.com
Web: www.3deepco.com
Contact # 1: Alex Cheimets

Business Description: Supplies Russian sil ver halide emulsions on plates. Green sensitive, red sensitive, and new full color emulsions available. Can also recommend appropriate developing and processing procedures,
SEE OUR ADVERTISEMENT
****

3-D Hologrammen
1012 GA Amsterdam
City: Holland
Country: Netherlands
Voice Phone: (31) 20 6247225
Fax Phone: (31) 20 6247225
email: j .kraak@inter.nl.net
Contact # 1: Erik S wetter
Business Description: Wholesale & Distribution - Retail - Art. Gallery/shop since 1987.
****

3-D Systems
P.O. Box 145
City: PI. Arena
State/Province: CA
Postal Code: 95468
Country: United States of America
Voice Phone: (1) 707 882 2066
email: Igcross@intercoastal.com
Contact # 1: Lloyd Cross

Business Description: Currently produces ""'virtual optics laboratory"" software for holographers and optics research, Pioneered integral holograms and other holographic techniques.
****

3D Holographics
Maple Lodge, Beavers Town
City: Donabate, Dublin
State/Province: Ireland
Postal Code: 11
Country: United Kingdom
Voice Phone: (353) I 843 6200
Business Description: Applied Holographics representative for Ireland
****

3D Images Ltd.
31 The Chine
Grange Park
City: London
State/Province: England
Postal Code: N21 2EA
Country: United Kingdom
Voice Phone: (44) 181 3640022
Fax Phone: (44) 1813641828
email: burder3d@aol.com
Web: http://www.tisco.coml3d-web/3d-images/
Contact # 1: David Burder

INTERNATIONAL DIRECTORY

Business Description: Manufacturer and distributor of 3D images. Also supplies lenticular 3D products. Producers of "Virtual Video". 3D supplies and glasses.
****

3D Optical Illusions
P.O. Box 765
City: Bayswater
State/Province: Victoria
Postal Code: 3153
Country: Australia
Voice Phone: (61) 39 729 6337
Voice Phone: (61) 18 776226
Fax Phone: (61) 39 729 6020
Contact # 1: Trevor McGaw
Business Description: Specializing in lenticular and holographic movement illusions with up to 20 different motion images.
****

3D Technologies & Arts
Scopolijeva 19
City: Ljubljana
Country: Slovenia
Voice Phone: (386) 61 558463
Fax Phone: (386) 61 1330189
email: holography&tripod.net
Web: http: //members.tripod.com/- holography
Contact # 1: Nikola Jelic
****

3D Vision
Hologramme-Laserprodukte
Ostertorsteinweg 1-2
City: Bremen
Postal Code: 0-28203
Country: Germany
Voice Phone: (49) (0)421 76797
Fax Phone: (+49) (0)421 76797
Contact # 1: Uwe Reichert
Business Description: Holograms, Holographic projects. General commercial holograms for sale and distribution.
****

30-40 Holographics
97 St John Street
City: London
State/Province: England
Postal Code: ECIM 4AS
Country: United Kingdom
Voice Phone: (44) 171 2503545
Fax Phone: (44) 171 2503566
email: graham@hologram.demon.co.uk
Contact #1: Graham Tunnadine
Business Description: Production house for all types of 3D visual displays. Large format, silver halide mastering facility on premises.
****

3DIMAGE
G.P.O. Box 95
City: Sydney
Postal Code: 2001
Country: Australia
Voice Phone: (61) 0148 804 905
Voice Phone: (61) 88 383 7255
Fax Phone: (61) 88 383 7244
email: info@3dimage.com.au
Web: http://www.mcm.com.au
Contact # 1: Simon Edhouse
Business Description: General Holographic design consultancy based on seven years experience in manufacturing. Specialising in design of large scale thematic exhibitions and alternative 3D effects.
SEE OUR ADVERTISEMENT
****

3M - Safety and Security Systems
3M Center, Bldg 225-4N-14
City: St. Paul
State/Province: MN
Postal Code: 55144- 1000
Country: United States of America
Voice Phone: (1) 800 328 7098
Fax Phone: (1) 800 223 5563
email: info@mmm.com
Web: www.mmm.com
Contact #1: Maureen Tholen
Business Description: Supplier of authenticating labels, including some holographic applications. Full range of origination services offered.
****

A.D. Tech (Advanced Deposition Technologies)
580 Myles Standish Blvd.
Myles Standish Industrial ·Park
City: Taunton
State/Province: MA
Postal Code: 02780
Country: United States of America
Voice Phone: (1) 508 823 0707
Voice Phone: (1) 800 767-5432
Fax Phone: (1) 508 823 4434
email: staff@adv-dep.com
Web: www.adv-dep.com
Contact # 1: Glenn J. Walters
Business Description: Security label business.
****

A.H. Prismatic, Inc.
139 Mitchell Avenue
Suite 102
City: South San Francisco
State/Province""t'A
Postal Code: 94080
Country: United States of America
Voice Phone: (1) 650 634 0500
Voice Phone: (1) 800 537 3631
Fax Phone: (1) 650 634 0545
Contact #1: Sheila Bagley
Business Description: Manufacturers of exclusive ranges of holographic gifts, toys, jewelry, photopolymers, and film holograms. Licensed products available: Star Trek; Star Trek: The Next Generation; Star Trek: Deep Space Nine; Star Wars. Call for U.S.A. wholesale catalog.
SEE OUR ADVERTISEMENT
****

A.H. Prismatic, Ltd.
New England House
New England Street
City: Brighton
State/Province:East Sussex, England
Postal Code: BNI 4GH
Country: United Kingdom
Voice Phone: (44) 1273686966
Fax Phone: (44) 1273 676692.
Contact # 1: Ian Dayus
Business Description: Manufacturers of exclusive ranges of holographic gifts, toys, jewelry, photopolymers, and film holograms. Licensed products avai lable: Star Trek; Star Trek: The Next Generation; Star Trek: Deep Space Nine; Star Wars.
SEE OUR ADVERTISEMENT
****

AB E,ueck Holoart
Roepers Weide 26
City: Hamburg
Postal Code: D-22605
Country: Germany
Voice Phone: (49) (0)40 8807151
Contact # 1: A. B. Rueck
Business Description: Holograms for conventional presentations, refection and transmission.
****

Abrams, Claudette
22 Bayview Avenue
City: Toronto, Wards Island
Postal Code: M5J IZI
Country: Canada
Voice Phone: (1) 416 203 7243
Fax Phone: (1) 416 203 7243
Contact #1 : Claudette Abrams
Business Description: Holographic artist and technician.
****

Academy of Media Arts Cologne
Peter-Welter Platz 2
City: Cologne
Postal Code: 0-50676
Country: Germany
Voice Phone: (49) (0)221 201 89 115
Fax Phone: (+49) (0)221 201 89 124
Contact # 1: Dieter Jung
Business Description: International Academy for Media Arts. Extensive holography lab and teaching.
****

Accuwave Corp.
1651 19th St.
City: Santa Monica
State/Province: CA
Postal Code: 90404
Country: United States of America
Voice Phone: (1) 310 4495540
Fax Phone: (1) 310 449 5539
email: staff@accuwave.com
Web: www.accuwave.com
Contact # 1: Jason Corwin
Business Description: Manufacture Holographic Optical elements (filters) for fiber optic industry.
****

Acme Holography
12 Sunset Road
City: West Somerville
State/Province: MA
Postal Code: 02144
Country: United States of America
Voice Phone: (1) 617 623 0578
email: bconn@media.mit.edu
Contact # 1: Betsy Connors
Business Description: Acme Holography is Boston's first private holography lab. We offer full service in reflection, transmission and computer generated holography, including design consultation and large-scale environmental holography.
****

Action Tapes
Unit 5
Boundary Road
City: Brackley
State/Province: England
Postal Code: NNI3 7ES
Country: United Kingdom
Voice Phone: (44) 1280700591
Fax Phone: (44) 1208 700590
email: action.tapes@dia1.pipex.com
Contact # 1: Alan J. Phillips .
Business Description: Adhesive tapes for industry: close tolerance custom slit roll s; varying length stroke diameter; high speed die cut; large and small volume runs. Fast and experienced service. Experienced staff to provide professional and technical advise.
****

AD 2000, Inc.
948 State Street
City: New Haven

State/Province: CT
Postal Code: 065 1 1
Country: United States of America
Voice Phone: (1) 203 624 6405
Voice Phone: (1) 800 334 4633
Fax Phone: (1) 203 624 1780
email: pscheir@ad2000.com
Web: ad2000.com/ad2000/
Contact # 1: Peter Scheir
Business Description: Fully custom & customized stock image embossed and photopolymer holograms. Our HOLOBANK contains the world's largest selection of stock image embossed holograms - available plain, as labels, foil, magnets, pins, roll stock, etc.
LOW RUN SECURITY OUR SPECIALITY
****

AD HOC Public Relations GmbH
Thesings Allee 10
City: Giitersloh
Postal Code: D-33332
Country: Germany
Voice Phone: (49) 5241 500990
Fax Phone: (+49) 5241 500999
Contact # 1: Wolfpeter Hocke
Business Description: pres agency, specialized on holography, public relation concepts, promotion, brochures,
****

Adlas G.M.B.H. & Co Kg.
See land Strasse 9
City: Luebeck
Postal Code: D-23569
Country: Germany
Voice Phone: (49) 451 390 9300
Fax Phone: (49) 451 390 9399
Business Description: Established 1986. Manufacturer of diode laser-pumped solid state lasers which operate in CW and pulsed mode with wavelengths in IR, visible and Uv. Branch office: 636 Great Road, Stow, MA 01775, USA. Owned by Coherent, Santa Clara, CA, USA
***"*

Advanced Holographic Laboratories
(a division of Astor Universal Ltd.)
Astor Road / Eccles New Road
City: Salford
State/Province: England
Postal Code: M5 2DA
Country: United Kingdom
Voice Phone: (44) (0)1617898131
Fax Phone: (44) (0)161 7878348
Contact # 1: Francis Tuffy
Business Description: Worldwide designers, originators and manufacturers of holographic pattern and image foils and film. See Astor Uuiversal advertisement for further details.
SEE OUR ADVERTISEMENT
****

Advanced Holographic Laboratories
(a division of Astor Universal Corp,)
3841 Greenway Circle
City: Lawrence
State/Province: KS
Postal Code: 66046-5444
Country: United States of America
Voice Phone: (1) 800 255 4605
Voice Phone: (1) 913 842 7674
Fax Phone: (1) 9138429748
email: astoruniversal@worldnet.att.net
Contact # 1: John Thoma
Business Description: Worldwide designers, originators and manufacturers of holographic pattern and image foils and films. See Astor Universal advertisement for further details.
SEE OUR ADVERTISEMENT
****

Advanced Optics, Inc.
110 N.E. Trilein Drive
Suite #4
City: Ankeny
State/Province: IA
Postal Code: 50021
Country: United States of America
Voice Phone: (1) 515 964 5050
Fax Phone: (1) 515 964 5050
email: heill@aol.com
Web: www.techexpo.com
Contact # 1: Wendy Heil
Business Description: Manufactures of custom and precision optics including first surface mirrors for use in holographic equipment.
****

Advanced Precision Technology, Inc.
4669 Hillside Drive
City: Castro Valley
State/Province: CA
Postal Code: 94546
Country: United States of America
Voice Phone: (1) 510 889 1650
Fax Phone: (1) 510 889 8195
Contact # 1: Bruce Pastorius
Business Description: Makes holographic optical element used in company built scaner for fingerprint recognition.
****

Advanced Technology Program
Bldg. 101- Room A430
City: Gaithersburg
State/Province: MD
Postal Code: 20899
Country: United States of America
Voice Phone: (1) 8002873863
Fax Phone: (1) 301 926 9524
Web: atp .nist.gov
Business Description: United States Government program providing funding for R&D projects.
****

Aerospatiale
Ets D'Aquitaine
Saint-Medard-En-Jalles
City: Bordeaux
Postal Code: F-33165
Country: France
Voice Phone: (33) 56 57 34 80
Fax Phone: (33) 56 57 30 70
Business Description: Scientific and industrial research, NDT testing
****

Aerotech Inc.
Electro-Optical Division
101 Zeta Drive
City: Pittsburgh
State/Province: PA
Postal Code: 15238
Country: United States of America
Voice Phone: (1) 4129637470
Fax Phone: (1) 412 963 7459
Contact #1: Steve A. Botos
Business Description: Manufacturers of helium neon tubes, power supplies and complete systems for OEM and end users. Other product lines include optical table positioners and precision rotary and linear positioning systems. Subsidiary Companies: Aerotech Ltd.- England, Aerotech GmbH-Germany.
****

Ag Electro-Optics Ltd.
Tarporley Business Centre
City: Tarporley, Cheshire
State/Province: England

Postal Code: CW6 9UY
Country: United Kingdom
Voice Phone: (44) 1829733 305
Fax Phone: (44) 1829733 679
email: sales@ageo.co.uk
Web: http://www.ageo.co.uk
Contact #1: J.A. Gibson
Business Description: Distributor of lasers, optics, lab equipment and fiber optics.
****

Agfa - Gevaert N. V.
(a division of Bayer)
Septestraat 27
City: Mortsel, Antwerp
Postal Code: B-2640
Country: Belgium
Voice Phone: (32) 3 444 8251
Fax Phone: (32) 3 444 8243
email: bebay2uk@ibmmail.com
Contact # 1: Frank Mortier
Business Description: Headquarters. Manufacturer of silver-halide recording materials.
****

Ahead Optoelectronics, Inc.
BI , No. 130, Section 3, Keelung Road
City: Taipei
Country: Taiwan
Voice Phone: (886) 27 369 1520
Fax Phone: (886) 27 362 0485
email: julie@ahead.com.tw
Web: http://www.ahead.com.tw
Contact # 1: Julie Lee
Business Description: A manufacturer specializing in holographic products, holographic systems, optomechanical systems. Can produce custom masters (conventional , 3D dot matrix, Lippmann) and holographic technology (recombining, hot-stamping, embossing, tamper-evident). Can offer two different dot matrix systems (one 200-400 dpi, one high security at 150-1300dpi and over 10 dots/sec).
SEE OUR ADVERTISEMENT
****

AHT 3D-Medien
Association for Hologram Techniques & 3D
Media (AH
Niederesch 28
City: Bad Rothenfelde
Postal Code: D-49214
Country: Germany
Voice Phone: (49) (0)542 5365
Fax Phone: (+49) (0)542 5359
Contact #1: Gunter Deutschmann
Business Description: Association of German hologram manufacturers. Association for hologram techniques and 3D media, engaged in the commercialization of holographic products. Realization of projects and campaigns in co-operation between the member companies. From consultation, design, mastering, hologram production to overprinting, converting, integration into the finished product, etc.
****

AKS Holographie-Galerie GmbH
Potsdamer Strasse 10
City: Essen
Postal Code: D-45145
Country: Germany
Voice Phone: (49) (0)201 756455
Fax Phone: (+49) (0)201 753582
email: akshol@aol.com
Web:http://members.aol.com/akshol/
hhome_d.htm
Contact #1: Detlev Abendroth
Business Description: producing all kinds of holograms encl. computer-generated up to a size

of 80 x 100 cm. One of the world's greatest film-hologram-edition. Internationally represented by distributing partners. Owns great holographic gallery.
****

AKS Holographie-Gallerie GmbH
Postsdamer Strabe 10
Eben
City: Ruhr
Postal Code: D-4145
Country: Germany
Voice Phone: (49) 20 I 756455
Fax Phone: (49) 20 I 753582
Business Description: Holography Gallerie - various types of holograms.
****

Alabama A&M University
Center for Applied Optical Sciences
P.O. Box 1268
City: Normal
State/Province: AL
Postal Code: 35762
Country: United States of America
Voice Phone: (1) 205 851 5870
Fax Phone: (1) 205 851 5622
email: marius@caos.aamu.edu
Web: www.caos.aamu.edu
Contact #1: Nicholai Kukhtarev
Business Description: R&D in all applications of holography including dynamic (real time) holography and photo-refractive materials.
****

Alfred Dirksen + Sohn, Modellwerkstiitten
Griiner Weg 8-10
City: Wesseling
Postal Code: D-50389
Country: Germany
Voice Phone: (49) 2236 42827
Contact # 1: Alfred Dirksen
Business Descriptiop: Consulting and design for holographic projects. Building of models for holography, hologram displays
****

Amagic Technologies Inc.
1652 Deere Ave.
City: Irvine
State/Province: CA
Postal Code: 92606
Country: United States of America
Voice Phone: (1) 7144743978
Voice Phone: (1) 800 262 4421
Fax Phone: (1) 714 474 3979
email: amagicl68@aol.com
Web: www.thomasregister.com/
Contact #1: Marilyn Huff
Business Description: Fully integrated manufacturer of holographic custom images, stock images, diffraction/digitized patterns as polyester, PVC/vinyl and polypropylene. Produces hot stamp foil, pressure sensitive labels, and film to 39.37 inches.
****

Amazing World Of Holograms
Corrigan's Arcade Foreshore Road
South Bay
City: Scarborough, North Yorkshire
State/Province: England
Postal Code: YOII IPB
Country: United Kingdom
Voice Phone: (44) 1723500696
Fax Phone: (44) 1482492 286
Business Description: Exhibitors and retailers of film, glass, embossed, dichromate and related products. Permanent display of 200 holograms

including mutliplexes which are updated and changed regularly. Main season May-October. Distributors of film & glass.
****

American Bank Note Holographies
399 Executive Blvd.
City: Elmsford
State/Province: NY
Postal Code: 10523
Country: United States of America
Voice Phone: (1) 914 592-2355
Fax Phone: (1) 914 592-3248
email: abnh@westnet.eom
Web: http ://www.abnh.com
Contact # 1: Keith Woodward
Business Description: World leader in development of embossed holography for security and commercial applications. Produces embossed holograms in a range of formats: foil, pressure-sensitive & tamper-evident labels, clear & wide-web laminates for packaging.
SEE OUR ADVERTISEMENT
****

American Holographic Inc.
60 I River Street
City: Fitchburg
State/Province: MA
Postal Code: 01420
Country: United States of America
Voice Phone: (1) 508 3403 0096
Fax Phone: (1) 508 348 1864
email: amerholo@tiac.net
Contact #1: Thomas Mikes
Business Description: Design , devdop and manufacture of components and instruments for use in industrial and medical measuring devices. We are using holographic diffraction grating design and manufacture capability to produce components for unique measurement instruments.
****

American Laser Corporation
1832 South 3850 West
City: Salt Lake City
State/Province: UT
Postal Code: 84104
Country: United States of America
Voice Phone: (1) 801 972 1311
Fax Phone: (1) 80 I 972 5251
email: sales@amlaser.com
Web: http://www.amlaser.com
Contact # 1: Dan Hoefer
Business Description: Established 1970. Manufacturer of Argon, Krypton and mixed gas laser systems and subsystems from 3 mw to 10w, in air or water-cooled configuration.
****

American Paper Optics Inc.
3080 Barlett Corporate Drive
City: Barlett
State/Province: TN
Postal Code: 38133
Country: United States of America
Voice Phone: (1) 901381-1515
Voice Phone: (1) 800 767 8427
Fax Phone: (I) 901381 1517
email: optics3d@lunaweb.net
Contact # 1: John Jerit
Business Description: American Paper Optics is the leading manufacturer of paper 3D glasses in the world. The products include Diffraction glasses, Polarized and Chroma Depth glasses. The latest holographic 3D g lass is called HoloSpex and creates an image of a logo when viewing a direct point of light.
****

American Propylaea Corporation
555 South Woodward, Suite 1109
City: Birmingham
State/Province: MI
Postal Code: 48009-6626
Country: United States of America
Voice Phone: (1) 248 642 7000
Fax Phone: (1) 248 642 9886
email: dddimage@aol.com
Business Description: Auto-stereoscopic realtime HOE based display devices for CAD, medicine, education, entertainment and defense applications. Capable of displaying full motion, projected imagery. Also color holograms and HOEs.
****

American Society for Nondestructive Testing
1711 Arlingate Lane
City: Columbus
State/Province: OH
Postal Code: 43228-0518
Country: United States of America
Voice Phone: (1) 800 222 2768
Voice Phone: (1) 614 274 6003
Fax Phone: (1) 614 274 6899
email: Itrask@asnt.org
Web: http://www.asnt.orgl
Contact #1: Larry Trask
Business Description: ASNT is a non-profit organization representing the field of non-destructive testing. They offer certification programs, Information services (Patent searches), and a Magazine/Journal called ""Materials Evaluation""
****

Ana MacArthur
P.O. Box 15234
City: Santa Fe
State/Province: NM
Postal Code: 87506
Country: United States of America
Voice Phone: (1) 505 438 8739
Fax Phone: (1) 505 438 8224
email: auroean@nets.com
Contact # 1: Ana MacArthur
Business Description: Holographic installation artist - unique installations. Also produces limited edition dichromate holograms and holographic sculptures. Experimental use of real-time interferometry.
****

Another Dimension Inc. (Spectore/AD!)
637 NW 12th Ave.
City: Deerfield Beach
State/Province: FL
Postal Code: 33442
Country: United States of America
Voice Phone: (1) 800 422 0220
Voice Phone: (1) 952 429 1017
Fax Phone: (1) 952 4212391
email: info@spectore.com
Web: http ://www.spectore.com
Contact #1: Mark Anoff
Business Description: Another Dimension, Inc. merged with Spectore Corp. in January 1997. Spectore/ AD! distributes and manufacturers holograms and 3DFX (Stereo-optics/Lenticular) both domestically and internationally. Primary focus is retailers.
****

Applied Holographies, Pic.
40 Phoenix Road
Crowther District 3
City: Washington, Tywe & Wear
State/Province: England
Postal Code: NE38 OAD

Country: United Kingdom
Voice Phone: (44) 1914175434
Fax Phone: (44) 1914163292
email: sales@appplied-holographics.com
Web: http://www.applied-holographics.com
Contact # 1: David Tidmarsh
Business Description: Holographic designers and manufacturers of secure optically variable devices in the form of hot stamping foils, tamperevident labels, semi-transparent laminates and securetransfer films for the protection of security printed documents and branded products from counterfeiting.
SEE OUR ADVERTISEMENT
****

Applied Optics
2662 Valley Drive
City: Ann Arbor
State/Province: MI
Postal Code: 48103-2748
Country: United States of America
Voice Phone: (1) 313 998 0425
Fax Phone: (1) 313 998 0425
email: upatnks@applopt.com
Contact # 1: Juris Upatnieks
Business Description: Applied Optics provides consulting services and laboratory breadboard testing of optical systems in coherent optics, holography, diffractive optical elements, and light control. System analysis using the ZEMAX optical design program.
****

Arbeitskreis Holografie B.Y.
Boeckelter Weg 47
City: Geldem
Postal Code: D-47608
Country: Germany
Voice Phone: (49) (0)283 3034
Contact # 1: Herman-Josef Bianchi
Business Description: Artistic holography
****

Armin Klix Holographie
Postfach 260218
City: Duesseldorf
Postal Code: D-40095
Country: Gemlany
Voice Phone: (49) (0)2 11 317775
Fax Phone: (+49) (0)211 317749
Contact # 1: Armin Klix
Business Description: Anfertigung von Displayhologrammen im Auftrag von Werbung und Industrie und Einzel-und Grosshandel. Katalog von Lieferbaren Hologrammen vorhanden. Einzelstucke und Grosserien.
****

Art Agentur Kaln
Venloer Stra13e 461
City: Koeln
Postal Code: D-50825
Country: Germany
Voice Phone: (49) (0)221 54 41 00
Voice Phone: (+49) 221 54 1400
Contact #1: Liesel Dr. Hollmann-Langecker
Business Description: mediation agency for modem art, focused on holography
****

Art Institute Of Chicago (The School of the ...)
Holography Department
112 South Michigan Ave.
City: Chicago
State/Province: IL
Postal Code: 60603
Country: United States of America
Voice Phone: (1) 312 345 3998
Fax Phone: (1) 312 345 3565

email: ewesly@artic.edu
Web: http://www.artic.edu
Business Description: The School of the Art Institute of Chicago offers an MFA degree with a concentration in holography, and is equipped with three tables, with one containing a stereogram printer for recording computer generated imagery or live subjects.
****

Art Lab
1000 Richmond Terrace
City: Staten Island
State/Province: NY
Postal Code: 1030 I
Country: United States of America
Voice Phone: (1) 7184478667
Fax Phone: (1) 718 447 8668
Contact #1: John Iovine
Business Description: The Art Lab is an art school which offers classes and workshops in the art of holography. Free brochure available.
****

Art, Science & Technology Institute (AST!)
2018 R Street, N.W.
City: Washington
State/Province: DC
Postal Code: 20009
Country: United States of America
Voice Phone: (1) 202 667-6322
Fax Phone: (1) 202 265-8563
Contact # 1: Laurent Bussaut
Business Description: Research and educational corporation dedicated to the advancement of art, science and technology in holography, laser, optics and photonics industries. The museum features a permanent collection of masterpiece holograms including one of the largest holograms in the world. The museum is visited by guided tour only, and reservations are required.
****

Artbridge Light Studios _
Madam Weg 77
City: Braunschweig
Postal Code: D-38120
Country: Germany
Voice Phone: (49) ~0)531 352816
Fax Phone: (+49) (0)531352816
Contact #1: Odile Meulien
Business Description: Design & Production of Holograms and Light Show - Organization of special happenings in coordination with multidisciplinary artists, engineers and scientists.
****

Artplay Holographika Studio
Ady Endre ut. 8
City: Budapest
Postal Code: H-1191
Country: Hungary
Voice Phone: (36) 1 281-9114
Fax Phone: (36) 1 282-4921
email: baloghho@caesar.elte.hu
Contact #1: Tibor Balogh
Business Description: A.P.Holografika provides mastering, whole process for custom embossed holograms (mainly in security applications), optical design using HOE-s. Wholesaler of holographic novelties and diffraction foils.
****

Asahi Glass Co.
R&D General Division,
2-1-2 Marunouchi Chiyoda-Ku
City: Tokyo
Postal Code: 100
Country: Japan

Voice Phone: (81) 3 3218 5825
Fax Phone: (81) 332145060
Contact # 1: Fumihiko Koizumi
Business Description: R&D on holography and single copy holograms.
****

Astor Universal Ltd.
Astor Road
Eccles New Road
City: Salford
State/Province: England
Postal Code: M5 2DA
Country: United Kingdom
Voice Phone: (44) 161 7898 131
Fax Phone: (44) 161 787-8348
Contact #1: Laurence Holden
Business Description: See listing for Advanced Holographic Laboratories.
SEE OUR ADVERTISEMENT
****

Atelier Holographique De Paris
13 pass Courtois
City: Paris
Postal Code: F-750 II
Country: France
Voice Phone: (33) I 43 79 69 18
Fax Phone: (33) I 40 09 05 20
Business Description: Artistic holography; Buying & Selling; Consulting
****

Australian Holographics
P.O. Box 160
City: Kangarilla
State/Province: SA
Postal Code: 5157
Country: Australia •
Voice Phone: (61) 88 383 7255
Fax Phone: (61) 88 383 7244
email: mcm@mcm.com.au
Web: http://www.mcm.com.au
Business Description: Specialists in Large Format holography. Official commercial partners to the South Australian Museum. Rainbow, laser-transmission, and Denisyuk holograms to 1.1 - 2.2 m. Reflections to 1.1 - 1.1 m. CW and Pulsed. Optical and laser sales. Studio rental. Stock image bank.
****

Automated Holographic Systems
II Stephanie Lane
City: Westfield
State/Province: MA
Postal Code: 0 I 085
Country: United States of America
Voice Phone: (1) 4135682897
email: ahologram@earthlink.net
Contact # 1: Jim Gibb
Business Description: Holographic consultant with 20 years of industry experience. Designs and manufacturers state of the art production equipment. Specializing in film DCG, embossed image ganging, and HOEs.
****

Avant-Garde Studio
34 North Rochdale Ave.
P.O. Box 296
City: Roosevelt
State/Province: NJ
Postal Code: 08555-0296
Country: United States of America
Voice Phone: (1) 609 448 6433
Fax Phone: (1) 609 448 6433
email : avant3d@aol.com
Contact #1: Amy E. Medford
Business Description: Model makers. Sculptors, excelling in relief (forced perspective, texture,

INTERNATIONAL DIRECTORY

fonn), for holographic models. We will design and/or work from, photographs, drawings or other material.
****

Baier Praegepressen
Maschinenfabrik Gebr. Baier KG
Lindenthaler Str. 78
City: Rudersberg
Postal Code: D-73635
Country: Gennany
Voice Phone: (49) 7183 532
Fax Phone: (+49) 7183 3481
Business Description: Manufacturer of machines for production of embossed holograms.
****

Barilleaux, Rene Paul
c/o Mississippi Museum of Art
20 I East Pascagoula St.
City: Jackson
State/Province: M S
Postal Code: 39201
Country: United States of America
Voice Phone: (1) 60 I 982 7898
Voice Phone: (1) 60 I 960 ISIS
Fax Phone: (1) 60 I 960 1505
email: LHOUND@worldnet.art.net
Business Description: Curator, consultant, writer, lecturer in area of artist-made holograms; organized "New Directions in Holography: The Landscape Reinvented," traveling through 1998; "Unfolding Light: The Evolution of Ten Holographers".
****

Batelle Pacific Northwest National Laboratory
P.O. Box 999
K8-09
City: Richland
State/Province: WA
Postal Code: 99352
Country: United ~tates of America
Voice Phone: (1) 509 375 4405
Fax Phone: (1) 509 372 6488
email: ma_lind@pnl.gov
Web: www.pnl.gov2080
Contact # 1: Michael Lind
Business Description: Theoretical and experimental research and development in applying holographic techniques to active and passive optical, radar, acoustic, ultrasound, EEG and EKG imaging. R&D for Dept. of Energy
****

Bauer, Josef
Augustenstra13e 121
City: Muenchen
Postal Code: D-80798
Country: Gennany
Voice Phone: (49) 89 2710248
Contact # 1: Josef Bauer
Business Description: Educational courses in holography
****

Beddis Kenley (Machinery) Ltd.
Unit 21
Moderna Business Park
City: Mytholmroyd, Halifax
State/Province: England
Postal Code: HX7 5QQ
Country: United Kingdom
Voice Phone: (44) 01422 886 550
Fax Phone: (44) 01422 886 614
Contact # 1: S.D. Smith
Business Description: Large sheet hot foil stamping machines for graphic enhancement and hologram application. Maximum sheet size 29 x 41 (73cm x 104cm).
****

Beijing Dongfang Laser Printing Tech. Co.
Yuqun Hu-Tong Jia- #18
Dong-Cheng-Qu
City: Beijing
Postal Code: 100010
Country: China
Voice Phone: (86) 10 64033223
Fax Phone: (86) 10 64034996
Contact # 1: Liu Haidong
****

Beijing Fantastic Hologram Products
Xue-Yuan-Lu #5
Hai-Dian-Qu
City: Beijing
Postal Code: 100083
Country: China
Voice Phone: (86) 10 62083373
Fax Phone: (86) 10 68378302
Contact #1: Wang Yao
Business Description: Sales of holograms A
****

Beijing Hologram Printing Tech. Co
PO Box 9622-#2
City: Beijing
Postal Code: 100086
Country: China
Voice Phone: (86) 10 68379223
Fax Phone: (86) 10 68378302
Contact # 1: Fu Ziping
Business Description: {folography sales and distribution.
****

Beijing Sanyou Laser Images Co.
Bei-San-Huan-Zhong-Lu #40
City: Beijing-
Postal Code: 100088
Country: China
Contact # 1: Li Qiang
Business Description: hologram distribution
****

Beijing Univ. of Posts & Telecommunications.
P.O.Box 163
City: Beijing
Postal Code: 100088
Country: China
Voice Phone: (86) 10 62281535
Fax Phone: (86) 10 62281774
email: xudx@bupt.edu.cn
Contact # 1: Hsu Da-hsiung
Business Description: Holography class
****

Beijing Xi Ji Wo Computer Graphic Working Co.
Dong-Huang-Cheng-Gen-Bei-Jie Jia-#20
City: Beijing
Postal Code: 100010
Country: China
Voice Phone: (86) 10 64255163
Fax Phone: (86) 10 64255163
Contact # 1: Chen Hui
Business Description: Holography classes.
****

Bellini, Victor
850 Howard Ave. # 5J
City: Staten Island
State/Province: NY
Postal Code: 1030 I
Country: United States of America
Voice Phone: (1) 718 442 4726
email: victor3D@webtv.net
Contact # 1: Victor Bellini
Business Description: Extensive collection and archive of holographic, lenticular, and other 3D collectibles. Appraisals for trading cards.
****

Bemrose Security and Promotional Printing
Wayz Goose Dr
City: Derby
State/Province: England
Postal Code: DE21 6XG
Country: United Kingdom
Voice Phone: (44) 133 229 4242
Fax Phone: (44) 133 2290366
****

Benyon, Margaret - Holography Studio
40 Springdale Avenue
City: Broadstone, Dorset
Postal Code: BH18 9EU
Country: United Kingdom
Voice Phone: (44) 1202698067
Fax Phone: (44) 1202 698 067
email: benyon@holography.demon.co.uk
Contact #1: Margaret Benyon
Business Description: Fine art holography, established in 1968. Limited edition and unique works available. Included in a large number of private art collections world-wide.
****

Berkhout, Rudie
223 West 21 st Street
City: New York
State/Province: NY
Postal Code: 10011
Country: United States of America
Voice Phone: (1) 212 255 7569
Fax Phone: (1) 212 727 0532
email: rudieberkhout@mindspring.com
Web: rudieberkhout.home.mindspring.com
Contact # 1: Rudie Berkhout
Business Description: Holographic Fine Artist who has had work ex hibited worldwide, including at the Whitney Museum of American Art (New York). Also teaches holography at the School of Visual Arts (NY).
****

Bernhard Halle Nachf. GmbH & Co.
Hubertusstr. 10-11
City: Berlin
Postal Code: 0-13589
Country: Gennany
Voice Phone: (49) 30 7 91 60 77
Fax Phone: (+49) 30 7 91 85 27
Contact #1: A. Frank
Business Description: Precision optics, polarization optics, catalog and custom made optics
****

BIAS
(Bremer Institute Applied Beam)
Klagenfurter Str 2
City: Bremen
Postal Code: D-28359
Country: Gennany
Voice Phone: (49) (0)421 218 5002
Fax Phone: (+49) (0)421 218 5059
Contact #1: Werner Prof. Juptner
Business Description: Industrial research; NOT
****

Bjelkhagen, Hans (see Lake Forest College)
****

Blue Ridge Holographics, Inc.
511 Stewart St.
City: Charlottesville
State/Province: VA
Postal Code: 22902
Country: United States of America
Voice Phone: (1) 804 296 I 110
Fax Phone: (1) 804 296 1182
email: steve@blueridgeholo.com
Web: http ://www.blueridgeholo.com
Contact # 1: Steve Provence

Business Description: Mastering facility for the production of embossed holograms. Consulting and design services offered. Extensive client list.
****

Bobst Group Inc.
146 Harrison Avenue
City: Roseland
State/Province: NJ
Postal Code: 07068
Country: United States of America
Voice Phone: (1) 973 226 8000
Fax Phone: (1) 973 226 6715
email: sales@bobstgroup.com
Web: http ://www.bobstgroup.com
Contact # 1: Doug Herr
Business Description: One of the world's largest manufacturers of hologram hot stamp machinery.
****

Booth, Roberta
5326 Sunset Blvd.
City: Los Angeles
State/Province: CA
Postal Code: 90027
Country: United States of America
Voice Phone: (1) 213 466 5767
Fax Phone: (I) 213 465 5767
Contact # 1: Roberta Booth
Business Description: I am a holographic artist working in transmission and reflection holography. I also work as a consultant for holographic projects and curate holography shows.
****

Boyd, Patrick
Archway House, Church Street
Easton On The Hill
City: Stamford
State/Province: England
Postal Code: PE9 3LL
Country: United Kingdom
Voice Phone: (44) 976 298 578 ....
Voice Phone: (44) 1780755 647
Fax Phone: (44) 181 6707810
email: boyd@atlas.co.uk
Contact #1: Patrick Boyd
Business Description: Holographic Fine Artist. Extensive portfolio.
****

Brainet Corporation - International Division
4F. Asset Bldg.
3-31 -5 Honkomagome Bunkyo-ku
City: Tokyo
Postal Code: I 13-0021
Country: Japan
Voice Phone: (81) 3 5395 7030
Fax Phone: (81) 3 5395 7029
email: brainet@bnn-net.or.jp
Contact # 1: Yutaka Inoue
Business Description: Distributor and producer of framed holograms and other processed holographic products from a variety of manufacturers. Sole agent distributor of Lightrix/USA and Laza/UK for Japan and Korea. Also provide consulting for custom work and exhibition services
****

Brandtjen & Kluge, Inc.,
539 Blanding Woods Road
City: St. Croix Falls
State/Province: WI
Postal Code: 55024
Country: United States of America
Voice Phone: (1) 71 5 483 3265
Voice Phone: (1) 800 826 7320
Fax Phone: (I) 715 483 1640

Contact # 1: John Edgar
Business Description: With over 75 years experience in manufacturing what are acknowledged as the industry's most reliable presses, Brandtjen & Kluge is uniquely qualified to de liver presses for efficient, trouble fre e application of hologram foils.
****

Bridgestone Technologies, Inc.
375 Howard Ave.
City: Bridgeport
State/Province: CT
Postal Code: 06605
Country: United States of America
Voice Phone: (1) 203366 1595
Fax Phone: (I) 203 366 1667
Contact # 1: Richard Zucker
Business Description: Product authentication systems. Anti-counterfeiting technology. Fully integrated provider of security printed products and services including holography, micro-tracers, biocoding and field investigation.
SEE OUR ADVERTISEMENT
****

British Aerospace Pic.
Sowerby Research Centre
Fpc: 267 PO Box 5
City: Filton, Bristol
State/Province: England
Postal Code: BSI 2 7QW
Country: United Kingdom
Voice Phone: (44) 1179 366 842
Fax Phone: (44) 11 79 363 733
Contact # 1: Steve Parker
Business Desc ri ption: Sheer Holography and NDT interferometry as it applies to material stress. Will work on proj ects that are of mutual benefit to British Aerospace and client.
****

Broadbent Consulting
1070 Commerce Street~uite A
City: San Marcos
S'tate/Province: CA
Postal Code: 92069
Country: United States of America
Voice Phone: (1) 760 752 1039
Fax Phone: (I) 1'60 752 1039
email: hologram@fi a.net
Contact # 1: Donald Broadbent
Business Description: An independent, privately owned, holographic fac ility producing HOEs and display holograms in various recording materials. Donald Broadbent has 36 years experience in holography.
****

Burgmer, Brigitte
Volksgartenstrabe 14
City: Koeln
Postal Code: D-50677
Country: Germany
Voice Phone: (49) 221 329472
Contact #1: Brigitte Burgmer
Business Description: artist, initiator of numerous holography projects
****

Burleigh Instruments, Inc.
Burleigh Park
City: Fishers
State/Province: NY
Postal Code: 14453
Country: United States of America
Voice Phone: (1) 716 924 9355
Fax Phone: (1) 716 924 9072
Web: www.burleigh.com
Contact # 1: Tim Klimasewski

Business Description: Burleigh Instruments, Inco is a manufacturer of electro-optical equipment including wavelength meters, laser spectrum analyzers, interferometers, and nanopositioning devices.
****

California Institute of the ArtS
School of Critical Studies
24700 McBean Parkway
City: Valencia
State/Province: Ca
Postal Code: 91355
Country: United States of America
Voice Phone: (1) 805 255 1050 x2406
Fax Phone: (1) 805 255 0 177
email: alschulr@muse.calarts.edu
Business Description: We teach introductory holography and Lippmann photography.
****

Cambridge Laser Labs
853 Brown Road
City: Fremont
State/Province: CA
Postal Code: 94539
Country: United States of America
Voice Phone: (1) 510 65 1 01 10
Fax Phone: (1) 510 651 1690
Contact # 1: Brian Bohan
Business Description: World renowned specialist in ion laser repair. Rental systems and used laser system sales. Price guide furnished on request.
****

Canon Inc. R&D Headquarters
890, Kawasaki-Shi 4
Saiwai-Ku Kawasaki
City: Kanagawa
Postal Code: 2 11
Country: Japan
Voice Phone: (81) 44 549 5424
Contact #1: Tetsuro Kuwayama
Business Description: Research. Courses in holography.
****

Capilano College
Physics Department - Holography Research Lab
2055 Purcell Way
City, Prov.: N. Vancouver, B.C.
Postal Code: V7J 3H5
Country: Canada
Voice Phone: (1) 604 983 7571
Fax Phone: (1) 604 983 7520
email : bsimson@capcoll ege.bc.ca
Web: hltp:l/www2.capcollege.bc.ca/- bsimson/
Contact # 1: Milessa Crenshaw
Business Description: Digital holography research. Synthesizing holograms from video (in tegrals).
****

Capitol Converting Equipment, Inc.
500 North Redfield Court
City: Park Ridge
State/Province: IL
Postal Code: 60068
Country: United States of America
Voice Phone: (1) 847 825 7891
Fax Phone: (1) 847 825 8661
Business Description : Manufacturers of Hot stamping presses. 29"x41 ", 33"x47", 39" x55"
****

Carl M. Rodia And Associates
13 Locust St.
City: Trumbull
State/Province: CT

INTERNATIONAL DIRECTORY

Postal Code: 06611
Country: United States of America
Voice Phone: (1) 203 261 1365
Fax Phone: (1) 203 2681619
email: carlrodia@aol.com
Contact # 1: Carl M. Rodia
Business Description: Comprehensive engineering consultation services in precision hologram manufacturing. Plant design and engineering, process engineering, troubleshooting and seminar training of manufacturing personnel.
****

Casdin-Silver Holography
99 Pond Avenue Suite D403
City: Brookline
State/Province: MA
Postal Code: 02146
Country: United States of America
Voice Phone: (1) 617 739 6869
Voice Phone: (1) 617 423 4717
Fax Phone: (1) 617 739 6869
Contact # 1: Harriet Casdin-Silver
Business Description: I have been creating holographic art and interactive holographic installations since 1968. Our company specializes in original holograms for advertising, architectural and theater settings, and expositions. We are also consultants and exhibition organizers/designers.
****

Cavomit
22 Pipinou Steet
City: Athens
Postal Code: GR-11257
Country: Greece
Voice Phone: (30) I 823 2355
Fax Phone: (30) I 231 4499
Contact # 1: Alkis Lembessus
Business Description: Hot-stamping equipment (cylinders - platen), hologram registration systems, foils and consumables. Local distributor for Astor Univ7sal, Kluge, Light Impressions, Applied Holographics, Revere Graphic Products.
****

Center for Applied Research in Art & Tech.
University of Gent
72 Lange Boongaardstraat
City: Gent
Postal Code: B-9000
Country: Belgium
Voice Phone: (32) 91 626384
Fax Phone: (32) 91 237326
Contact # 1: Prof. Pierre Boone
Business Description: research
****

Central Glass Co., Ltd .
Kowa-Hitosubashi Bldg
7-1 Kanda-Nishikicho 3-Chom
City: Tokyo (Chiyoda-Ku)
Postal Code: 101
Country: Japan
Voice Phone: (81) 3 3259 7354
Contact # 1: Chikara Hashimoto
Business Description: Research - Heads Up Display
****

Centre d'Art Holographique et Photonique
College de Maisonneuve (c-2200)
3800 Sherbrooke est
City, Prov: Montreal, Quebec
Postal Code: HIX 2A2
Country: Canada
Voice Phone: (1) 514 254 7131 ex4509
Fax Phone: (1) 514253 8909
email: holostar@cmaisonneuve.qc.ca

Web: http ://www.cmaisonneuve.qc.ca/
holostar.html
Contact # 1: Eric Bosco
Business Description: We are situated in a CEGEP (post-secondary school). We have 2 fully equipped tables with HeNe's and an argon (with etalons). We do all types of holograms up to 2 feet by 3 feet. We give courses, workshops, production work and rent lab space.
****

Centro de Investigaciones en Optica, A.C.
Loma del Bosque 115
Col. Lomas del Campestre
City, Prov.: Leon, Gto.
Country: Mexico
Voice Phone: (52) 47731017
Voice Phone: (54) 47731018
Fax Phone: (52) 47 175000
email: dfa@riscl.cio.mx J:
Contact #1: Fernando Mendoza, Ph.D. 1
Business Description: Our Optical Research Center offers: optical design and construction of a wide variety of optical components (lenses, mirrors, prisms, etc.); high standards in R&D in optical NDT for industrial applications, optical fiber sensors, rare earth doped fibers and optical shop testing.
****

CFC Applied Holographies
500 State St.
City: Chicago Heights
State/Province: IL
Postal Code: 60411
Country: United States of America ·
Voice Phone: (1) 800 438 4656
Voice Phone: (1) 708 891 3456
Fax Phone: (1) 708 758 5989
email: 104347.1654@compuserve.com
Web: http ://www.applied-holographics.com/
Contact # 1: Dave Beeching
Business Description: For security, decoration and packaging applications; we offer a full range of tamper evident and authentication labels, hot stamping foils, and release and size coat combinations. Print treatments and custom metallizing available.
SEE OUR ADVERTISEMENT
****

Checkpoint Security Services Limited
Export Dept
IIS Chatham Street
City: Reading, Berkshire
State/Province: England
Postal Code: RGI 7JX
Country: United Kingdom
Voice Phone: (44) 118925 8250
Voice Phone: (44) 1189258251
Fax Phone: (44) 1189583749
email: lcarmichael@checkpoint.co.uk
Web: http ://www.checkpoint.co.uk
Contact # 1: Lisa Carmichael
Business Description: Point of issue hologram applicators for continuous and sheet documents, to authenticate or protect high value, negotiable documents. Available with metallised, and demetallised foil in various die block styles.
****

Chengdu Xinxing Inst. for Development of Tech.
Bin-Jiang-Lu-Wai Dong-Xia-He-Ba # 60
City, Prov.: Chengdu, Sichuan
Postal Code: 610061
Country: China
Contact # 1· Zeng Xuzhang
Business Description: Holography classes.
****

Cherry Optical Holography
2047 Blucher Valley Road
City: Sebastopol
State/Province: CA
Postal Code: 95472
Country: United States of America
Voice Phone: (1) 707 823 7171
Fax Phone: (1) 707 823 8073
Contact # 1: Greg Cherry
Business Description: Highest quality display holography available. Stock and custom reflection/transmission holograms on glass plates or film up to 40" x 72" in size. Open and Limited edition fine art holograms. Custom mastering services offered for silver halide and photopolymer replication.
****

Chiba University - Faculty Of Engineering
1-33 Yayoi-Cho
City: Chiba
Postal Code: 260
Country: Japan
Voice Phone: (8 1) 472 511 111
Contact #1: Jumpei Tsujiuchi.
Business Description: research; holographic.
****

China Ann Arbor Holographical Institute
Rm. 403 Bldg. 22
Zhong Lou Xing Chen
City, Prov.: Jiangsu, Suzhou
Postal Code: 215006
Country: China
Voice Phone: (86) 512 227 461
Contact # 1: Yaguang Jiang
Business Description: Hologram sales, classes.
****

CHIRON Technolas GmbH
Max-Planck Strasse 6
City: Domach
Postal Code: D-85609
Country: Germany
Voice Phone: (49) (0)89 945514 0
Fax Phone: (+49) (0)89 945514 70
Contact # 1: Mr. Junger
Business Description: Ophthalmologic Systems.
****

Chongqing Yinhe Laser Products Ltd.
Shi-Ma-He Xia-Hua-Yuan #6,
Jiang-Bei-Qu
City: Chongqing
Postal Code: 630021
Country: China
Voice Phone: (86) 81 1 5312071
Fax Phone: (86) 811 5312050
Contact # 1: Zhang Zheng
Business Description: Laser distributor.
****

Chromagem Inc.
573 South Schenley
City: Youngstown
State/Province: OH
Postal Code: 44509
Country: United States of America
Voice Phone: (1) 330 7933515,
Fax Phone: (1) 330 793 3515
email: chromagem@aol.com
Contact # 1: Thomas J. Cvetkovich
Business Description: Established in 1981 . Hologram mastering facility specializing in shooting photoresist and photopolymer masters for use in commercial mass-production of embossed holograms: 2D, 3D, stereogram and dot matrix. Extensive experience working with major corporate accounts. Design and consultation services.
SEE OUR ADVERTISEMENT
****

Chronomotion
424 Ninth St.
City: Santa Monica
State/Province: CA
Postal Code: 90402
Country: United States of America
Voice Phone: (1) 310 393 9859
Fax Phone: (1) 310 458 6269
email: mburney@ix.netcom.com
Contact # 1: Michael Burney
Business Description: Developed and patented a general process for producing "electronic holograms" with a real image projected into the room with the viewer.
****

Cifelli, Dan
712 Bancroft Road # 332
City: Walnut Creek
State/Province: CA
Postal Code: 94598
Country: United States of America
Voice Phone: (1) 510 930 8033
Fax Phone: (1) 510 930 8033
email: 74544.1630@compuserve.com
Contact # 1: Dan Cifelli
Business Description: Holography consulting and brokering since 1979 for stock and custom products. Evaluate, match, and develop your product for holography mastering and production techniques (photopolymer, embossed, and dichromate); sales potential and market positioning (ASI, premiums, specialty, security/anticounterfeiting etc.); manufacturing and converting processes/materials; cost analysis; and patent! licensing potential.
****

City Chemical
100 Hoboken Ave.
City: Jersey City
State/Province: NJ
Postal Code: 07310
Country: United States of America
Voice Phone: (1) 20 I 653 6900 ...
Voice Phone: (1) 800 248 2436
Fax Phone: (1) 20 I 653 4468
Business Description: Photo-chemicals and chemical supplies. .
****

Coburn Corporation
1650 Corporate Road West
City: Lakewood
State/Province: NJ
Postal Code: 0870 I
Country: United States of America
Voice Phone: (1) 908 367 5511
Fax Phone: (1) 908 367 2908
email: coburncorp@aol.com r
Web: http://www.coburn.com
Contact # 1: John White ·r
Business Description: Offers a wide range of stock repeating geometric holographic designs in pressure sensitive film ; conventionally printable substrate; various traditional and designer oriented colors avai lable. Introductory sample kits available.
****

Coherent Luebeck GmbH
Seelandstrasse 9
City: Luebeck
Postal Code: D-23569
Country: Germany
Voice Phone: (49) (0)451 3909300
Fax Phone: (+49) (0)451 395725
Business Description: Manufacturer of diode laser-pumped solid state lasers which operate in CW

and pulsed mode with wavelengths in IR, visible and Uv. Branch office: 636 Great Road, Stow, MA 01775, USA. Owned by Coherent, Santa Clara, CA, USA
****

Coherent, Inc. - Laser Group
5100 Patrick Henry Drive
City: Santa Clara
State/Province: CA
Postal Code: 95054
Country: United States of America
Voice Phone: (1) 408 764 4000
Voice Phone: (1) 800 527 3786
Fax Phone: (1) 408 7644800
email : tech_sales@cohr.com
Web: http://www.cohr.com
Contact # 1: Paul Ginouves
Business Description: Coherent is the world leader in high-power ion and diode-pumped solid-state (DPSS) lasers. Our products for professional holography include argon ion lasers (up to 8 W at 488.0 nm, krypton lasers (up to 3.5 W at 647. 1 nm), and DPSS lasers (up to 5 W at 532 nm).
SEE OUR ADVERTISEMENT
****

Colour Holographics
Unit 7a, 1-2 Domingo Street
City: London
State/Province: England
Postal Code: EC IY OTA
Country: United Kingdom
Voice Phone: (44) 171 251 0511
Fax Phone: (44) 171 736 4710
Contact # \: Michael Medora
Business Description: We produce multiplex embossed stereogram masters. We also produce HOE's as well as DCG fu ll colour multiplex reflection masters. We have both have been Holographers since 1982 working with Third Dimension, Richmond Holographics, Raven Holographics, Medora Waves and now Colour Holographics.
****

Concordia University
Communications Studies
7141 Sherbrooke St. W
City, Prov: Montr~al, Quebec
Postal Code: PQ H4B I R6
Country: Canada
Voice Phone: (1) 514 848 2539
Voice Phone: (1) 514 848 2424
Fax Phone: (1) 514 848 3492
email: hal@vax2 .concordia.ca
Web: http://www.concordia.ca/
Business Description: 3 Dimension Research Center
****

Continental Optical
15 Power Drive
City: Hauppauge
State/Province: NY
Postal Code: 11788
Country: United States of America
Voice Phone: (1) 516 582 3388
Fax Phone: (1) 516 582 1054
Business Description: Optics and custom orders.
****

Control Module Inc.
227 Brainard Road
City: Enfield
State/Province: CT
Postal Code: 06082
Country: United States of America
Voice Phone: (1) 860 745 2433
Voice Phone: (1) 800 722 6654

Fax Phone: (1) 860 741 6064
email: rbilleri@controlmod.com
Web: http://www.controlmod.com
Contact # I : Ralph Billeri
Business Description: Experience in the design and manufacture of automatic data collection equipment and systems, CMI offers exciting innovations in 1996, including Holonetics (TM), a machine-readable hologram, offering the highest security protection avai lable for Access Control and Labor Management.
****

Control Optics
13111 Brooks Drive, Unit J
City: Baldwin Park
State/Province: CA
Postal Code: 91706
Country: United States of America
Voice Phone: (1) 626 813 1991
Fax Phone: (1) 626 813 1993
email: liucoc@interserv.com
Web: http://www.controloptics.coml
Contact # 1: Wai-Min Liu
Business Description: Provides full-service optical engineering supporting industry and education. Offers full range of holographic table top optics, positioning devices and mounts. New products include holography and fiberoptic experimenter's kits.
SEE OUR ADVERTISEMENT
****

Corion Corp.
8 East Forge Parkway
City: Franklin
State/Province: MA
Postal Code: 02038 •
Country: United States of America
Voice Phone: (1) 508 528 4411
Fax Phone: (1) 508 520 7583
email: sales@corion.com
Web: http ://www.Corion.com
Contact # 1: Don McLeod
Business Description: Corion Corp. manufactures volume and one-of-a-kind custom and stock optical components including coatings, filters, optics and optical assemblies for use in the UV Visible-IR spectrum. Mostly biomedical applications.
****

Corning Incorporated
City: Coming
State/Province: NY
Postal Code: 14831
Country: United States of America
Voice Phone: (1) 800 222-7740
Voice Phone: (1) 800 492 1110
email: info@corning.com
Web: http://www.coming.comlindex.html
Business Description: On April 24, 1997 Corning announced the acquisition of Optical Corporation of America who produces Precision, large aperture (to 36 inch diameter) aspheric mirrors for holographic production systems.
****

Corp. Mexicana De Impresion S.A. de c.v.
General Victoriano Zepeda 22, Col. Observatorio
Delegacion Miguel Hidalgo, c.P.
City: Mexico D.F.
Postal Code: 11840
Country: Mexico
Voice Phone: (52) 5 273-5583
Fax Phone: (52) 5 272-2916
email: bety@fenix.ifisicocu .unam.mx
Contact # 1: Salvador Nava-Calvillo
Business Description: Comisa is Mexico City government's printing company. We have in-

INTERNATIONAL DIRECTORY

stalled a holographic production line in order to use holograms for high security purposes in all out official documents and certificates.
****

Courtauld Chemicals
PO. Box 5, Station Road
Spondon
City: Derby
State/Province: England
Postal Code: DE21 7BP
Country: United Kingdom
Voice Phone: (44) 01332 661422
Fax Phone: (44) 0 1332 660178
Contact # 1: Richard ford
Business Description: die acetate film
****

Creative Holography Index, The
The International Catalog for Holography
46 Crosby Road
City, Prov.: West Bridgford, Nottingham
State: England
Postal Code: NG2 5GH
Country: United Kingdom
Voice Phone: (44) 7050 133 624
Fax Phone: (44) 7050 133 625
email: index@monand.demon.co.uk
Web: http ://www.holo.com/peper/search. html
Contact # I : Andrew Pepper
Business Description: The Creative Holography index is an international cata logue, in colour, and includes an artist produced hologram. Available as the complete collection. It features artists working with holography as a creative medium and includes critical essays, biographies, statements and a hologram. Cost 89.95 Sterling.
****

Creative Label
2450 Estes Drive
City: Elk Grove Vi llage
State/Province: IL
Postal Code: 6000'7
Country: United States of America
Voice Phone: (1) 847 956 6960
Fax Phone: (1) 847 956 8755
Contact #1: Jerry Koril
Business Description: Full range decorative graphic finishers. Large volume capability. Bindery application of holograms to paper, cardboard, and plastics. Kluge (2 stream) and Bobst (4 stream) machines. Call for more information.
****

Crown Roll Leaf, Inc. / Holo-grafx
91 Illinois Ave.
City: Paterson
State/Province: NJ
Postal Code: 07503
Country: United States of America
Voice Phone: (1) 973 742 4000
Voice Phone: (1) 800 63 1 3831
f ax Phone: (1) 973 742 0219
email: sa les@crownrollleaf.com
Web: http://www.crownrollleaf.com
Contact # 1: Kathy Kassover
Business Descri ption: Crown Roll Leaf is a major manufac turer of hot stamp foils suited for holographic applications. In addit ion, our in house production facil ities are capable of full origination through fi nishing and converting.
SEE OUR ADVERTISEMENT
****

Curt Abramzik
Goethestr. 67
City: Offenbach
Postal Code: D-63067
Country: Germany

Voice Phone: (49) (0)69 884911
Contact # 1: Curt Abramzik
Business Description: Manufacturer, importer and exporter of magician equipment and novelties. Holograms up to 100 x 100 cm, reflection and transmission holograms on film and glas, multiplex holograms, dichromate- and embossed-holograms, diffraction foils.
****

Customer Service Instrumentation
7 Meadowfield Park South
City: Stocks field, Northumberland
State: England
Postal Code: NE43 7QA
Country: United Kingdom
Voice Phone: (44) 1661 842741
Fax Phone: (44) 1661 842288
email: ghscott@netcom.co.uk
Contact # 1: G.H. Scott
Business Description: Manufacture front surface mirrors and optics for holography. 4.
****

CVI Laser Corporation
200 Dorado Place
City: Albuquerque
State/Province: NM
Postal Code: 87 192
Country: United States of America
Voice Phone: (1) 505 296 9541
fax Phone: (1) 505 298 9908
email: cvi@cvilaser.cofu
Web: www.cvilaser.com
Contact # 1: Bob Soales
Business Description: Manufactures holographic quality single and multiple element lenses, mirrors, windows and beamsplitters for all standard holographic laser sources. free 104-page catalog available.
****

D. Brooker & Associates
Rt. I, Box 12A
City: Derby
State/Province: Iowa
Postal Code: 50068
Country: United States of America
Voice Phone: (1) 515 533 2103
fax Phone: (I) 515 533 2104
email: dbrooker@netins.net
Contact # 1: Dennis Brooker
Business Description: NEW kit - Enables user with inkjet or laser printer to apply images and text to holographic vinyls. Kit includes materials, instructions and patterns - make bus & greeting cards, labels, ornaments and much more! Call for details!
****

Dai Nippon Printing Co., Ltd.
Central Research Institute
250-1 Aza-Kahasawa
City, Prov.: Kashiwa-City, Wakashiba
State: Chiba
Postal Code: 277
Country: Japan
Voice Phone: (81) 4 7134 0512
fax Phone: (8 1) 4 71332540
Contact # J: Takashi Wada
Business Description: Central research center. Embossed holography research. Embossed holograms done for customers.
****

Daimler Benz Aerospace
Dornier Medizintechnik GmbH
Industriestrasse 15
City: Germering
Postal Code: D-82110

Country: Germany
Voice Phone: (49) 89 84108 0
Fax Phone: (49) 89 84108 575
Contact # J: Ms. Thiemon
Business Description: Industrial Research; holographic non-destructive testing. HOE research.
****

Dan Han Optics
188-261 An Nyeong-Ri Tean-Eup
City, Prov: Hwasong-Gun, Kyung Ki-Do
Country: Korea
Voice Phone: (82) 0331 351 030
Fax Phone: (82) 0331 351 031
Contact # 1: Chung Song
Business Description: General optical supplies.
****

Datacard Corporation
II1II Bren Road West
City: Minneapolis
State/Province: MN
Postal Code: 55440
Country: United States of America
Voice Phone: (1) 612 933 1223
Voice Phone: (1) 800 621 6972
fax Phone: (1) 612 9310418
email: info@datacard.com
Web: http ://www.datacard.coml
Contact # 1: Mark Iverson
Business Description: Security and authentication applications utilizing holographic technologies. Capable of high volume runs for government and commercial users.
****

Datasights Ltd.
Alma Road
Ponders End
City: Enfield, Middlesex
State: England
Postal Code: EN3 7BB
Country: United Kingdom
Voice Phone: (44) 181 8054151
Fax Phone: (44) 181 805 8084
Contact # 1: Frank Sharpe
Business Description: Manufacture mirrors for use in holography. Beamsplitters and gratings also available.
****

David Dann Modelmaking Studios
PO Box 396, 4 East Hill Rd.
City: White Sulphur Springs
State/Province: NY
Postal Code: 12787
Country: United States of America
Voice Phone: (1) 914 292 1679
fax Phone: (1) 914 292 1679
email: davidann@zelacom.com
Contact #1: David Dann
Business Description: A maker of holographic models for more than a decade, David Dann's work has been seen on the covers of National Geographic, Omni, Marvel and Malibu comics, and many, many other places. Clients have included American Bank Note, Polaroid, HoloGrafx, Blue Ridge Holography, and Bridgestone Tech.. Brochure and samples upon request.
****

De La Rue Holographics
Stroudley Road
Daneshill Industrial Estate
City: Basingstoke, Hampshire
State: England
Postal Code: RG24 8FW
Country: United Kingdom
Voice Phone: (44) 1256463000
Fax Phone: (44) 1256460800

Contact # 1: Annette Kiely
Business Description: De La Rue Holographics, a division of De La Rue International Ltd., offers customers high technology protection against product counterfeit and tampering though security optical microstructures. It operates in two main markets: security products and brand protection.
****

Decolux GmbH
Verdistra13e 7
City: Muenchen
Postal Code: D-81247
Country: Germany
Voice Phone: (49) 89 8112044
Voice Phone: (+49) 89 8112045
Fax Phone: (+49) 89 8118582
Contact # 1: Horst Mairiedl
Business Description: 28 years self adhesive foils, 16 years effect-foils, 12 years diffraction foils, 6 years holography, 5 years holographic stamping foils, contacts to master hologram manufacturers and all important companies in this field
****

Deem, Rebecca
709 1/2 West Glen Oaks Blvd
City: Glendale
State/Province: CA
Postal Code: 91202
Country: United States of America
Voice Phone: (1) 818 549 0534
Fax Phone: (1) 818 549 0534
Contact #1: Rebecca Deem
Business Description: Holographic artist. Originates masters for mass production holograms in embossed, DCG and photopolymer materials. Both pulsed and CW lasers available.
****

Deep Space Holographics
1070 Moss Street # I 05
City, Prov.: Victoria, B.C.
Postal Code: V8V 4P3
Country: Canada
Voice Phone: (1) 250 384 3927
email: eyetrek@islandnet.com
Contact # 1: Marc de Roos
Business Description: Exotic fine art/commercial sculpture/animation, conceptual/industrial design, display merchandising, exhibits and special effects. Since 1980 secured worldwide distribution of our DCG designs via Holocrafts, including Star Trek holograms design.
****

Dell Optics Company, Inc.
25 Bergen Blvd.
City: Fairview
State/Province: NJ
Postal Code: 07022
Country: United States of America
Voice Phone: (1) 201 941 101O
Fax Phone: (1) 201 941 9524
Contact # 1: Belle Steinfeld
Business Description: Custom working of precision optical components. Established 1950. 15 Employees at this address.
****

Denisyuk, Yuri N.
A.F.loffe Physicotechnical Institute
Politechnicheskaya 26
City: St. Petersburg
Postal Code: 194021
Country: Russia
Voice Phone: (7) 812 247 9384
Contact #1: Yuri N. Denisyuk

Business Description: Holography teacher. One of the pioneers of holography.
****

Deutsche Gesellschaft fur Holografie e. V.
Geschaeftsstelle
Bergstr. 6
City: Halle
Postal Code: D-06108
Country: Germany
Voice Phone: (49) 345 2026751
Fax Phone: (+49) 345 2026752
email: nimoe@burg-halle.de
Web: http ://www.burg-halle.de/dgh
Contact # 1: Niklas Moeller
Business Description: The society was founded to promote awareness of holography, and its members are mainly holographers and artists. To this end, the group intends to organize exhibitions. Interferenzen is a periodical published by this organization.
****

Deutscher Drucker Verlagsgesellschaft mbH & Co. KG
Senefelderstra13e 12
City: OstfildemlRuit
Postal Code: D-73760
Country: Germany
Voice Phone: (49) 711 442096
Voice Phone: (+49) 711 442098
Fax Phone: (+49) 711 442099
Contact # I : Theodor J. Anton
Business Description: publisher of worldwide journals for the printing industry. "Deutscher Drucker" 42 issues/year, distributed in more than 50 countries, "World-Wide Printer" 6 issues/year, distributed in 158 countries, supplement "EI Arte Tipografico" in 34 countries, some holography related articles with real holograms
****

Diamond Images, Inc.
P.O. Box 1701
City: Miami
State/Province: FL
Postal Code: 33133
Country: United s.tates of America
Voice Phone: (1) 305 443 2310
Voice Phone: (1) 305 323 8406
Fax Phone: (1) 305 443 8346
email: mark@diamondimages.com
Web: http://www.Diamondlmages.com
Contact # 1: Mark Diamond
Business Description: Holographer Mark Diamond brings 24 years experience to full color stereograms. Work is featured in museums and collections in 15 countries. Founding member of Museum of Holography New York. Specializing in portraiture and animated digital compositing.
****

Diavy sri
Via Vivaldi 108
City: so liera (Modena)
Postal Code: 1-41019
Country: Italy
Voice Phone: (39) 59 565758
Fax Phone: (39) 59 566074
Business Description: Subsidiary of Diaures; producer of holographic metallic paper as part of a venture with Scharr Industries, USA.
****

Die Dritte Dimension
Frankfurter Strasse 132-134
City: Neu fsenburg
Postal Code: D-63263

Country: Germany
Voice Phone: (49) (0)610 33367
Fax Phone: (+49) (0)610 326709
Contact # 1: Elke Hein
Business Description: Greatest specialized shop for holography in Germany. Always over 1,000 different holograms in stock. Very comprehensive fine art section. Branch office: Nordwest-Zentrum, Tituscorso, 60439 FrankfurtiM. Germany.
****

Dietmar Oehlmann
Bortfelder Stieg 4
City: Braunschweig
Postal Code: 0-38116
Country: Germany
Voice Phone: (49) (0)531352816
Fax Phone: (+49) (0)531 352 816
Contact #1: Dietmar Oehlmann
Business Description: Master of Arts in Holography from the Royal College of Arts, with own light creation lab to design and produce special effects in holography for artworks, performances and stage decoration.
****

Diffraction Ltd.
P.O. Box 909
Route 100
City: Waitsfield
State/Province: VT
Postal Code: 05673
Country: United States of America
Voice Phone: (1) 8024966642
Fax Phone: (1) 802 496 6644
email: hologram@madriver.com
Contact # 1: Bill Park@r
Business Description: Products and services relating to diffractive optics and holographic optical elements (HOEs) including micro fabrication and photomask production.
****

Digital Matrix Co.
67 Whitson Street
City: Hempstead
State/Province: NY
Postal Code: 11550
Country: United States of America
Voice Phone: (1) 516 481 7990
Fax Phone: (1) 5164817320
Web: http ://www.galvanics.com
Contact #1: Alex Greenspan
Business Description: Manufacturers of high precision computerized electro forming systems for the production of nickel embossing shims for the holography industry. Turnkey systems, training and consultation.
****

Dimension 3
3380 Francis-Hughes St.
City, Prov.: Laval, Quebec
Postal Code: H7L 5A 7
Country: Canada
Voice Phone: (l) 514 662 0610
Fax Phone: (1) 514 662 0047
email: pierre@dimension3.net
Web: http ://www.dimension3.net
Contact # 1: Pierre Gougeon
Business Description: We offer creative solutions to holographic projects. We are a full holographic production house (DCG, foil, transmission, photopolymer), large format and micro embossed with animation and colour control Photograms ™ (lenticular photography/ printing).
SEE OUR ADVERTISEMENT
****

INTERNATIONAL DIRECTORY

Dimensional Arts
40 I Carver Road
City: Las Cruces
State/Province: NM
Postal Code: 88005
Country: United States of America
Voice Phone: (1) 505 527 9183
Fax Phone: (I) 505 527 9927
email: arts@holo.com
Web: www.holo.com
Contact # 1: Ken Harris
Business Description: Exclusive manufacturer of the Light Machine, a patent protected digital origination system. Custom stock Dotz(r) dot matrix patterns available. Capable of 2D, 3D and full color stereogram work. Can transfer technology worldwide.
****

Dimensional Foods Co.
8 Faneuil Hall Market Place
City: Boston
State/Province: MA
Postal Code: 02109
Country: United States of America
Voice Phone: (1) 617 973 6465
Fax Phone: (1) 617 973 6406
email: holo@lightvision.com
Web: http://lightvision.coml
Contact # 1: Erich Begleiter
Business Description: Scientific and artistic research. Licensing a proprietary "micro relief" process to food manufacturers for producing chocolate and hard candy holograms.
****

Dimensions
Taj Pura
City: Sialkot
Country: Pakistan
Voice Phone: (92) 432 85197
Voice Phone: (92) 432 66006
Fax Phone: (92) 43;558336
Contact # 1: Mr. Shahjahan
Business Description: International agents and importers of holograms, diffraction foils and other holographic products.
****

Dimuken (GB)
33 Stapledon Rd
Orton Southgate
City: Peterborough
State/Province: England
Postal Code: PE2 6TD
Country: United Kingdom
Voice Phone: (44) 1733 230 044
Fax Phone: (44) 1733 230 012
email: sales@dimuken.demon.co.uk
Web: http://www.dimuken.co.uk
Contact # 1: John Bentley
Business Description: Manufactures holographic hot stamping machinery which can do hot stamping or blind embossing by switching machinery components.
****

Direct Holographies
PO Box 295
City: Strasburg
State/Province: PA
Postal Code: 17579
Country: United States of America
Voice Phone: (1) 717 687 9422
Voice Phone: (1) 888 43Dspex
Fax Phone: (1) 717 687 9423
email: directholo@cpcnet.com
Contact # 1: Jacque Phillips
Business Description: Manufacturer of SHOCK SPEX (holographic sunglasses), SHOCK MUGS, SHOCK STEINS and SHOCKFOBS. Distributor of a large range of silver-halide, embossed and photopolymer holograms. Line includes embossed stickers and magnets. Custom inquires invited.
SEE OUR ADVERTISEMENT
****

Directa GmbH
Hammer Stral3e 40
City: Muenster
Postal Code: D-48153
Country: Germany
Voice Phone: (49) 251 521551
Voice Phone: (+49) 251 5214 11
Contact # 1: Ute Schulze
Business Description: Shop (holography and gifts), wholesale of holograms and accessories.
****

Doris Vila Holographics
445 Grand Street
City: Brooklyn
State/Province: NY
Postal Code: I 1211
Country: United States of America
Voice Phone: (1) 718 388 6533
Fax Phone: (1) 718 388 6533
email : vila@dorsai.org
Contact # 1: Doris Vila
Business Description: Custom holography in state-of-the-art in-house lab, silver halide limited editions, architectural-scale & fine-art originals, mastering and transfers, consultations and classes available by appointment.
****

Dornier Medizintechnik GmbH
Industriestrasse 15
City: Germering
Postal Code: D-82110
Country: Germany
Voice Phone: (49) (0)89 84108 0
Fax Phone: (+49) (0)89 84108 575
Contact # 1: W. Langer
Business Description: Industrial Research; holographic non-destructive testi ng. HOE research. Medical systems in Lithotripsy, Surgery, Orthopaedics.
****

Dr. Steeg & Reuter GmbH
Berner Str. 109
City: Frankfurt
Postal Code: D-60437
Country: Germany
Voice Phone: (49) 69 500010-0
Fax Phone: (+49) 69 5083673
Business Description: Manufacturer of optics and crystaloptics. Main activities: Polarization, laser, interference-optics, surface mirrors, lenses for UVIVIS/IR, X-ray monochromators. The company was founded in 1855, 142 years ago.
****

Dri-Print Foils
329 New Brunswick Ave.
City: Rahway
State/Province: NJ
Postal Code: 07065
Country: United States of America
Voice Phone: (1) 732 382 6800
Fax Phone: (1) 732 382 8760
Business Description: Provider of hot stamp foils. We will produce hot stamp holographic foils for you from start to finish.
****

IF YOUR COMPANY PROVIDES GOODS OR SEVICES TO THE HOLOGRAPHY INDUSTRY, GET YOUR FREE LISTING!

DuPont (E.I. DuPont De Nemours & Co.)
Holographic Materials Division
P. O. Box 80352
City: Wilmington
State/Province: DE
Postal Code: 19880-0352
Country: United States of America
Voice Phone: (1) 302 695 4893
Fax Phone: (1) 302 695 9631
email: staff@dupont.com
Web: http://www.dupont.com
Contact #1: Paula Bobeck
Business Description: Manufacturer of photopolymer emulsions for sale to holography businesses.
SEE OUR ADVERTISEMENT
****

Dutch Holographic Laboratory BV
Kanaaldijk Noord 61
City: Eindhoven
Postal Code: NL-5642JA
Country: Netherlands
Voice Phone: (31) 40 281 7250
Fax Phone: (31) 40 281 4865
email: walter@iaehv.nl
Web: http://www.euroweb.com/DHL
Contact # 1: Walter Spierings
Business Desc ription: Manufacturer of Holoprinter and Holotrack equipment. Production of holograms on silver halide, photoresist and photopolymer. Computer-generated holograms and multiple photo-generated holograms (MPGH). Also traditional recording techniques.
SEE OUR ADVERTISEMENT
****

E.C.S.C.
1490 W Artesia Blvd
City: Gardena
State/Province: CA
Country: United States of America
Voice Phone: (1) 310 217 8021
Fax Phone: (1) 310 217 0950
email: ecsc@eio.com
Web: http://www.eio.com
Contact # 1: Barry Gott
Business Description: Electronics and Computers Information. Holography discussion group.
****

Ealing Electro-Optics Inc.
89 Doug Brown Way
City: Holliston
State/Province: MA
Postal Code: 01746
Country: United States of America
Voice Phone: (1) 508 429 8370
Voice Phone: (1) 800 343 4912
Fax Phone: (1) 508 429 7893
email: staff@ealing.com
Web: http://www.ealing.com
Business Description: Ealing manufactures and distributes optical tables and benches, subminiature manual and controlled positioners, optical mounts, custom optics, pinholes, filters, HeNe Lasers, spati al filters , laser mounts, Tungsten sources, interferometers, textbooks, and off-theshelf lenses and mirrors.
****

Eastman Kodak Company
343 State St.
City: Rochester
State/Province: NY
Postal Code: 14650-0811
Country: United States of America
Voice Phone: (1) 800 242 2424
Voice Phone: (1) 800 823 4474
Fax Phone: (1) 800 4979

Business Description: Manufacturer of silver halide recording materials. Glass plates & film. See chapter ""Recording Materials"" in this book for further information.
****

Edmund Scientific Company
101 East Gloucester Pike
City: Barrington
State/Province: NJ
Postal Code: 08007
Country: .United States of America
Voice Phone: (1) 609 547 3488
Fax Phone: (1) 609 573 6295
email: scientifics@edsci .com
Web: http ://www.edsci.com
Business Description: Mail-order catalogue, wholesale, and retail. We offer one of the largest selections of precision optics and optical components and accessories for the optical lab. Holography products for schools, science fairs, etc.
****

EI Don Engineering
4629 Platt Rd.
City: Ann Arbor
State/Province: MI
Postal Code: 48408
Country: United States of America
Voice Phone: (1) 313 973 0330
email : eldonlaser@aol.com
Contact #1: Don Gillespie
Business Description: Surplus and refurbished lasers for the holographer. Full warranty. Technical request calls welcomed.
****

Electro Optical Industries, Inc.
859 Ward Drive
City: Santa Barbara
State/Province: CA
Postal Code: 9311 I
Country: United States of America
Voice Phone: (1) 805 964 670 I
Fax Phone: (1) 805 967 8590
Contact # I : Joseph Lansing -
Business Description: Manufacturer of infrared test and calibration instrumentation including; collimators, choppers, blackbody sources, differential temperature sources, FUR test equipment, radiometers and LLL-TV target simulators.
****

Electro Optics Developments Ltd.
Howards Chase
Pipps Hill Industrial Estate
City, Prov: Basildon, Essex
State: England
Postal Code: SS14 3BE
Country: United Kingdom
Voice Phone: (44) 1268531 344
Fax Phone: (44) 1268 531 342
Contact # 1: Chris Varney
Business Description: Made to order optics such as HOE items, gratings, etc.
****

Electro-Optics Lab, NECTEC
King Mongkut's Institute of Technology
Chalongkrung Road, Ladkrabang
City: Bangkok
Postal Code: 10520
Country: Thailand
Voice Phone: (66) 2 326 9045
Fax Phone: (66) 2 326 9045
email: fkh@nwg.nectec.or.th
Business Description: A national lab that is also Thailand's first hologram manufacturer. Produces embossed holograms and photopolymer holo-

grams. Provides service in training, consulting and hologram mastering. Also conducts academic research in holograpgy, photonics and optoelectronics.
****

Elusive Image
603 Munger Street, # 213
City: Dallas
State/Province: TX
Postal Code: 75202
Country: United States of America
Voice Phone: (1) 2 14 720 6060
Fax Phone: (1) 214 754 7009
Contact # 1: Fred Wilbur
Business Description: Holography gallery and store located in Dallas's "West End Marketplace". Extensive collection of fine art and educational holograms avail able for exhibition rental and sale.
****

Embossing Technology Ltd
Steep marsh, Nr Petersfield
City: Hants
State/Province: England
Postal Code: GU32 2BN
Country: United Kingdom
Voice Phone: (44) 1730 895 390
Fax Phone: (44) 1730 894 383
Business Description: Wide web embossing by contract. Also stock images. Also for sale is complete system for originating embossed holograms, including laser.
****

Engineering Animation, Inc.
2321 N. Loop Drive
City: Ames
State/Province: IA
Postal Code: 500 I 0
Country: United States of America
Voice Phone: (1) 515 296 9408
Fax Phone: (1) 515 296 704
Contact # 1: Brad Shafer
Business Description: EAI develops, produces and sells 3D animation products that address visualization, animation and graphics needs of its customers. Prodqcts include: 3 D interactive software titles on CD'ROM; animation software (UNIX); and custom 3D computer animations.
****

EPA - Elektro-Physik Aachen GmbH
Julicher Strabe 336-340
City: Aachen
Postal Code: D-52070
Country: Germany
Voice Phone: (49) 241 53 I 778
Fax Phone: (+49) 241 1822100
Contact # 1: Eva Schulze Brockhausen
Business Description: EPA delivers HOEs, dichromated holograms for advertising, processing to jewelry.
****

ETA-Optik Gmbh
Niethausener Strasse 15
City: Heinsberg
Postal Code: D-52525
Country: Germany
Voice Phone: (49) (0)245 66654
Fax Phone: (+49) (0)245 64433
Contact #1: Wilbert Dr. Windeln
Business Description: DCG pendants, diffraction gratings and custom HOE
****

Evolution Design, Inc.
570 SW 181stWay

City: Pembroke Pines
State/Province: FL
Postal Code: 33029
Country: United States of America
Voice Phone: (1) 305 534 9808
Fax Phone: (1) 305 534 9808
email: pliberato@aol.com
Contact # 1: Pablo Liberato
Business Description: A design, production and management firm that is dedicated to the integration and conceptualization of product design/packaging and holography (embossed or photopolymer) from the beginning up to completion of the process.
****

Excitek Inc.
277 Coit Street
City: Irvington
State/Province: NJ
Postal Code: 07111
Country: United States of America
Voice Phone: (1) 973372 1669
Fax Phone: (1) 973 372 855 1
Contact # 1: George Cubberly
Business Description: Supplier of re-manufactured argon and klypton ion laser tubes, and used laser systems. Established in 1984. 10 Employees at this address.
****

Expanded Optics Limited
Moon Lane
City, Prov.: Barnet, Hertfordshire
State: England
Postal Code: EN5 5ST
Country: United Kingdom
Voice Phone: (44) 181 441 2283
Fax Phone: (44) 181 4496143
Contact # 1: T.R. Hollinsworth
Business Description: Manufacturer of medical and industri al endoscopes; HOEs used in microprecision optics for medical viewing.
****

Fantastic Holograms
P.O. BOX 492026
8400 Pen a Blvd. (DIA Terminal - Level 5)
City: Denver
State/Province: CO
Postal Code: 80249
Country: United States of America
Voice Phone: (1) 303 342 3440
Fax Phone: (1) 303 342 3440
Contact # I : RB Osada
Business Description: Well stocked holography store selling a variety of unique giftware including holographic pictures, jewelry, executive gifts, books, and optical novelties.
****

Far East Holographics
12/F Hang Wai Commercial Bldg
231-233 Queen's Road East
City, Prov.: Wanchai, Hong Kong
Country: China
Voice Phone: (852) 2 893 9773
Fax Phone: (852) 2 893 0640
Contact # I : Adrian 1. Halkes
Business Description: Finisher and distributor of holograms and holographic products.
****

Feofaniya Ltd.
P.O. Box 164
City: Kiev
Postal Code: 252 191
Country: Ukraine
Voice Phone: (380) 44 261 4343
Voice Phone: (380) 044 261 4343

INTERNATIONAL DIRECTORY

Fax Phone: (380) 044 261 4343
email: eeic@gluk.apc.org
Contact # 1: Sergey Kornienko
Business Description: Non destructive testing, pulse holography, embossing & shim making, production of holography stickers. Holography portraits studio.
****

Feroe Holographic Consulting
1420 45th Street #33
City: Emeryville
State/Province: CA
Postal Code: 94608
Country: United States of America
Voice Phone: (1) 510 658 9787
Fax Phone: (1) 510 658 9788
Contact # 1: James F eroe
Business Description: Consultant with 20 years hands-on experience in holography: silver-halide reflection and transmission, photoresist and embossed. Will travel to your site.
****

Fielmann-Verwaltung KG
Weidestrasse I 18a
City: Hamburg
Country: Germany
Voice Phone: (49) (0)40 2707 60
Fax Phone: (+49) (0)40 2707 6399
Contact # 1: Uta Kerpen
Business Description: Optician, 159 stores, Collector of Holograms, Exhibitions.
****

Fisher Scientific
Educational Materials Division
485 South Frontage Road
City: Burr Ridge
State/Province: IL
Postal Code: 60521
Country: United States of America
Voice Phone: (1) 800 955 1177
Voice Phone: (1) 610 655 4410
Fax Phone: (1) 630 655 4335
Web: fisheredu.com
Business Description: Supplies science lab equipment including holography kits, lab manuals, lasers and laser related equipment.
****

FLEXcon
I FLEX con Industrial Park
City: Spencer
State/Province: MA
Postal Code: 01562-2642
Country: United States of America
Voice Phone: (1) 508 885 8200
Fax Phone: (1) 508 885 8400
email: staff@flexcon.com
Web: http://www.flexcon.com
Business Description: Manufacturer of holographic and prismatic materials used for authentication and decoration. Holograms can be combined with overt and covert security features to provide unique so lutions to graphic films, packaging and security applications. Wide web embossing in excess of 60 inch width.
****

Flight Dynamics
16600 SW 72nd Ave.
City: Portland
State/Province: OR
Postal Code: 97224
Country: United States of America
Voice Phone: (1) 503 684 5384
Fax Phone: (1) 503 684 0169
Business Description: Manufacturer of HOEs and Head-Up Displays.
****

Floating Images, Inc.
95 Post Avenue
City: Westbury
State/Province: NY
Postal Code: 11590
Country: United States of America
Voice Phone: (1) 516 338 5000
Fax Phone: (1) 516 338 5008
email: foryst@floatingimages.com
Web: http://www.floatingimages.com
Contact #1: Carole Foryst
Business Description: Floating Images, Inc. has developed the software and hardware for a new, patent pending, ""floating 3D, off-the-screen experience"" display technology. Floating Images actually produces images at different depths on any display, such as CRT and LCD, for television, computer, rojection, and other formats.
****

Focal Image Ltd
20 Conduit Place
City: London
State/Province: England
Postal Code: W2 1HZ
Country: United Kingdom
Voice Phone: (44) 171 7062221
Fax Phone: (44) 171 7062223
email: kaveh@focal.demon.co.uk
Web: http://www.focalimage.com
Contact # 1: Kaveh Bazargan
Business Desc riptien: Re search and developement in holographic recording and di splay systems.
****

Foil Stamping Embossing Association
P.O. Box 12090
City: Portland
State/Province: OR
Postal Code: 97212
Country: United States of America
Voice Phone: (1) 503 33 I 6221
Fax Phone: (1) 503 331 6928
email: fsea@aol.com
Web: http ://www.fsea.com
Contact # 1: Mary Fuller
Business Description: A non-profit international trade association of the foil stamping, embossing, die cutting and other graphic finishing industries. It's purpose is to develop a cohesive alliance within the trade for the advancement of the entire finishing industry.
****

FoilMark Holographic Images
(a division of Foilmark, Inc.)
5 Malcolm Hoyt Drive
City: Newburyport
State/Province: MA
Postal Code: 01950
Country: United States of America
Voice Phone: (1) 508 462 7300
Voice Phone: (1) 800 468 7826
Fax Phone: (1) 508 462 0831
email: staff@foilmark.com
Web: http://www.foilmark.com
Contact # 1: David Dion
Business Description: FoilMark Holographic Images, a division of FoilMark, Inc. , is a manufacturer of diffraction embossed films. These films are printable and are available in many different mediums; for example, unsupported film, film laminated to paper or board, pressure sensitive material, static cling products, and hot stamping foils.
****

Fong Teng Technology
No 41 , Lane 63 , Hwa Chen Road

City: Hsin Chuang, Taipei
Country: Taiwan
Voice Phone: (886) 2 2 998 4760
Fax Phone: (886) 2 2 992 1240
Contact # 1: Mark Chiang
Business Description: 60 inch hologram and dot-matrix pattern foil manufacturer, service from origination to finished product.
****

Foreign Dimension Ltd., The
190 I Manley Commercial Bldg.
367-375 Queen's Road, Central
City, Country: Hong Kong, China
Voice Phone: (852) 2 542 0282
Fax Phone: (852) 2 541 6011
email: schvarzy@netvigator.com
Web: http://www.dimension.com.hkl
Contact:#1: FS - PLEASE ENTER TEXT
Business Description: Specialists in manufacturing all kinds of holographic and illusion products (watches, keyrings, etc.). If you are a hologram manufacturer, we can also make top quality products at unbeatable prices using your holograms.
SEE OUR ADVERTISEMENT
****

Foreign Dimension Ltd., The
The Peak Galleria
Level 2, Shops 29 & 42, The Peak
City, Country: Hong Kong, China
Voice Phone: (852) 2 8496361
Fax Phone: (852) 2 5416011
email: schvarzy@netvi gator.com
Web: www.dimension.com.hk
Contact # 1: Frederic Schvartzman
Business Description: Holography shop/showroom offering all varieties of holograms for sale to the public. We also offer holograms for sale wholesale to other businesses.
SEE OUR ADVERTISEMENT
****

Fornari, Arthur David
195 Garfield Place
City: Brooklyn
State/Province: NY
Postal Code: 11215
Country: United States of America
Voice Phone: (1) 718 965 3956
Contact # 1: Arthur David Fornari
Business Description: Artistic holographer; silver halide transmission & reflection holograms.
****

Forth Dimension Holographics
36 East Franklin Street
City: Nashville
State/Province: IN
Postal Code: 47448
Country: United States of America
Voice Phone: (1) 812 988 8212
Fax Phone: (1) 812 988 8212
Contact # 1: Rob Taylor
Business Description: Holographic gallery and retail shop. Hologram mastering/consulting. Small run silver halide transmission, reflection holograms.
****

Foshan Holosun Packaging Co. Ltd
Zhang-Cha-Zhen
Zhang-Cha-Yi-Lu, Yu-Dai-Kai-Fa-Qu
City, Prov.: Foshan, Guangdong
Postal Code: 528000
Country: China
Voice Phone: (86) 757 2212368
Fax Phone: (86) 757 2211228
Contact # 1: Qin Yijun
****

Frank DeFreitas Holography Studio
8 15 Allen Street
City: Allentown
State/Province: PA
Postal Code: 18102
Country: United States of America
Voice Phone: (1) 800 458 3525
Fax Phone: (1) 8004583525
email: director@holoworld.com
Web: http://www.holoworld.com
Contact # 1: Frank DeFreitas
Business Description: A full service holography studio family owned and operated since 1983.
****

Frank 1. Deutsch Inc.
17 Spielman Road
City: Fairfield
State/Province: NJ
Postal Code: 07004
Country: United States of America
Voice Phone: (1) 800 394 7713
Fax Phone: (1) 973 808 1168
Business Description: High speed precision web presses for hot stamping of holographic images, die cutting, lamination sheeting and rewinding of holograms.
****

Frank M. Schenker's Aquarius-Vertrieb
Crailsheimer Stra13e I
City: Kirchberg/Jagst
Postal Code: 0-74592
Country: Germany
Voice Phone: (49) 7954 222
Contact # 1: Frank M. Schenker
Business Description: Retailer and wholesaler.
****

Free University Of Brussels
Faculty Of Applied Sciences
Alna-Tw Pleinlaan 2
City: Brussels
Postal Code: B-I050
Country: Belgium _
Voice Phone: (32) 2 629 3452
Fax Phone: (32) 2 629 3450
Contact # 1: Erik Styns
Business Description: DOE and HOE research.
****

Fresnel Technologies Inc.
101 West Morningside Drive
City: Fort Worth
State/Province: TX
Postal Code: 76110
Country: United States of America
Voice Phone: (1) 817 926 7474
Fax Phone: (1) 817 926 7146
email: info@fresneltech.com 1
Web: http://www.fresneltech.com
Contact # 1: Linda H. Claytor J
Business Description: Manufactures plastit Fresnel lenses & lens arrays from its POLY IR plastics for use into the infrared; also other optical products for use into the ultraviolet from acrylic & other plastics.
****

Fringe Research Holographics
Interference Hologram Gallery
1179A King Street West, Suite 010
Ciry, Prov.: Toronto, Ontario
Postal Code: M6K 3C5
Country: Canada
\ 'oice Phone: (1) 416 535 2323
Contact # 1: Michael Sowdon
Business Description: Gallery of Artistic holography; silver halide holograms; pulse portraits; gallery; workshops; traveling exhibit.
****

Fuji Electric Co. Ltd - Mecatronics Division
1-12-1 Yuraku-Cho Chiyoda-Ku
City: Tokyo
Postal Code: 100
Country: Japan
Voice Phone: (8 1) 3 2 11 7 111
Business Description: C02 laser manufacturer.
****

Fujitsu Laboratories Ltd.
Peripheral Systems Laboratories
10-1 Morinosato-Wakamiya
City, Prov.: Atugi, Kanagawa
Postal Code: 2430124
Country: Japan
Voice Phone: (81) 462 50-8821
Voice Phone: (81) 462 48-3111
Fax Phone: (81) 462 48-3233
email: nakashim@flab.fujitsu.co.jp
Web: www.fujitsu.com
Contact # 1: Masato Nakashima
Business Description: Research and development applications for computer i/o systems.
****

G.M. Vacuum Coating Lab, Inc.
882 Production Place
City: Newport Beach
State/Province: CA
Postal Code: 92663
Country: United States of America
Voice Phone: (1) 714 642 5446
Fax Phone: (1) 714 642 7530
Contact # 1: Dan Coursen
Business Description: Custom manufacturing only, usually on your substrate. Will do coatings for front surface mirrors, beam splitters, etc. for holographic use.
****

Galerie 3D
Goldbacher Stra13e 31
City: Aschaffenburg
Postal Code: 0-63739
Country: Germany
Voice Phone: (49) 6021 26447
Contact #1: Stefan Merget
Business Description: Retailer of holograhic jewelry, wall decor and giftware.
****

Galerie 7000
Kleiner Schlo13platz
City: Stuttgart
Postal Code: 0-70173
Country: Germany
Voice Phone: (49) 71 I 223727
Contact #1: Martin Hofmann
Business Description: Hologram Gallery
****

Galerie Illusoria
Schwarztorstrasse 70
City: Bern
Postal Code: CH-3007
Country: Switzerland
Voice Phone: (41) 3138 1 7731
Fax Phone: (41) 3 1 381 773 1
Contact # 1: Sandro del-Prete
Business Description: Hologram Gallery.
****

Galerie WesterlandiSylt
Strandstra13e
City: WesterlandiSylt
Postal Code: 0-25980
Country: Germany
Voice Phone: (49) 4651 21313
Contact # 1: Martin Hofmann
Business Description: Gallery
****

Galvoptics Ltd.
Harvey Road
Burnt Mills Industrial Estate
City: Basildon, Essex
State: England
POM~ Code: SSl3 IES
Country: United Kingdom
Voice Phone: (44) 1268 728 077
Fax Phone: (44) 1268590445
Contact # 1: R. D. Wale
Business Description: Optics; mirrors, lenses.
****

GEHOL sarI
28 quai des Bateliers
City: Strasbourg
Postal Code: F-67000
Country: france
Voice Phone: (33) 88 52.17.16
Fax Phone: (33) 88 52.17.44
email: gehol@calvanet.calvacom.fr
Contact # 1: Jean-Luc Perreau
Business Description: I have an holography shop and a laboratory with a Denisyuk table.
****

General Design
2005 - 18th Street
City: San Francisco
State/Province: CA
Postal Code: 94107
Country: United States of America
Voice Phone: (1) 4155509193
email: bk@sfo.com
Web: sfo.com/-bk
Contact # 1: Brian Kane
Business Description: Q-eative services - Computer graphics for print, video and holography. 3D Computer Modeling and 20 Computer Composition. General image design and construction.
****

General Holographics, Inc.
P.O. Box 82247
City, Prov.: Burnaby, B.c.
Postal Code: V5C 5P7
Country: Canada
Voice Phone: (1) 604 685 6666
Voice Phone: (1) 800 667 9669
Fax Phone: (I) 604 685 6678
email: bsimson@capcollege.bc.ca
Web: http://www2.capcollege.bc.ca/-bsimsonl
Contact #1: Paula Simson
Business Description: Distributor of dichromate & embossed gift and jewelry items, silver halide wall and desk decor, and photopolymer for the Canadian market. Custom and stock.
****

GeoIa
P.O. Box 343
City: Vilnius
Postal Code: 2006
Country: Lithuania
Voice Phone: (370) 2 232737
Fax Phone: (370) 2 232838
email: geola@post.omnitel.net
Web: http://www.geola.com/
Business Description: Manufacturer of Pulsed Neodymium Lasers for Holography and compact automated Holoportraiture systems. International Sales of Soviet Holographic materials. Optics Sales. Hologram Studio Rental. Reflection Hologram Stock Images. Pulsed Holography jobs to I x2m. Equipment and Hologram rental.
SEE OUR ADVERTISEMENT
****

INTERNATIONAL DIRECTORY

Gerhard Winopal Forschungsbedarf
Echtemfeld 25
City: Hannover
Postal Code: 0-30657
Country: Germany
Voice Phone: (49) 51 I 65444
Contact # 1: Gerhard Winopal
Business Description: Isolation tables for holography, Holopal system
****

Gigahertz-Optik
Fischerstr. 4
City: Puchheim
Postal Code: 0-82178
Country: Germany
Voice Phone: (49) 89 8002143
Fax Phone: (+49) 89 806724
Contact # 1: Wolfgang G. O. Diihn
Business Description: Optics retailer
****

Glaser - Technical Consulting
24 Hashnayim Street
City: Givatayim
Postal Code: 53239
Country: Israel
Voice Phone: (972) 3 673 2734
Fax Phone: (972) 3 6732734
email: feglaser@weizmann.ac.il
Web: http://www.weizmann.ac.ill
Contact # 1: Shelly Glaser
Business Description: Technical consulting on holography (HOE, display, etc.), diffractive optics (DOE and systems containing DOEs), nonconventional optical systems (lenslet array based etc.), and image processing (specifically imaging optics for image processing). Services include feasibility studies, system design and evaluation, courses, etc.
****

Glass Mountain Optics
9517 Old McNeil l.oad
City: Austin
State/Province: TX
Postal Code: 78758-5225
Country: United States of America
Voice Phone: (1) 512 339 7442
Fax Phone: (1) 512 339 0589
email: hardyhar@ix.net.com
Web: www.glassmountain.com
Contact #1: Don Conklin
Business Description: Specialize in custom manufacturing front surfaced collimating mirrors. Emphasis on massive optics. See our web site for surplus mirrors.
****

Global Images
1 Northumberland Ave
City: London
State/Province: England
Postal Code: WC2N 5BW
Country: United Kingdom
Voice Phone: (44) 171 872 5452
Fax Phone: (44) 171 872 5611
Contact # 1: Walter Clarke
Business Description: Specialists in high volume, low cost, quality embossing equipment. ISO 9002 Qualification.
****

Gorglione, Nancy
2047 Blucher Valley Road
City: Sabastopol
State/Province: CA
Postal Code: 95472
Country: United States of America
Voice Phone: (1) 707 823 7171

Fax Phone: (1) 707 823 8073
emai l: gorglione@aol.com
Contact # 1: Nancy Gorglione
Business Description: Holographic artist specializing in architectural installations and public art environments. Extensive portfolio of one-of-akind fine artworks.
****

Grau, G., Prof. Dr. techno
Institut f. Hochfrequenztechnik der Uni Karlsruhe
Kaiserstral3e 12
City: Karlsruhe
Postal Code: 0-7613 1
Country: Germany
Voice Phone: (49) 721 6082480
Voice Phone: (49) 721 6082492
Contact # 1:" G. Grau, Prof. Dr. techno
Business Description: Research on synthetic (computergenerated) holograms 1
****

Gresser, E. , KG
An Der Warth 10
City: Ochsenfurt
Postal Code: 0-97199
Country: Germany
Voice Phone: (49) (0)933 22 77
Fax Phone: (+49) (0)93378 41
Contact # 1: Joachim Mueller
Business Description.: Laser measurement techniques,Lasers, medical
****

Guangdong Dongguan South Holoprint Co.
Dongguan ChW-Qu Qi-Feng-Lu #77
City, Prov.: Dongguan, Guangdong
Postal Code: 511700
Country: China
Contact # 1: Fan Cheng
Business Description: Hologram distribution.
****

Guangzhou Chuntian Industrial Techniques Inc.
Dong-Shan-Qu Jiang-Ling-Dong # 14
1st Floor
City, prov.: Guangzhou, Guangdong
Postal Code: 510080
Country: China
Contact #1: Xu Chuntian
****

Guangzhou Inst.of Electronics
Lab for Holography & Optoelectronic Tech
Xian-Lie-Zhong-Lu # 1 00
City, Prov.: Guangzhou, Guangdong
Postal Code: 510070
Country: China
Voice Phone: (86) 20 7668176
Fax Phone: (86) 20 7668176
Contact # 1: Wang Tianji
Business Description: Holography research and classes
****

H & W, Holographie & Werbung Petra Abramzik
Fachagentur flir Holographie & Lasertechnik
Goethestral3e 67
City: Offenbach/Main
Postal Code: 0-63067
Country: Germany
Voiee Phone: (49) 69 815112
Contact # 1: Petra Abramzik
Business Description: More than 500 holograms on glass and film, laser-display holograms, multiplex, integral-, dichromate-, and embossed holograms
****

H. Kallenbach - H.M.V.
Holographie-Marketing-Vertrieb
Endenicher Stral3e 14
City: Bonn
Postal Code: 0-53115
Country: Germany
Voice Phone: (49) 228 651332
Contact # 1: Heinz Kallenbach
Business Description: Wholesale of holograms on film, glass, and embossed holograms, holographic gifts.
****

Hallmark Capital Corp.
230 Park Avenue Suite 2430
City: New York
State/Province: NY
Postal Code: 10169
Country: United States of America
Voice Phone: (1) 212 249 9634
Fax Phone: (1) 212 249 9537
Contact # 1: Patricia M. Hall
Business Description: New York based investment banking firm, specializing in raising debt and equity financing for privately-held companies. Mergers & acquisitions are a strong secondary activity. Hallmark has raised capital for both public and private holography companies.
****

Harvard Apparatus, Canada
6010 Vanden Abeele
City: St. Laurent
Postal Code: H4S lR9
Country: Canada
Voice Phone: (1) 514 335 079
Fax Phone: (1) 514 335 3482
email: harvarda@ultranet.com
Web: http://www.harvardapparatus.com!
Contact #1: Elle Massuda
Business Description: Mirrors, prisms, optical items for holography.
****

Haus der Technik e. V
Hollestral3e 1
City: Essen
Postal Code: 0-45127
Country: Germany
Voice Phone: (49) 20 I 18031
Voice Phone: (+49) 201 1803242
Fax Phone: (+49) 201 1803269
Contact # 1: E. Prof. Dr.-lng. Steinmetz
Business Description: Seminars for engineers and scientists, focus on holography in engineering, printing and graphics.
****

HOI Panama
Holographic Dimensions [nco
5201 Zone 5
City: Panama
Country: Panama
Voice Phone: (507) 212 0177
Voice Phone: (507) 212 0173
Fax Phone: (507) 212 0177
email: jimmywoo@sinfo.net
Web: http://www.sinfo.net/holographic
Contact # 1: Jimmy Woolford _
Business Description: Security Holograms.
SEE OUR ADVERTISEMENT
****

Heil3, Peter, Dr., Priv.-Doz.
An der Bleiche 2
City: Korschenbroich
Postal Code: 0-41352
Country: Germany
Business Description: Hobby holographer, teacher, books and seminars on holography.
****

Hellenic Institute Of Holography
28 Dionyssou Street
City: Chalandri
Postal Code: GR-I5234
Country: Greece
Voice Phone: (30) I 684 6776
Fax Phone: (30) I 685 0807
Contact #1: Alkis Lembessis
Business Description: Established in 1987, the Institute aims at the overall introduction and promotion of holography in Greece. Exhibitions, courses, vocational training and mastering laboratory.
****

Heptagon Oy
Otaniemi Science and Technologry Park
Tekniikantie 12
City: Espoo
Postal Code: FIN-02150
Country: Finland
Voice Phone: (358) 9 4354 2041
Fax Phone: (358) 9 4354 2041
email: info@heptagon.fi
Web: http ://www.heptagon.fi
Contact #1: Jyrki Saarinen
Business Description: Heptagon provides complete design services for designing diffractive optical elements (DOEs) to customer requirements. Heptagon also offers consulting services and engineering assistance to the fabrication and exploitation of DOEs.
****

Hiat Image Technology Group, Inc.
2F, No. 16 Lane 6, Sec. I, Hang Chou S. Rd.
City: Taipei
Country: Taiwan
Voice Phone: (886) 2 393 0306
Fax Phone: (886) 2 3958122
Contact #1: Billy Chou
****

Highlite
Alexanderstra13e 63-65
City: Aachen
Postal Code: D-52062
Country: Germany
Voice Phone: (49) 241 35402
\'oice Phone: (+49) 241 31407
Contact # 1: Hans J. Bose
Business Description: Manufacturer of dichromate holograms up to 30x40cm
****

H:vI S-Elektronik
Hans M. Strassner GmbH
Tannenweg 7
Ci ty: Leverkusen
Postal Code: 0-51381
Country: Germany
Voice Phone: (49) 21713814
Voice Phone: (+49) 21713815
Contact # 1: Hans M. Strassner
Business Description: Manufacturing and sale of high-tech electronics. Lock-In amplifier, pre-amplifier, optical chopper. optical components, ultrasonic devices, broadband amplifiers
****

HODIC Holographic Display Artists & Enginiers Club
Engineering Department Chiba University
1-33. Yayoi-cho
City: Chiba
Postal Code: 263
Counrry: Japan
Voice Phone: (81) 472 511 III
Fax Phone: (8 1) 472 517 337

Contact # 1: Miss Tomoko Sakai
Business Description: Regular meetings 4 times a year with oral presentations on holography given. Publishes HOOIC circular. Membership open to all.
****

Hofmann-Lange, Brigitte, Dr.
Kaiserstr. 12
City: Weinheim
Postal Code: D-69469
Country: Germany
Voice Phone: (49) 6201 58571
Contact # 1: Brigitte Hofmann-Lange, Dr.
Business Description: Seminars on holography
****

HOL 3, Galerie fur Holographie GmbH
Europa Center
City: Berlin
Postal Code: 0-10789
Country: Germany
Voice Phone: (49) 30 261 4490
Fax Phone: (49) 30 344 6379
Contact # 1: Valeska Cordner-Guled
Business Description: Exhibition of mainly stock holograms of various producers. Sometimes feature one man shows. Sales of holograms and all sorts of related holographic items.
****

Holar Seele KG
Wasserwerksweg 16a
City: Aurich
Postal Code: 0-26603
Country: Gemlany
Voice Phone: (49) 4941 10005
Fax Phone: (+49) 4941 63644
Contact # 1: Gerd Dipl.-Ing. See Ie
Business Description: Large format holograms up to 1 x3m for use in architecture, display holograms also possible. Portrait holograms with pulsed laser.
****

Holarium
Museum flir Holografie
Kirchplatz
City: Esens .
Postal Code: D-26427
Country: Germany
Voice Phone: (49) 4971 3088
Contact #1: Hardo Sziedat
Business Description: Museum for holography.
****

Holicon Corporation.
3312 Belle Plain Ave. #2
City: Chicago
State/Province: IL
Postal Code: 60618
Country: United States of America
Voice Phone: (1) 312 267 9288
Fax Phone: (1) 312 267 9288
Contact #1: Richard Bruck
Business Description: Holograms produced with pulsed lasers. Holographic portraits. Large format reflection and rainbow holograms. Mass production of silver-halide holograms.
****

Holo 3D S.p.A.
AREA Science Park
Padriciano, 99
City: Trieste
Postal Code: 1-34012
Country: Italy
Voice Phone: (39) 40 226 327
Fax Phone: (39) 40 226 431
email: hol03d@specialnet.cmt.it

Contact # 1: Alhbrante Mano
Business Description: Origination of holograms for security and promotional purposes. Primarily image origination, embossing and printing of holographic labels
****

Holo Art
Island Holographics
City: Northport
State/Province: NY
Postal Code: 11768
Country: United States of America
Voice Phone: (1) 516 757 3866
email: pgdb@aol.com
Contact # 1: Dave Battin
Business Description: Supplier of educational diffraction / holographic art kits.
****

Holo GmbH
Lutterdamm 82
City: Bramsche
Postal Code: D-49565
Country: Germany
Voice Phone: (49) (0)546 91123
Fax Phone: (+49) (0)546 91122
Contact # 1: Thomas Lucy
Business Description: Holograms up to 1 x 1 m; embossed holography. Holo-design.
****

Holo Images Tech Co. , Ltd.
17, Alley 20, Lane 7, Jong Hwa Road
City: Yung Kang City, Tainan County
Country: Taiwan
Voice Phone: (886) 66 237 3896
Fax Phone: (886) 66 23il 4641
Contact #1: Craig Chiou
Business Description: Embossed holograms and products.
****

Holo Impressions Inc
47-1 Wu Chuan Rd
Wu-Ku Industrial Park
City: Taipei Shein
Country: Taiwan
Voice Phone: (886) 2 299 7576
Fax Phone: (886) 2 299 7050
Contact # 1: Jonathan Hsu
Business Description: Embossed holography.
****

Holo Sciences, LLC
480 East Rudasill Road
City: Tucson
State/Province: AZ
Postal Code: 85704
Country: United States of America
Voice Phone: (1) 520 293 9393
Fax Phone: (1) 520 696 0773
email: deck_O@azstamet.com
Contact # 1: Chuck Hassen
Business Description: H I Mastering and stereo-gram creation from video or computer graphic source imagery. Special capabilities to produce full-parallax, animated stereo grams. Custom 3-D computer modeling and animation services.
****

Holo Service/Service-Druck
Morsestra13e 5
City: Duesseldorf
Postal Code: 0 -40215
Country: Germany
Voice Phone: (49) 221 370917
Contact #1: Klaus Kleinheme
Business Description: Retailer of holograms.
****

INTERNATIONAL DIRECTORY

Holo Time Gericke
SchertlinstraBe 32
City: FreiberglNeckar
Postal Code: D-71691
Country: Germany
Voice Phone: (49) 7141 76850
Fax Phone: (+49) 7141 73050
Contact #1: W. K. Gericke
Business Description: Hologram full service: all kinds of holograms, consulting, engineering, sale, mediation.
****

Holo-Idee Reiner Kleinheme
ChopinstraBe 2a
City: Meerbusch-Struemp
Postal Code: D-40670
Country: Germany
Voice Phone: (49) 2159 8733
Contact # 1: Reiner Kleinheme
Business Description: Consultant for industry and advertising, organizer of exhibitions, manufacturing of sample holograms on film before embossing process.
****

Holo-Laser
12 rue de Vouille
Ecole
City: Paris
Postal Code: F-75015
Country: France
Voice Phone: (33) I 45 31 52 75
Fax Phone: (33) I 48 33 17 07
Contact # 1: Dr.. Jean Louis Tribillon
Business Description: Embossed holography and equipment; artistic holography; buying and selling; education.
****

Holo-Or Ltd
PO Box 1051
Kiryat Weiznvnn
City: Rehovot
Country: Israel
Voice Phone: (972) 89 469 687
Fax Phone: (972) 8 9466 378
Contact #1: Uri Levy
Business Description: Manufactures computer-generated diffractive optical elements by VLSI techniques. Catalogue elements and custom designs. Substrates include AnSe, GaAs, various glasses. DOE work station - dedicated workstation for element design, mask generation.
****

Holo-Spectra
7742 Gloria Ave.
City: Van Nuys
State/Province: CA
Postal Code: 91406
Country: United States of America
Voice Phone: (1) 818 994 9577
Fax Phone: (1) 818 994 4709
Contact # 1: Bill Arkin
Business Description: Embossed hologram production. Embossed mastering equipment, laser repairs. Optical table and holographic equipment resold.
****

Holo/Source Corporation
11930 Farmington Rd
City: Livonia
State/Province: MI
Postal Code: 48150
Country: United States of America
Voice Phone: (1) 313 427 1530
Fax Phone: (1) 313 525 8520
email: sales@holo-source.com
Web: http://www.holo-source.coml
Contact # 1: Lee Lacey
Business Description: Paperboard sheets of holographic film and paper. Holographic image mastering of all types and finished flexo printed holographic labels.
****

Holoart Studio
No.13 II Lane
Shin-Fu Street ,Shihlin
City: Taipei
Country: Taiwan
Voice Phone: (886) 2 8323843
Fax Phone: (886) 2 8339445
email: linyow@tcts.seed.net.tw
Contact # 1: Yow-Snin Lin
Business Description: Artistic reflection hologram up to 30 x 40cm, stock and custom design.
****

HoloCom
40 I Carver Road
City: Las Cruces
State/Province: NM
Postal Code: 88005
Country: United States of America
Voice Phone: (1) 505 527 9183
Fax Phone: (1) 505 5279927
email: arts@holo.com
Web: http://www.holo.com
Contact # 1: Ken Marris
Business Description: Holographic web site and web provider. Holographic technology transfer and holographic photoresist training. Custom originations in photoresist.
****

Holocrafts
Canadian Holographic Developments Ltd.
Box 1035
City, Prov.: Delta, B.C.
Postal Code: V4M 3T2
Country: Canada
Voice Phone: (I) 604 946 1926
Fax Phone: (1) 604 946 1648
Contact #1: Karoline Cullen
Business Description: Holocrafts manufacturers dichromate holograms. Offering stock and custom production in a variety of formats such as plain discs, watches, keychains, pendants and 3 "" x 3 "" plates. Providing a tradition of excellence since 1979.
SEE OUR ADVERTISEMENT
****

Holocrafts Europe Ltd.
Barton Mill House
Barton Mill Road
City: Canterbury, Kent
State: England
Postal Code: cn !BY
Country: United Kingdom
Voice Phone: (44) 1227463223
Fax Phone: (44) 1227450399
Contact # 1: Chris Luton
Business Description: Specialists in manufacture of dichromate reflection holograms. Also produce holographic gift products as well as selling photopolymer.
SEE OUR ADVERTISEMENT
****

Holodesign
Fiichtenbusch & Hufmann GbR
Am Forst 38
City: Wesel
Postal Code: D-46485
Country: Germany
Voice Phone: (49) 28152837
Contact # 1: Annette Fuechtenbusch
Business Description: Manufacturing of holograms.
****

Holografia Polska
ul. sw.Mikolaja 16117
City: Wroclaw
Postal Code: PL-50-128
Country: Poland
Voice Phone: (48) 71 343 46
Fax Phone: (48) 71 33948
Contact # 1: Boguslaw Stich
Business Description: Practical application of holography and importers of holographic products.
****

Holografica
30 I South Light Street
City: Baltimore
State/Province: MD
Postal Code: 21202
Country: United States of America
Voice Phone: (1) 410 685 3331
Contact # 1: Renee Fee
Business Description: Retail store with full range of holographic products.
****

Holografie-Hofmann
Labor und Galerien
C.H.-GaiserstraBe 20
City: Goeppingen
Postal Code: D-73033
Country: Germany
Voice Phone: (49) 7161 12200
Contact # 1: Martin Hofmann
Business Description: Gallery, manufacturer of holograms up to 30x40cm, custom made up to 1x1m
****

Hologram Center Holmby
Holmby gamla skola
City: Flynge
Postal Code: S-24032
Country: Sweden
Voice Phone: (46) 46 524 30
****

Hologram Company RAKO GmbH
Moellner Landstrasse 15
City: Witzhave
Postal Code: D-22969
Country: Germany
Voice Phone: (49) (0)410 693 250
Fax Phone: (+49) (0)410 693 249
Contact #1: Wilfried Schipper
Business Description: Specializing in production of embossed holograms and the sale of embossing equipment.
****

Hologram Development Corp.
37 Standish Ave.
City, Prov.: Toronto, Ontario
Postal Code: M4W 3B2
Country: Canada _
Voice Phone: (1) 416 925 5569 -
Contact #1: Ed Burke
Business Description: General hologram services.
****

Hologram Fantastic
368 West Market
Mall of America
City: Bloomington
State/Province: MN

Postal Code: 55425
Country: United States of America
Voice Phone: (1) 612 858 9416
Voice Phone: (1) 612 722 4423
Web: mallofamerica.com/direct/home.htm
Contact # 1: Nina Bourque
Business Description: Retail store in high traffic mall selling a full range of holographic giftware and related optical novelties.
****

Hologram Industries
22 Ave de l'Europe
Parc Gustave Eiffel, Bussy Saint-Georges
Marne La Vallee Cedex 3
Postal Code: F-77606
Country: France
Voice Phone: (33) I 64763100
Fax Phone: (33) I 64763570
Contact #1: Hughes Souparis
Business Description: Embossed holography. Security holograms.
****

Hologram Land
284 E. Broadway
'via ll Of America
City: Bloomington
State/Province: MN
Postal Code: 55425
Country: United States of America
Voice Phone: (1) 612 854 9344
Fax Phone: (1) 612 854 7857
Web: mallofamerica.com/direct/home.htm
Contact # 1: Sue Rickert
Business Description: Retail store specializing in everything holographic. Product range includes: artwork, watches, jewelry, T-shirts, small gift items and optical novelties. Framing and lighting accessories provided. Also carry various scientific novelties and glow-in-the-dark merchandise.
****

Hologram Research, Inc.
P. O. Box 377
City: Locust Valley
State/Province: NY
Postal Code: 11560
Country: United States of America
Voice Phone: (1) 5169222560
Voice Phone: (1) 8889016655
Fax Phone: (1) 5166248687
email: hologram@lihti.org
Web: wwwhologramres.com
Contact # 1: Joseph Bums
Business Description: Exclusive source for Ilford silver halide plates and film. Custom hologram to 42" x 72", stereograms and limited edition in silver halide, photoresist, embossed and in. jection -molded from our argon and he-ne laser labs. Exhibition services available from our extensive hologram collection; brokering services for the New York Metro area. Fully equipped studio for rent.
SEE OUR ADVERTISEMENT
****

Hologram Varga Miklos
Ki raly u.1 02.
City: Budapest
Postal Code: H-I068
Country: Hungary
\ 'oice Phone: (36) I 351 4725
\ 'oice Phone: (36) 20 342 076
Fax Phone: (36) I 227 354
Contact # 1: Miklos Varga
Business Description: Stock and custom made holograms on best quality photopolymer.
****

Hologram World, Inc.
1860 Berkshire Lane North
City: Plymouth
State/Province: MN
Postal Code: 55441
Country: United States of America
Voice Phone: (1) 612 559 5539
Voice Phone: (1) 800 882 4656
Fax Phone: (1) 6125592286
Contact # 1: Jim Paletz
Business Description: One of the largest wholesale distributors of holographic novelties. We represent over 50 holographic manufacturers. Specialize in helping the new retail store owner in all stages of development from start to finish
SEE OUR ADVERTISEMENT.
****

Hologram, etc.
1640 Camino del Rio North
City: San Diego
State/Province: CA
Postal Code: 92108
Country: United States of America
Voice Phone: (1) 619 683 3159
Contact # 1: Michel Nguyen
Business Description: Retail holograms. Have large mobile call and sell photopolymers, embossed and various types of holograms.
****

Hologramas S.A. de c.Y. (Mexico)
Pino 343-3 Col. Atlampa
City: Mexico City
State/Province: D.E
Country: Mexico
Voice Phone: (52) 5 5411791
Voice Phone: (52) 5 5479046
Fax Phone: (52) 5 5474084
email: holomex@holomex.com.mx
Web: http ://www.holomex.com.mx
Contact # 1: Dan Liebennan
Business Description: Manufacturer of embossed holograms since 1984. Our services include -security labels, overlaminates, hot-stamping foil, wide web materials for flexible, and rigid packaging up to 43" wide, and technology transfers.
SEE OUR ADVERTISEMENT
****

Hologramm
Mendelssohnstraf3e 49
City: Frankfurt
Postal Code: 0-60325
Coqntry: Germany
Voice Phone: (49) 69 746116
Contact #1: Juergen Koerner
Business Description: Hologram Gallery.
****

Holograms 3D
286 Earl 's Court Road
City: London
State/Province: England
Postal Code: SW5 9AS
Country: United Kingdom
Voice Phone: (44) 171 3702239
Fax Phone: (44) 171 3732511
Contact # 1: Jonathan Ross
Business Description: Jonathan Ross has a personal holography collection available for touring shows. He also deals privately in holographic art and consults on commercial applications.
****

Holograms and Lasers International
Retail Merchandise Division
1200 McKinney, Suite 433
City: Houston
State/Province: TX

Postal Code: 770 I 0
Country: United States of America
Voice Phone: (1) 713 650 9204
Fax Phone: (1) 713 650 9204
email: felix@electrotex.com
Web :// www.holoshop.com/
Contact #1: Perry Felix
Business Description: Holograms and Lasers International operates the largest retail hologram shop in Houston, Texas with a complete library of over 750 hologram images and products to view on the Internet 's World Wide Web.
****

Holograms and Lasers International
Hologram Production Facilities
1200 McKinney, Suite 433
City: Houston
State/Province: TX
Postal Code: 77010
Country: United States of America
Voice Phone: (1) 7136509204
Fax Phone: (1) 713 650 9204
email: felix@electrotex.com
Web: http://www.holoshop.coml
Contact # 1: Perry Felix
Business Description: A full service Ruby Pulse and CW laser hologram production lab providing complete origination services for reflection, transmission and mass production holograms, specializing in fine quality large format hologram portraits and trade show exhibits.
****

Holograms Fantastic and Optical Illusions
P.O. Box 765
City: Bayswater 4
State/Province: Victoria
Postal Code: 3153
Country: Australia
Voice Phone: (61) 18 776 226
Voice Phone: (61) 18 776226
Fax Phone: (61) 3 9729 6020
Contact # 1: Trevor McGaw
Business Description: Glass, film and foil (opp, PET & PVC) 20, 2D/3D, 3D & multi images & patterns. Services to printers, hot-stampers, packaging, label, security marketing, sales promotion and advertising. Specialists in foil holography.
****

Holograms International
211 18th Street
City: Huntington Beach
State/Province: CA
Postal Code: 92648
Country: United States of America
Voice Phone: (1) 714 536 0608
Fax Phone: (1) 714 536 0608
email: holograms@worldnet.att.net
Contact # 1: Dave Krueger
Business Description: Distributor of all kinds of holograms to retail stores and wholesale accounts. We are known for our fast delivery, friendly consulting and factory-direct prices. Call or write for quote or catalogue.
****

Holograms Unlimited (U.K. Gold Purchasers)
110 Central Park Mall
City: San Antonio
State/Province: TX
Postal Code: 78216
Country: United States of America
Voice Phone: (I) 210 530 0045
Voice Phone: (1) 800 722 7590
Fax Phone: (1) 210 530 0048
Web: eden.com/~mainlink/tx/sat/art/rai/ index.htm

INTERNATIONAL DIRECTORY

Contact # 1: Marvin Uram
Business Description: Full line distributor of hologram and related products for specialty retailers - representing more than 80 firms. One stop shopping at competitive prices.
****

Holographic Applications
21 Woodland Way
City: Greenbelt
State/Province: MD
Postal Code: 20770-1728
Country: United States of America
Voice Phone: (1) 30 I 345 4652
Fax Phone: (I) 30 I 345 4653
Contact # 1: Suzanne St. Cyr
Business Description: Design and product engineering services for consumer products and censed promotions using 3-D imaging technologies. Product specifications and quality assurance. Vendor selection and product management General contractor delivering finished, packaged product.
****

Holographic Consulting Agency
Jampot Cottage, WestHill,
Elstead, Godalming,
City: Surrey
Postal Code: GU8 6DQ
Country: United Kingdom
Voice Phone: (44) 1252702781
Fax Phone: (44) 973763121
email: Holoconsul@aol.com
Contact # 1: Mark Dicker
Business Description: Equipment & Technology
****

Holographic Design Systems
1134 West Washington Blvd.
City: Chicago
State/Province: IL
Postal Code: 60607
Country: United States of America
Voice Phone: (1) 312 829 2292
Fax Phone: (1) 312 829 9636
email: museumh@concentric.net
Web: http://www.concentric.netl-Museumh/
Contact # 1: Robert Billings
Business Description: Unrivaled creativity, combining artistic imagination with complete technical mastery of all forms of holography resulting in a worldwide reputation for excellence. The most complete labs in the industry with the most powerful and advanced lasers and computers. Our clients include the most innovative and sophisticated companies in the US and abroad.
SEE OUR ADVERTISEMENT
****

Holographic Dimensions
7503 N.W 36th Street
City: Miami
State/Province: FL
Postal Code: 33166
Country: United States of America
Voice Phone: (1) 305 994 7577
Fax Phone: (1) 305 994 7702
email: Sales@hgrm.com
Web: http ://www.hgrm.com
Contact # 1: Kevin Brown
Business Description: Holographic Dimensions, Inc. is a vertically integrated manufacturer of holographic imagery, with extensive experience in high volume security and authentication holograms. A recently, traded Public Company, it has complete in-house facilities from artwork origination to embossing.
SEE OUR ADVERTISEMENT
****

Holographic Dimensions, Poland S.A.
UI. Traugutta 25
City: Lodz
Postal Code: PL-90-950
Country: Poland
Contact # 1: Grace Golen, VP
Business Description: Holographic Dimensions, Inc. is a vertically integrated manufacturer of holographic imagery, with extensive experience in high volume security and authentication holograms. A recently, traded Public Company, it has complete in-house facilities from artwork origination to embossing.
SEE OUR ADVERTISEMENT
****

Holographic Finishing
501 hendricks Street
City: Ridgefield
State/Province: NJ
Postal Code: 07657
Country: United States of America
Voice Phone: (1) 201 941 4651
Fax Phone: (1) 201 941 4453
Contact #1: Charlie Vulcano
Business Description: Holographic Hot Stamping, Embossing, Die Cuting Security Seals.
****

Holographic Images Inc.
521 Michigan Ave,
City: Miami Beach
State/Province: FL
Postal Code: 33139
Country: United States of America
Voice Phone: (1) 305 531 5465
Fax Phon"!'\l) 305 5324090
Contact # 1: Matthew Schrieber
Business Description: Mastering and replication facility dedicated to the production of multi-color/ full color limited-edition holographic artworks produced on silver halide film .
****

Holographic Impressions
96 North Almaden Blvd.
City: San Jose
State/Province: CA
Postal Code: 95110-2490
Country: United States of America
Voice Phone: (1) 408 292 890 I
Fax Phone: (1) 408 292 0417
email: smithmckay@aol.com
Contact #1: Dave McKay
Business Description: Manufacturer and distributor of unique line of greeting cards, stationary and business cards. Stock products available.
****

Holographic Industries, Inc.
P.O. Box 1109
City: Libertyville
State/Province: IL
Postal Code: 60048
Country: United States of America
Voice Phone: (1) 847 680 1884
Fax Phone: (1) 847 680 0505
Contact # 1: Robert Pricone
Business Description: Parent company of "Lightwave" retail. Consulting and distribution for giftware industry.
****

Holographic Label Converting (HLC)
7669 Washington Avenue South
City: Edina
State/Province: MN
Postal Code: 55439
Country: United States of America
Voice Phone: (1) 612 944 7408

Fax Phone: (1) 612 944 7210
email: hlc@pclink.com
Contact #1: Scott Labelle
Business Description: Full service capabilities, 2D/3D holography, designing, embossing, hotstamping, precision die-cutting, wide variety of foils. Custom holographic labeling, magnetic holograms, packaging and more ... You think of it, and we can put it together.
****

Holographic Laserdesign
Kunst- und Geschenkartikel
Trendelbuscher Weg I
City: Ganderkesee
Postal Code: D-27777
Country: Germany
Voice Phone: (49) 4221 89298
Contact # 1: Tilmann Eimers
Business Description: Custom made holograms
****

Holographic Optics Inc.
358 Saw Mill River Rd
City: Millwood
State/Province: NY
Postal Code: 10546
Country: United States of America
Voice Phone: (1) 914 7621774
Fax Phone: (1) 914 762 2557
Contact #1: Jose R. Magarinos
Business Description: Manufacturer of holographic optical elements, particularly holographic filters, holographic mirror and beamsplitters. Design and manufacture of prototypes.
****

Holographic Products
1711 St. Clair Ave.
City: St. Paul
State/Province: MN
Postal Code: 55105
Country: United States of America
Voice Phone: (1) 612 698 6893
Fax Phone: (1) 612 6981619
email: sugarOO I@tc.umn.edu
Contact # 1: Stephen Sugarman
Business Description: Holographic Products is actively pursuing new product development in educational toys, intermedia print design, ad specialties, promotions, and premiums. Specialists in security laminates (for ID card application) and tamper evident labeling. Also conducts hands-on elementary school workshops, and instructional presentations.
****

Holographic Studios
240 East 26th Street
City: New York
State/Province: NY
Postal Code: 100 I 0
Country: United States of America
Voice Phone: (1) 212 686 9397
Fax Phone: (1) 212 481 8645
email: drlaser@interport.net
Web: hmt.com/ holography/holostudios/
index.html -
Business Description: New York's only gallery and commercial holographic lab. Custom and stock holograms. Single or mass-produced. Integral portrait cinematography, mastering, and scan copies from small to large format. Computer generated holograms. Classes.
****

Holographic Studios Rosowski
Lindenau 23
City: Issum

Postal Code: 0-47661
Country: Gennany
Voice Phone: (49) 2835 1684
Contact #1: Ralf Rosowski
Business Description: Custom made white-light reflection holograms and white-light transmission holograms
****

Holographic Systems Muenchen GmbH
Melchior-Huber-Strasse 25
City: Ottersberg
Postal Code: 0-85652
Country: Gennany
Voice Phone: (49) (0)812 9925-0
Fax Phone: (+49) (0)812 9925-99
Contact #1: Gunther Dausmann
Business Description: Holographic production, including machines.
****

Holographics (Uk) Ltd.
12 Whidborne St
City: London
State/Province: England
Postal Code: WCIH 8EU
Country: United Kingdom
Voice Phone: (44) 171 8332236
Fax Phone: (44) 171 8332237
Contact #1: Jon Vogel
Business Description: Holographic & 3-D multimedia, design origination and production specialists (est. 1982) providing comprehensive service for the corporate, retail, & leisure sectors.
****

Holographics Inc.
44-0 I Eleventh Street
City: Long Island City
State/Province: NY
Postal Code: 1110 1
Country: United States of America
Voice Phone: (1) 718 784 3435
Fax Phone: (1) 718 7060813
Contact # 1: Fred Nicholas -
Business Description: Company involved in 3 main areas of research and development primarily involving pulsed holography: multi/full color artistic portraiture (Ana Maria Nicholson); NOT for government and corporate applications (Dr. John Webster, Tim Schmidt); and laser development (Peter Nicholson).
****

Holographics North Inc.
444 South Union Street
Ci ty: Burlington
State/Province: VT
Postal Code: 05401
Country: United States of America r
Voice Phone: (1) 802 658 2275 \'
Fax Phone: (1) 802 658 5471 ..
email: perryjf@vbi.champlain.edu
Web: http://www.holonorth.com/
Contact # 1: John Perry
Business Description: Designers/manufacturers of large fonnat holograms for commercial, museum and fine art applications. Multicolor, animated holograms up to 44x72 inches (1.1 m x 1.8 m). Design, model building, production, installation and consulting services.
****

Holographie & Design
Kalk-Miilheimer-StraBe 124
City: Koeln
Postal Code: 0-51103
Country: Gennany
Voice Phone: (49) 221 857386
Contact #1: Peter Ludwig

Business Description: Custom made holograms, exhibition for rent, modelling, large collection
****

Holographie Anubis
Oberer Kaulberg 37
City, Prov.: Bamberg, Barvaria
Postal Code: 0-96049
Country: Gennany
Voice Phone: (49) 951 57951
Fax Phone: (49) 951 59529
email: holographie.anubis@t-online.de
Contact #1: M.T. Frieb
Business Description: We are producer and distributor of all formats of holograms. Import and export. Consulting and education. Mass production. Full pulse laser facilities. Fully pictured wholesale catalog (122 pages).
****

Holographie Fachstudio Bad Rothenfelde
Postfach 1304
Niederesch 28
City: Bad Rothenfelde
Postal Code: 0-49214
Country: Gennany
Voice Phone: (49) (0)542 5365
Fax Phone: (+49) (0)542 5359
Contact # 1: Gunter Deutschmann
Business Description: Expert consultancy for integration of holographic products into finished advertising media, including application, overprinting etc. Founder and office of the AHT (Arbeitskreis Hologramm-Techniken & 3D Medien).
****

Holographie Konzept GmbH
KiirberstraBe 3
City: Frankfurt
Postal Code: 0-60433
Country: Gennany
Voice Phone: (49) 69 531073
Contact # 1: Gabriele Chmielewski
Business Description: Delivers all kinds of holograms.
****

Holographie Labor .,
Bertelsmann AG
Auf dem Eickholt 47
City: Gutersloh
Postal Code: 0-33334
Country: Gennany
Voic,e Phone: (49) 5241 580192
Fax Phone: (49) 5241 580549
Contact #1: Saurda Uwe
Business Description: Holograms, Holographic projects.
****

Holographle Roth
AispachstraBe 16
City: Reutlingen
Postal Code: 0-72764
Country: Gennany
Voice Phone: (49) 712 1 45344
Contact # 1: Ulrich G. Roth
Business Description: Hologram production, sale, rent, exhibition.
****

Holographie-Labor Mielke
GeorgenstraBe 61
City: Muenchen
Postal Code: 0-80799
Country: Germany
Voice Phone: (49) 89 2712989
Contact # 1: H. M. Mielke
Business Description: Manufacturer of holo-grams

and HOEs. Format: IOx12 cm up to 80x80cm. Holograms on glass or film, silver-halide, dichromate, photo polymer and photoresist.
****

Holography and Media Institute of Quebec
I139 Ave des Laurentides
City, Prov.: Quebec City, Quebec
Postal Code: GIS 3C2
Country: Canada
Voice Phone: (1) 4186872985
Voice Phone: (1) 418 656 3095
Fax Phone: (1) 418 687 2985
email: Marie-Andree.Cossette@arv.ulaval.ca
Web: http ://www.ulaval.cal
Contact # 1: Marie-Andree Cossette
Business Description: Established in 1990 as an international centre for the study and production of holographic art. Workshops and private tutorials. Residency programs offered on an invitation only basis. Director Marie-Andree Cossette is Associate Professor of Visual Art at Laval University, Quebec.
****

Holography Center of Austria (Holo,Trend)
Kahlenbergstrasse 6
City: Wurmla
Postal Code: A-3042
Country: Austria
Voice Phone: (43) 22758210
Fax Phone: (43) 2275 82105
Contact # 1: Irmfried Wober
Business Description: Our holography laboratory, founded in 1985, is the first in Austria and the most comprehensive in the region. We offer high quality pulsed and.CW laser origination, mastering and production (transportable pulsed laser systems for portrait, stereograms up 1x1 meter and computer generated holograms available). In addition, we organize exhibitions in Austria and Germany and sell embossed holograms.
SEE OUR ADVERTISEMENT
****

Holography Development Co.Ltd.
News building No.2 Room622
Shen Nan Zhong Road #2
City, Prov.: Shenzhen, Guangdong
Postal Code: 518027
Country: China
Voice Phone: (86) 755 2271973
Fax Phone: (86) 755 2271973
email: Holohot@nenpub.szptt.net.cn
Web: http://www. holoworld.comlwww/hdc/
Contact # 1: Hunter Wang
Business Description: The Holography Development Company is located in the P.R. China. Our aim is to serve holographic companies throughout the world. We can make contact and find outlets for your product all over China. Some of the potential customers would be the department stores, hotels and restaurants, museums, advertising and ex hibition companies, etc. We are not a holographic manufacturer. Our job is to market yo ur product as your agent.
****

Holography Group TEMoo
Nordala 3031P
City: Angelholm
Postal Code: S-262 73
Country: Sweden
Voice Phone: (46) 431 82577
Fax Phone: (46) 431 82577
email: nordala@algonet.se
Business Description: 10 years old organization for enthusiasts. Lab for holographers.
****

INTERNATIONAL DIRECTORY

Holography Institute of San Francisco
PO Box 24-153
City: San Francisco
State/Province: CA
Postal Code: 94124-0153
Country: United States of America
Voice Phone: (1) 4158227123
emai l: hologram@well.com
Contact # 1: Jeffrey Murray
Business Description: Specializing in one-onone
instruction in display holography (design, optics, recording, processing, etc.). Courses can be
customized to the required balance of theory and
practice for artistic, technical or production applications. Classes for artists, scientists, young and
old. No prerequisites for beginners. Call for class
schedule. Research facilities include a dedicated
holography studio with 5 isolation tables, HeNe
lasers, and optics for image plane holography up
to 50x60 cm.
SEE OUR ADVERTISEMENT
****

Holography Israel
21 Hakomemiut Str.
City: Herzlia
Postal Code: 46683
Country: Israel
Voice Phone: (972) 09 572 387
Voice Phone: (972) 09 559 766
Fax Phone: (972) 09 570 569
Contact #1: Hameiri Shimon
Business Description: Holography Israel special-
izes in exhibitions-lectures and demonstrations
to pupils and students-advertising, commission,
sales and production of art holograms.
****

Holography Marketplace
(c/o Ross Books)
P.O.Box 4340
City: Berkeley
State/Province: CPr
Postal Code: 94704
Country: United States of America
Voice Phone: (1) 510 8412474
Voice Phone: (1) 800 367 0930
Fax Phone: (1) 510 841 2695
email: staff@rossbooks
Web: http:\\www.holoinfo.com
Contact # 1: Alan Rhody
Business Description: The world's most compre-
hensive and informative book for, and about, the
holography industry. Published annually, each edi-
tion of THE HOLOGRAPHY MARKETPLACE
includes: a worldwide corporate directory; chap-
ters of useful reference material; interviews with
industry experts; and a sample kit of actual holograms produced by major manufacturers. Seven
editions available.
ORDER YOUR COPY TODAY
****

Holography Presses On (HPO)
20 I North Fruitport Road, Box 193
City: Spring Lake
State/Province: MI
Postal Code: 49456-0193
Country: United States of America
Voice Phone: (1) 616 842 5626
Fax Phone: (1) 616 842 5653
Contact # 1: Jan Bussard
Business Description: Holographic stock or cus-
tom shapes and sizes applied with heat or pressure
for adhesion to all substrates. Specialize in stick-
ers and textile applications. Sealed edges prevent
delamination in all weather; washable. Worldwide
distributors sought.
SEE OUR ADVERTISEMENT
****

Holographyx inc.
99 Ronald Court
City: Ramsey
State/Province: NJ
Postal Code: 07446
Country: United States of America
Voice Phone: (1) 201 3274414
Voice Phone: (1) 954 345 500 I
Fax Phone: (1) 201 327 2606
email: sdhgx@sprynet.com
Contact # 1: Scott Devens
Business Description: Full service creators and
producers of holographic promotional programs
and materials for the consumer product market..
****

Holographyx Inc.
10661 NW 43rd Court
City: Pompano Beach
State/Province: FL
Postal Code: 33065-2320
Country: United States of America
Voice Phone: (1) 954 345 500 I
Contact # 1: Bob Arnott
Business Description: Full service creators and
producers of holographic promotional programs
and materials for the consumer product market.
****

Hololand S.c.
Batumi 6 m 43
City: Warszawa -.
Postal Code: PL-02-760
Counlly: Poland
Voice Phone: (48) 22 427 463
Fax Phone: (48) 2 625 5567
email: stepin@if.pw.edu.pl
Contact # 1: Pawel Stepien
Business Description: Low volume holographic
labels, consultancy, security CGH, R&D.
****

Hololaser Gallery
PO Box 23386
City: Dubai
Country: United Arab Emerates
Voice Phone: (97 1) 4 518 989
Fax Phone: (971) 4 528 015
Contact # 1: Abdul Wahab Baghdadi
Business Description: Holography Gallety and ho-
lographic items. We are the first and only gallery
in the Gulf Countries. Also do laser shows.
****

HoloMedia Ab/Hologram Museum
PO Box 45012
City: Stockholm
Postal Code: S-10460
Counlly: Sweden
Voice Phone: (46) 8 411 1108
Fax Phone: (46) 8 107638
Contact # 1: Mona Forsberg
Business Description: Broker for embossed and
custom made artistic holography; buying & seil-
ing holograms; holography education; gallery.
Display unit available. Hologram center.
****

Holomedia France
16 rue Maurice Fontvielle
City: Toulouse
Po~tal Code: F-31000
Country: France
Voice Phone: (33) 62 27 1704
Fax Phone: (33) 62 27 17 04
Business Description: Wholesale and distribu-
tion of silver halide, jewelry and fine art holo-
grams. Two retail shops in Toulouse and Lyon,
France.
****

Holomedia France
4 r St Jean
City: Lyon
Postal Code: F-69005
Country: France
Voice Phone: (33) 4 7240 2840
Business Description: Wholesale and distribution
of silver halide, jewelry and fine art holograms.
Two retail shops in Toulouse and Lyon, France.
****

Holomex
4 Borrowdale Avenue
City, Prov.: Harrow, Middlesex
State: England
Postal Code: HA3 7PZ
Country: United Kingdom
Voice Phone: (44) 181 427 9685
Contact # 1: Mike Anderson 4. Business Descrip-
tion: Supplier of film processing kits and safe-
lights. Designs and manufacturers a ""holographic
camera and viewer"".
****

Holonix
Box 45577
City: Seattle
State/Province: WA
Postal Code: 98145
Country: United States of America
Voice Phone: (1) 206 689 6966
email: jkollin@holonix.com
Web: holonix.com
Contact # 1: Joel Kollin
Business Description: Optical systems design and
consultation and for 3-D imaging, holography,
scanning systems, lighting and image projection.
****

Holophile, Inc.
56 Abner Lane
City: Killingworth
State/Province: CT
Postal Code: 06419
Counlly: United States of America
Voice Phone: (1) 860 663 3030
Fax Phone: (1) 860 663 3067
email: info@holophile.com
Web: http ://www.connix.coml-barefootl
Contact # 1: Paul D. Barefoot
Business Description: Founded in 1975, Holo-
phile provides consulting services in holography
and ""spectral imagery"" (3-D projection of mov-
ing images) to corporations, museums and display
builders. Our company is a premier producer of
holography exhibitions for museums, science centers and children 's museums.
****

Holoptics
Heidjerweg 13
City: Oldenburg
Postal Code: D-26133
Country: Germany
Voice Phone: (49) (0)44145166
Fax Phone: (+49) (0)441 4860928
Contact # 1: Joerg Schweer
Business Description: Production of silver halide
holograms, origination of SHH (F.E.: Eye in Pyra-
mid), mass production of small size SHH, distri-
bution of holograms and holographic articles, so-
phisticated display-systems. Mobile exhibitions,
listed supplier of holograms for fas tidious depart-
ment stores.
****

IF YOUR COMPANY PROVIDES GOODS
OR SERVICES TO THE HOLOGRAPHY
INDUSTRY, GET YOUR FREE LISTING!

Holopublic Unbehaun
Hirschstrasse 84
City: Wuppertal
Postal Code: D-42285
Country: Germany
Voice Phone: (49) (0)202 8411 8
Contact #1 : Klaus Unbehaun
Business Description: Consulting, education, newsletters "Holography 3D Software" and "AHT Reflexionen", fine arts (Holofotografik) book "Holo Show International", founding member "AHT-Association for Holography and New Media".
****

Holos Art Galerie
4 Place Grenus
City: Geneva
Postal Code: CH-1201
Country: Switzerland
Voice Phone.: (41) 221 732 855 1
Fax Phone: (41) 221 7325191
Contact # 1: Pascal Barre
Business Description: Gallery, retail sales.
****

Holosco, Ernest Barnes
Bajada de Viladecols, 2
City: Barcelona
Postal Code: 08002
Country: Spain
Yoice Phone: (34) 33107113
Fax Phone: (34) 3 319 1676
Business Description: Holography Lab. Reflection and transmission, transfer to photoresist - embossing facilities . Consulting services.
****

Holosta Holographie-Galerie
Hofkamp 51
City: Wuppertal (Elberfeld)
Postal Code: 0-42103
Country: Germany
Contact # 1: Brigitta Staiger
Business Description: First gallery for holography in Wuppertal.
Retailer of holograms on glass or film, multiplex holograms, embossed holograms, dichromate, holographic jewellery
****

Holostar
College de Maisonneuve
3800 Sherbrooke Est
City, Prov.: Montreal, Quebec
Postal Code: HIX 2A2
Country: Canada
\'o ice Phone: (1) 514254-7131 #4509
Fax Phone: (1) 514 253-8909
email: holostar@cmaisonneuve.qc.ca
Web: cmaisonneuve.qc.calholostar.html
Contact #1: Eric Bosco
Business Description: Center dedicated to the promotion of sciences through holography. Education, services, workshops, research and production. Since 1988.
****

Holostik India Pvl. Ltd.
50. Adhchini
Sri Aurobindo Marg.
City: New Delhi
Poslal Code: 11 00 17
Country: India
Voice Phone: (91) 11 665 690
Voice Phone: (91) 11 669725
Fax Phone: (91) 11 6868828
Contact # 1: Govind Sharma
Business Description: We are among the first to have set up a fully automated plant for manufacture of security and promotional holograms and films in India. Soon setting up master lab and 40 inch wide web machine for holographic packaging.
****

Holostudio Beate Krengel
Mariawalder Strafle 2
City: HeimbachlEifel
Postal Code: 0 -52396
Country: Germany
Voice Phone: (49) 2446 3592
Contact #1: Beate Krengel
Business Description: Distribution of optics and components for holography
****

Holotec
Heilwigstrafle 19
City: Muenchen
Postal Code: 0-81825
Country: Germany
Voice Phone: (49) 89 429741
Contact # 1: M. Wagensonner
Business Description: Activities:
- Display holography (max. 30x40 cm).
- Portrait holography (Pulse-Master).
- Holography seminars.
****

HOLOTECH -Texel
Boodtlaan 41
City: De Koog - Texel, Holland
Postal Code: 1796
Country: Netherlands
Voice Phone: (31) 22 202 7352
Fax Phone: (3 1) 22 202 7429
Contact # 1: Dave Platts
Business Description: Gallery
****

Holotek
(a division of ECRM Inc.)
205 Summit Point Drive
City: Henrietta
State/Province: NY
Postal Code: 14467
Country: United Stat~ of America
Voice Phone: (1) 716 321 6000
Voice Phone: (1) 888 465 6832
Fax Phone: (1) 716 3216001
Business Description: Engineering and design of sub systems for laser optic scanning devices. Commercial and industrial applications.
****

HoloVision
Tumblingerstr. 32
City: Munich
Postal Code: D-80337
Country: Germany
Voice Phone: (49) (0)89 746 9336
Fax Phone: (+49) (0)89 746 9382
email: 101625.3552@comp
Contact #1: Julian Fischer
Business Description: Production of holograms on silver halide up to 100xi00 cm. Holographic stereograms, computer generated holograms, multiple color holograms, pulse holograms, traditional recording techniques, design, model making, installation and consulting services.
****

Holovision AB
Box 70002
City: Stockholm
Postal Code: S-10044
Country: Sweden
Voice Phone: (46) 8 331 186
Fax Phone: (46) 8 331 186
Contact #1: Jonny Gustafsson.
Business Description: Specializing in silver halide holography with pulsed lasers. Denisyuk and transferred-type reflection holograms up to 30 x 40 cm. Rainbow holograms with pulsed laser up to 2 x I m.
****

Holovision Systems Inc.
119 South Main SI.
City: Findlay
State/Province: OH
Postal Code: 45840
Country: United States of America
Voice Phone: (1) 419 422 3604
Fax Phone: (1) 419 422 4270
email: alt@brighl.net
Contact # 1: Ronald L. Kirk
Business Description: Holovision Systems, Inc. specializes in technological development for holographic and 3-dimensional image display. Holovision currently manufactures and markets Real Image (TM) displays which produce live full color 3-dimensional video projections into 3-D space for point of sale, trade show and exhibit applications. It is also developing a higher level of technology called Holoview(TM) or real time holographic displays of medical CADCAM and cinemagraphic applications.
****

Holoworld.com
Internet Webseum of Holography
815 Allen Street
City: Allentown
State/Province: PA ..
Postal Code: 18102
Country: United States of America
Voice Phone: (1) 800 458 3525
Voice Phone: (1) 610 770 0341
email: director@holoworld.com
Web: http ://www.holoworld.coml
Contact # 1: Frank DeFreitas
Business Description: Designer of the Internet Webseum of Holography - a multi-award winning web site dedicated to amateur and hobbyist holography.
****

Holtronic GmbH
Melchior-Huber-Str. 25
City: OttersbergiPliening
Postal Code: 0-85652
Country: Germany
Voice Phone: (49) 8121 81005
Contact #1: G. Dausmann
Business Description: - display-holography (custom made and series, editions, portraits)
- embossed holography (self adhesive und hot stamped holograms)
- holographic-cameras (Holomatic, Holomatic-Dental, Secumatic, Micromatic)
- security (entrance control systems, copysafe holograms)
- holographic interferometry
- holographic-optical elements (head-up-displays, gratings etc.)
****

Honeywell Technology Center
MN 65 - 2500
3660 Technology Drive
City: Minneapolis
State/Province: MN
Postal Code: 55418
Country: United States of America
Voice Phone: (1) 612 951 7738
Fax Phone: (1) 612 9517438
email: cox@src.honeywell.com

Web: http://www.honeywell.com
Contact # 1: Dr. 1. Allen Cox
Business Description: Diffractive optics. Micro-machining.
****

HopSec/dii
P.O. Box 765
City: Bayswater
State/Province: Victoria
Postal Code: 3153
Country: Australia
Voice Phone: (61) 39 729 6337
Voice Phone: (61) 18 776226
Fax Phone: (6 1) 39 729 6020
Contact # 1: Trevor McGaw
Business Description: Security, Optical Holography and Lenticulars.
****

HRT
Holographic Recording Technologies GmbH
Am Steinaubach 19
City: Steinau
Postal Code: 0-36396
Country: Germany
Voice Phone: (49) (0)666 7668
Fax Phone: (+49) (0)666 7463
Contact # 1: Richard Birenheide
Business Description: We produce and distribute a line of high quality silver-halide emulsions with low noise and high diffraction efficiency. We also have di stributors in the USA.
SEE OUR ADVERTISEMENT
****

Hyogo Prefectual Museum of Modem Art
Art Curator
Kobe-3-8-3 Harada-Dori
City: Nada-Ku Kobe 657, Hyogo Ken
Country: Japan
Voice Phone: (81) 7~ 80 I 1591
Fax Phone: (81) 78 8614731
Contact #1: Hitoshi Yamazaki
Business Description: 20th century Art, History of Art and Holography, Art and Optics, curating a exhibition of holography into Art.
****

I.S. Gill
214 Kailash Hills
East of Kailash
City: New Delhi
Postal Code: 110065
Country: India
Voice Phone: (91) II 1684 0377
Voice Phone: (9 1) II 6847 0377
Business Description: Bindry - application of holographic foil and stickers.
****

IBM Almaden Research Center
K03/G2
650 Harry Road
City: San Jose
State/Province: CA
Postal Code: 95120
Country: United States of America
Voice Phone: (1) 408 927 1283
Fax Phone: (1) 408 927 30 II
email: mikeross@almaden.ibm.com
Contact # 1: Michael Ross
Business Description: Scientific holography research; holographic storage.
****

Ibsen Micro Structures AlS
CAT, Frederiksborgvej 399
P.O. Box 30
City: Roskilde

Postal Code: DK-4000
Country: Denmark
Voice Phone: (45) 46 75 40 14
Fax Phone: (45) 4675 40 12
email: ibsen@risoe.dk
Web: http://www. ibsen.dk/
Business Description: Producer of photoresist plates for holography and diffractive optic.
****

iC Holographies
8 Flitcrot St.
City: London
State/Province: England
Postal Code: WC2H 8DJ
Country: United Kingdom
Voice Phone: (44) 171 240 6767
Fax Phone: (44) 171 240 6768
email: 100413.3406@compuserve.com
Contact #1: Chris Levine
Business Description: Hlographic design and digital mastering
****

ICI Polyester
(a division of ICI America)
Concord Plaza - 3411 Silverside Road
City: Wilmington
State/Province: DE
Postal Code: 19850
Country: United States ~f America
Voice Phone: (1) 800 635 4639
Fax Phone: (1) 302 887 5365
email: icipet.com
Web: http://www.icipolyester.com/home.htm
Business Description: Supplies polyester film, polyethylene naphthalate and polyester resins for metalizing and/or direct embossing. Recently acquired by DuPont.
****

Illinois Institute Of Technology
MechanicallMaterials & Aerospace Engineering
Engineering Building #1 Rm 252-B
City: Chicago
State/Province: IL
Postal Code: 60616
Country: United States of America
Voice Phone: (1) 312 567 3220
Fax Phone: (1) 312 567 7230
email: mesciamrnarella@mimna.iit.ezll
Contact # 1: Cesar Sciammarella
Business Description: Holographic interferometry; industrial holographic research; non-destructive testing.
****

Illuminations
1252 7th Avenue
City: San Francisco
State/Province: CA
Postal Code: 94122
Country: United States of America
Voice Phone: (1) 4156640694
email: Ibrill@slip.net
Contact # 1: Louis Brill
Business Description: Involved in developing & expanding market & sales efforts for holographic retail/wholesale product lines. Assist in preparation of promotions and collateral sales materials, identify potential sales markets & implementation of sales.
****

Image Technical Development Co.
Huazhong University of Science & Technology
City, Prov.: Wuhan, Hubei
Postal Code: 430074
Country: China
Voice Phone: (86) 27 7547655

Fax Phone: (86) 27 7547655
email: Megpei@sever20.hust.edu.cn
Contact # 1: Zhang Zhaoqun
Business Description: Holography classes.
****

Imagen Holography, Inc.
10 Park Ave. Suite 500
City: Basalt
State/Province: CO
Postal Code: 81621
Country: United States of America
Voice Phone: (1) 970 927 0360
Fax Phone: (1) 970 927 0359
Contact #1: Alan P. Morterud
Business Description: Specialized holographic products for mainstream marketing applications, including HOLOTEX and Advanced Holographic Textiles - a soft, supple, fully washable rendering of reflection holograms in both 2D & 3D, to a variety of fabrics.
****

Imagenes Holograficas De Columbia
Raul Delgado Z - Manager
AVDA 5 Norte # 17 - 23
City: Cali
Country: Columbia
Voice Phone: (57) II 572 6684032
Fax Phone: (57) II 572 6685450
****

Images Company
39 Seneca Loop
City: Staten Island
State/Province: NY
Postal Code: 10314
Country: United States of America
Voice Phone: (1) 718 698 8305
Fax Phone: (1) 718 982 6145
email: imagesco@he.net
Web: http://www.he.net/-imagesco/
Business Description: Sells holographic equipment and materials to educational institutions, students and private holographers. Equipment And materials availab le: lasers, film, development kits, mounting kits for lenses, beam splitters, mirrors, spatial filters, safelights, filters and display lights.
****

Imagination Plantation
2650 18th street, 2nd floor
City: San Francisco
State/Province: CA
Postal Code: 94110
Country: United States of America
Voice Phone: (1) 4154870841
Fax Phone: (1) 415 487 2103
email: ipd@iplant.com
Web: www.iplant.com
Contact #1: Noah Hurwitz
Business Description: 3D imaging and content creation for all media. Experienced in modeling for holographic applications, including direct output to master.
****

ImEdge Technology
Eastview Technology Center
350 Main St.
City: White Plains
State/Province: NY
Postal Code: 1060 I
Country: United States of America
Voice Phone: (1) 914 946 5536
Fax Phone: (1) 914 946 5460
email: mmetz@imedge.com
Web: http://eastview.orgilmEdge/
Contact # 1: Michael Met

Business Description: Research, development and manufacturing of edge-lit holograms; creative and innovative holography and optics problem solving; custom display volume holograms; edge-lit holographic optical elements; consulting; edge-lit fingerprint imaging device and other industrial holographic products.
****

Imperial College Of Science
Optics Section
BlackeirLaboratory
City: London
State/Province: England
Postal Code: SW7 2BZ
Country: United Kingdom
Voice Phone: (44) 171 5895111
Business Description: Courses in holography; holography research; particle measurement.
****

Industrial Technology Institute
290 I Hubbard Road
City: Ann Arbor
State/Province: MI
Postal Code: 48105
Country: United States of America
Voice Phone: (1) 313 769 4156
Voice Phone: (1) 800 292 4484
Fax Phone: (1) 313 769 4064
email: inquiry@iti .org
Web: http://www.iti.org
Contact #1: Kevin Harding
Business Description: ITI is a not-for-profit contract R&D organization. We help manufacturers identify and apply technology to solve production problems. ITI provides technical consulting, R&D, testing (including holographic NOT) and training programs.
****

Industrial Technology Research Inst.
Holography Department
Bldg 44, 195 Chung Hsing Road, Section 4
City, Prov.: Chutung, Hsinchu
Postal Code: 310 I 5
Country: Taiwan
Voice Phone: (886) 35 917 482
Fax Phone: (886) 35 917 479
Contact # 1: Dr. 1.1. Su
Business Description: Research in HUD (Head Up Display) and Dot Matrix Hologram systems
****

INETI - Institute of Information Technologies
LAER - Aerospace Laboratory
Estrada do Paco do Lumiar
City: Lisboa Codex
Postal Code: P-1699
Country: Portugal
Voice Phone: (351) I 7 165181
Fax Phone: (351) I 7166067
email: xana@laer.ineti.pt
Web: http://www.laer.ineti.pt/
Contact #1: Ana Alexandra Andrade
Business Description: R&D in Holography and OVDs. Special interests in security features. Consultants on teclmology and customized production projects of embossed holograms. Inhouse design and origination.
****

Infinity Laser Laboratories
681 I Flanders Station
City: Polk City
State/Province: FL
Postal Code: 33868
Country: United States of America
Voice Phone: (1) 941 984 3108
Fax Phone: (1) 941 9844244

email: infinity@digital.net
Contact #1: Thad Cason
Business Description: Full service company from concept and design to finished product. Laboratories include photopolymer capability with pulsed, continuous wave and solid state lasers including Argon, Krypton, Nd:YAG:KTP and HeCd.
****

Infox Corporation
3rd floor No 283
Sec2 Fu-hsing South Road
City: Taipei
Country: Taiwan
Voice Phone: (886) 2 7056699
Fax Phone: (886) 2 7551800
Contact # 1: Alex C.T. Chen
Business Description: Maker of injected-molded holograms.
****

Infrared Optical Products, Inc.
PO Box 292
City: Farmingdale
State/Province: NY
Postal Code: I 1735-0664
Country: United States of America
Voice Phone: (1) 516 694 6035
Fax Phone: (1) 516 694 6049
Contact # 1: Barry Bassin
Business Description: Manufacturer of infrared lenses, windows, reflectors, beamsplitters, computer-designed IR lens systems. Front surface optical coatings for mirrors.
****

Ingenieurbiiro Geiger
Dieding 7
City: Ebersberg
Postal Code: 0-85560
Country: Germany
Voice Phone: (49) 8092 6383_
Voice Phone: (+49) 8092 6583
Fax Phone: (+49) 8092 31658

Contact # 1: Thomas Dipl.-Ing. Geiger
Business Description: Holgraphy-Lab for manufacturing of large format white-light holograms, stamp masters for emobssing
****

INNOLAS (UK) Ltd.
67 Somers Road
City: Rugby, Warwickshire
State: England
Postal Code: CV22 7DG
Country: United Kingdom
Voice Phone: (44) 1788550 777 1
Fax Phone: (44) 1788550888
email: InnoLasUK@aol.com
Contact #1: William Brown
Business Description: Innolas Ltd. is a newly formed company and ongmates from Laser Techniques, United Kingdom and InnoLas GmbH., Planegg-Steinkirchen, (Munich) Germany. It is largely created on the sale of the Lumonics Ltd. Solid State Scientific Products business to both InnoLas (UK) Ltd. and InnoLas GmbH. Innolass Ltd. Continues to manufacture scientific Q switched ND:YAG lasers and ruby holographic lasers. Laser Technical Services is the agent for InnoLas Ltd. in North & South America.
****

Innovative Technology Associates
3639 East Harbor Blvd. # 203 E
City: Ventura
State/Province: CA
Postal Code: 9300 I
Country: United States of America

Voice Phone: (1) 805 650 9353
Fax Phone: (1) 805 984 2979
Contact # 1: Joseph Gaynor
Business Description: Technical and business consultants specializing in materials and processes relating to deformable films , photoplastics and photorefractive polymers. Special knowledge of holographic memory storage technologies.
****

Inrad, Inc.
181 Legrand Avenue
City: Northvale
State/Province: NJ
Postal Code: 07647
Country: United States of America
Voice Phone: (1) 2017671910
Fax Phone: (1) 20 I 767 9644
Contact #1: Maria Murray
Business Description: Manufacturer of nonlinear materials, harmonic generation systems, electro-optic and acousto-optic devices and drivers. Provide optical components, assemblies and optical coatings for the UV, visible, IR.
****

Inside Finishing Magazine
P.O. Box 12090
City: Portland
State/Province: OR
Postal Code: 97212
Country: United States of America
Voice Phone: (1) 503 3316221
Fax Phone: (1) 503 331 6928
email: fsea@aol.com
Contact # 1: Jeff Peterson ·
Business Description: Only trade publication specifically targeting the graphic finishing industry. Editorial focus is on foil stamping, embossing, holograms, die cutting, folding/gluing, and coatings. Published quarterly by the Foil Stamping & Embossing Association.
****

Inspeck, inc
360 rue Franquet, Suite 20
City, Prov.: St. Foy, Quebec
Postal Code: G I P 4N3
Country: Canada
Voice Phone: (1) 418 6502112
Fax Phone: (1) 4186502141
email: inspeck@riq.qc.ca
Contact # 1: Li Song
Business Description: Manufacturers a 3-D color digitizer (scans into computer the coordinates of object and color). Data is imported as DXF or other system independent data format. Works on PC and can be ported to SGl, SUN, etc. for use with any rendering program.
****

Institut fur Angewandte Phys ik
Schlossgartenstr. 7
TH Dannstadt
City: Darmstadt
Postal Code: 0-64289
Country: Germany
Voice Phone: (49) 6 151 162786
Fax Phone: (49) 6151 164534
emai l: Andreas Billo@Physik.TH-Darmstadt.de
Web: http://www.physik.th-darmstadt.de/andreas
Contact # 1: Andreas Billo
Business Description: Investigation of cavitation bubbles, sprays and droplets with high speed holographic particle image velocimetry (HPIV).
****

Institute for Holographic Tech.
Cenh·al Univ. of Nationalities

INTERNATIONAL DIRECTORY

Dept. of Physics
City: Beijing
Postal Code: 100081
Country: China
Voice Phone: (86) 769 2467775
Fax Phone: (86) 769 2478473
Contact # 1: Zhu Weili
Business Description: Holography classes.
****

Institute of Applied Optics
National Academy of Sciences of Ukraine
10-G Kudryavaskaya St.
City: Kiev
Postal Code: 254053
Country: Ukraine
Voice Phone: (380) 44 212 21 58
Fax Phone: (380) 44 212 48 12
email: vmarkov@iao.freenet.kiev.ua
Contact # 1: Vladimir B. Markov
Business Description: R&D in the areas of: display holography (reflection, transmission, mastering) and colour holography for museum items reproduction; holographic interferometry and its application in industry and cultural heritage protection; diffraction on the volume grating; laser physics, including properties of the cavity, tunable lasers, etc.; non-linear optical effects and multibeam interaction
****

Institute Of Optical Science
Central University
City: Chung-Li
Postal Code: 32054
Country: Taiwan
Voice Phone: (886) 3 425 7681
Fax Phone: (886) 3 425 8816
Contact # 1: Tang Yaw Tzong
Business Description: HOEs, academic research.
****

Integraf
P.O. Box 586
745 N. Waukegan Rd.
City: Lake Forest
State/Province: IL
Postal Code: 60045
Country: United States of America
Voice Phone: (1) 847 234 3756
Fax Phone: (1) 847 615 0835
email: jeong@lfc.edu
Contact #1: T.H. Jeong
Business Description: Distributor of holographic films and plates, including Russian emulsions. Expert consultation on HOEs, NDT, system designs and other holographic projects. Also, educational materials and stock holograms.
SEE OUR ADVERTISEMENT
****

Interactive Industries, Inc.
40 Todd Rd
City: Shelton
State/Province: CT
Postal Code: 06484
Country: United States of America
Voice Phone: (1) 203 929 9000
Fax Phone: (1) 203 929 9001
Contact # 1: Ron Phillips
Business Description: Custom, Stock, 3-D, 2-D, Key Chains, Bookmarks, Mugs, Posters, Calendars, Signs, Rulers, Gambling Specialties, Labels, Tags, Souvenirs, Buttons, Portfolios
****

Interferens Holografi D.A.
Museum , Gallery, Studio
Halvor Hoels Gt 6
City: Hamar

Postal Code: N-2300
Country: Norway
Voice Phone: (47) 62 25050
Voice Phone: (47) 62 30659
Fax Phone: (47) 62 30659
Contact # 1: Olav Skipnes
Business Description: Ongoing exhibition of Norway/Es largest co llection of holograms. Makes glass (mainly reflection) holograms of museum exhibits. Continuous wave laser. Norway's largest collection of holograms. Our specialty: museum exhibits.
****

International Data Ltd.
Units 5 & 6
Station Industrial Estate, Oxford Rd
City: Wokingham, Berkshlre
State: England
Postal Code: RG41 2YQ
Country: United Kingdom
Voice Phone: (44) 1189772 255
Fax Phone: (44) 1189772296
email: 101 763.1273@compuserve.com
Contact # 1: Dawn Dreelaw
Business Description: Manufacturer of plastic cards. Capable of hot stamping holographic materials.
****

Internl. Hologram ManufacturersAssociation
Runnymede Malthouse
Runnymede Road
City: Egham
State/Province: England
Postal Code: DY20 9BD
Country: United Kingdom
Voice Phone: (44) 1784 497 008
Fax Phone: (44) 1784 497 001
email: 100142.1164@compuserve.com
Business Description: IMHA was founded in 1993 to promote the interests of hologram manufacturers and the holography industry worldwide. It is a non profit membership organization open to all producers of holograms, suppliers of equipment and material for the manufacture of holograms and hologram converters and finishers. Annual meeting in November.
SEE OUR ADVERTISEMENT
****

Intrepid World Communications
(a subsidiary of American Propylaea Corp.)
555 South Woodward, Suite 1109
City: Birmingham
State/Province: MI
Postal Code: 48009
Country: United States of America
Voice Phone: (1) 248 642 9885
Fax Phone: (1) 248 642 9886
Contact # 1: Ann Marie Harrison
Business Description: Distributor of display holograms, in particular true-color holograms. Currently working on archiving Vatican treasures. Subsidiary of American Propylaea Corp.
****

Ion Laser Technology, Inc.
3828 South Main Street
City: Salt Lake City
State/Province: UT
Postal Code: 84115
Country: United States of America
Voice Phone: (1) 801 262 5555
Fax Phone: (1) 801 2625770
Contact # 1: Don Gibb
Business Description: Manufacturer of industrial, scientific and medical lasers, including air and water cooled Argon lasers suitable for holography.
****

Ishii, Ms. Setsuko
#404,
1-23 26 Kohinata,Bunkyo-Ku
City: Tokyo
Postal Code: 102
Country: Japan
Voice Phone: (81) 03 3945 9017
Fax Phone: (81) 03 3945 9068
Contact # 1: Setsuko Ishii
Business Description: Holographic Fine Artist.
****

Island Graphix
22 Bayview Ave.
Ward's Island
City, Prov.: Toronto, Ontario
Country: Canada
Voice Phone: (1) 416 203 7243
; Fax Phone: (416) 2037243
email: page@astral.magic.ca
Contact # 1: Michael Page
Business Description: Production of all formats Computer Generated Holograms, Mass Production, Consulting, Stock images
****

James River Products
800 Research Road
City: Richmond
State/Province: VA
Postal Code: 23236
Country: United States of America
Voice Phone: (1) 804 378 1800
Fax Phone: (1) 804 378 5400
email: jrp@richmond.infi.net
Web: http://hmt.comlholography/jrplindex.html
Contact # 1: Mike Florence
Business Description: World leader in embossed hologram machinery. Products include: origination lab equipment, photoresist plate spinning, electroform facilities, embossing machines, die cutting equipment, supporting technology and training - plus custom embossed holograms.
SEE OUR ADVERTISEMENT
****

Japan Communication Arts Co.
Yonezawa Bldg2F
2-37 Suehirocho, Kita-Ku
City: Osaka
Postal Code: 530
Country: Japan
Voice Phone: (81) 06 314 1919
Fax Phone: (81) 06 3151900
Contact # 1: Mineko Fukuma
Business Description: Sales of cards with hologram.
****

Jayco Holographics
29-43 Sydney Road
City: Watford, Herts
State: England
Postal Code: WD I 7PY
Country: United Kingdom
Voice Phone: (44) 1923246760
Fax Phone: (44) 1923 247 769
Contact # 1: Rohit Mistry
Business Description: Complete prodtlction service for embossed holograms. Embossing masters through to fully finished product. Our years of experience enables Jayco to offer outstanding quality of product and service at competitive prices.
****

Jeffery Murray Custom Holography
P.O. Box 24 - 153
City: San Francisco
State/Province: CA

Postal Code: 92124
Country: United States of America
Voice Phone: (1) 415 822-7123
email: hologram@well.com
Contact # 1: Jeffery Murray
Business Description: Museum displays, artists collaborations, commercial advertising, image research, HOE custom optics for visual display. Research specialties: high quality display holography, holographic optics, holographic recording systems. One-offs, ltd. editions and unusual projects. Research facilitie s: holography darkrooms, silver halide processing, 5 isolation tables with HeNe lasers. Able to produce H I masters, reflection and transmission image plane transfers, and Denisyuk holograms up to 50x60 cm.
****

Jiangsu Sida Images, Inc.
Nanjing Normal University
City, Prov.: Nanjing, Jiangsu
Postal Code: 210024
Country: China
Voice Phone: (86) 25 3318848
Fax Phone: (86) 25 7718174
Contact #1: Sun Hangjia
Business Description: Holography classes.
****

Jodon Inc.
62 Enterprise Drive
City: Ann Arbor
State/Province: MI
Postal Code: 48103
Country: United States of America
Voice Phone: (1) 313 7614044
Voice Phone: (1) 800 989 jodon
Fax Phone: (1) 313 761 3322
email: johng@wwn.com
Contact #1 : John Wernenski
Business Description: Manufacturer of HeNe lasers, laser systems, specialty laser tubes, optical and electro-optical instruments and systems. Holographic films, plates and chemicals.
****

John, Pearl
Beverly Hyrst - Flat 4
Addiscombe Road
City: Croydon, Surrey
State: England
Postal Code: CRO 6SL
Country: United Kingdom
Voice Phone: (44) 181 6541420
Contact #1: Pearl John
Business Description: Artist with pOl1folio, including pulsed and multiplex holograms. M.A. degree awarded from Royal College of Art, London. Also provides educational, design, and consulting services.
****

Jurewicz, Arlene
Box 4235 RRI
City: Lincolnville
State/Province: MA
Postal Code: 04849
Country: United States of America
Voice Phone: (1) 207 763 3182
email: holo@midcoast.com
Web: .midcoast.coml-holo/holo I.html
Contact #1: Arlene Jurewicz
Business Description: Introductory level presentations on holography for all age groups. Research on on the use of holography in education and as a perceptual tool. Colaborations welcomed.
****
K. Thielker & T. Rost
Holographievertrieb

Vermeerweg 15
City: Wesseling
Postal Code: D-50389
Country: Gellllany
Voice Phone: (49) 2236 43138
Contact # 1: Klaus Thielker
Business Description: Wholesale of holograms, traveling exhibitions, holograms for rent..
****

K.C. Brown Holographics
17 Salisbury Road
New Malden
City: Surrey
State/Province: England
Postal Code: KT3 3HZ
Country: United Kingdom
Voice Phone: (44) 181 942 8294
Fax Phone: (44) 181 877 3400
email: 100306.2015@compuserve.com
Contact # 1: Kevin Brown
Business Description: Pulse portraits; artistic holography.
****

Kaiser Optical Systems, Inc.
PO Box 983
371 Parkland Plaza
City: Ann Arbor
State/Province: MI
Postal Code: 48106
Country: United States of America
Voice Phone: (1) 313 665 8083
Fax Phone: (1) 313 665 8199
Web: http ://www.kosi.com/Products/Raman/5slcover.htm
Business Description: Compact fiber coupled Raman spectrometers for routine quality control and remote, real time, in line process monitoring applications with microscope accessory for line applications. Fast f/ 1.8 volume transmission grating based Holographic Imaging Spectrographs for visible, fluorescence and Raman applications. Holographic Notch and Laser Bandpass filters for Raman laser induced fluorescence spectroscopy.
****

Ka-Lor Cubicle & Supply Co. Inc.
P.O. Box 804 .,
City: Fair Lawn
State/Province: NJ
Postal Code: 07410
Country: United States of America
Voice Phone: 201 891 8077
Fax Phone: 201 891 6331
Contact # 1: D.Brett
Business Description: Stay in the dark! Darkroom curtains and track system for your holography studio or laboratory. Easily installed. Shipped worldwide. Call for details and consultation.
SEE OUR ADVERTISEMENT
****

Karas Studios Holografia
Ave Maria, 46
City: Madrid
Postal Code: 28012
Country: Spain
Voice Phone: (34) I 530 89 88
Fax Phone: (34) I 530 89 88
email: karasrb@ddnet.es
Web: http ://www.webvent.com/karas
Contact # 1: Ramon Benito
Business Description : Established in 1988, Karas owns a collection of holographic pieces selected under a so le criteria: art. Among others, the most relevant works of the Spanish art holographers are in the Karas Collection. Three Galleries. Publications. Art curators.
****

Kauffman, John
Box 477
City: Point Reyes Station
State/Province: CA
Postal Code: 94956
Country: United States of America
Voice Phone: (1) 415663 1216
Fax Phone: (1) 415663 1216
Contact # 1: John Kauffman
Business Description: Holographic Fine Artist. Specializes in multi color reflection holograms. Extensive portfolio.
****

Keio University
Dept Of Electrical Engineering
3-1 4-1 Hiyoshi Kohoku-Ku
City: Yokohama
Postal Code: 223
Country: Japan
Voice Phone: (81) 045 563 1141
Fax Phone: (81) 045 563 3421
Contact # 1: Dr. Masato Nakajima
Business Description: Research using HNDT
****

Kendall Hyde Ltd.
Kingsland Industrial Park
Stroudley Road
City: Basingstoke, Hants
State: England
Postal Code: RG24 8UG
Country: United Kingdom
Voice Phone: (44) 125 684 0830
Fax Phone: (44) 125 6840443
Contact # 1: Clive Birch
Business Description: Optical coating specialists. Front surface mirrors, and any type of film coating.
****

Keystone Scientific Co.
PO Box 22
City: Thorndale
State/Province: PA
Postal Code: 19372-0022
Country: United States of America
Voice Phone: (1) 610 269 9065
Fax Phone: (1) 610 269 4855
Business Description: Distributor for Agfa, Slavich and Kodak holographic films, plates and chemicals. Manufacturer of holography kits and automatic processors.
****

Kimmon Electric Co., Ltd.
TM21 Building
1-53-2 ltabashi, Itabashi-ku
City: Tokyo
Postal Code: 173
Country: Japan
Voice Phone: (81) 3 5248 4811
Fax Phone: (81) 3 5248-0021
email: lasers@kimmon.com
Web: http://www.k:immon.com
Contact #1: Shinichi Fukuda
Business Description: Manufacturer of Helium Cadmium lasers which are used in holography. Kimmon manufacturers the highest powered HeCd laser in the world. The model IK4171 I-G is the laser of choice for holographers because of its 180m W @442 TEMoo specified output power and long lifetime.
****

Kinetic Systems, Inc.
20 Arboretum Road
City: Boston
State/Province: MA

INTERNATIONAL DIRECTORY

Postal Code: 02131
Country: United States of America
Voice Phone: (1) 617 522 8700
Voice Phone: (1) 800992 2884
Fax Phone: (1) 617 522 6323
email: sales@kineticsystems.com
Web: http://www.kineticsystems.com
Contact # 1: Moss Blosvern
Business Desc ripti on: Manufacturers of Vibr plane standard and special Honeycomb optical tables in four grades up to 5' x 16' x 24'. Larger sizes available by butt splicing. Also vibration isolation support systems.
****

Kolbe-Druck mit Tochtergesellschaften
Im Industriegelaende 50
City: Versmold
Postal Code: 0-33775
Country: Germany
Voice Phone: (49) 5423 9670
Fax Phone: (+49) 5423 41 230
Contact # 1: Roland Pahnke
Business Description: Bindry: Hot stamping holographic foil , Lenticular-Printing, 3-0, motion.
****

Krystal Holographies International Inc.
U.S. Holographics Division
365 North 600 West
City: Logan
State/Province: UT
Postal Code: 84321
Country: United States of America
Voice Phone: (1) 435 753 5775
Voice Phone: (1) 800 998 5775
Fax Phone: (1) 435 753 5876
email: krystal@sunrem.com
Web: http://www.khiinc.com
Contact # 1: Dave Rayfield
Business Desc~iption: Marketing/manufacturing for mass-produced and custom photopolymer and dichromate holograms for retail and ad specialty needs. Stock and custom products available. KHI is also a manufacturer of mastering and mass replication equipment.
SEE OUR ADVERTISEMENT
****

Krystal Holographies International Inc.
555 West 57th Street
City: New York
State/Province: NY
Postal Code: 10019
Country: United States of America
Voice Phone: (1) 212 261 0400
Voice Phone: (1) 80 I 753 5775
Fax Phone: (1) 212 262 0414
email: hologram@krystaltech.com
Web: http://www.khiinc.com
Contact # 1: Marion Baker
Business Description: World headquarters of Krystal Holographics International Inc., a manufacturer and marketer of photopolymer holograms produced by proprietary technology. Experienced with large volume orders of custom holograms for commercial applications including packaging, giftware, security/authentication, advertising premiums, etc.
SEE OUR ADVERTISEMENT
****

Krystal Holographies Vertriebs-GmbH
Birnenweg 15
City: Reutlingen
Postal Code: 0-72766
Country: Germany
Voice Phone: (49) 7121 9461 58
Fax Phone: (+49) 7121 946 1 55

email: R.Stooss@krystaltech.de
Web: http: //www.khiinc.com
Contact # 1: Richard Stooss
Business Description: Eye-Catching and brand enhancement at its best! Distribution of innovative KrystalGram products in Western Europe: High quality 3D-Holograms on DuPont Omnidex photopolymer film, aimed for high volume promotional & OEM applications. Custom and stock designs available. Krystalmark security systems for high level security. You need to see it!
SEE OUR ADVERTISEMENT
****

Kurz Foils
3200 Woodpark Blvd.
City: Charlotte
State/Province: NC
Postal Code: 28206
Country: United States of America
Voice Phone: (1) 800 950 3645
Voice Phone: (1) 704 596 9091
Fax Phone: (1) 704 596 3321
Contact # 1: Donald Tomking
Business Description: Provides holographic hot stamp foil to the industry. Custom designs done to customer specifications. Also provide bindry hot stamping service.
****

Laboratories of Image Information
Science and Technology
LC Bldg. 10F 1-4-2 Shin-senri Higashi-machi
City: Toyonaka, Osaka
Postal Code: 565
Country: Taiwan
Voice Phone: (81) 06 873 2053
Fax Phone: (81) 06 873 2056
Business Description: R&D on Holographic Display, Dynamic Holography, Holographic Optical Elements
****

Laboratory for Optical Data Processing
Carnegie Mellon University
Dept. of Electrical And Computer Engineering
City: Pittsburg
State/Province: PA
Postal Code: 15213
Country: United States of America
Voice Phone: (1) 412 268 2464
Fax Phone: (1) 412 268 6345
email: marlene@ece.cmu.edu
Web: http://www.ece.cmu.edul
Contact # 1: David Casasent
Business Description: Research in optical data processing, pattern recognition, product inspection, neural nets. Processors, filters and feature extractors using computer generated holograms.
****

Laboratory Vinckiner
Holography Workshop Univ Gent
41 St Pietersnieuwstraat
City: Gent
Postal Code: B-9000
Country: Belgium
Voice Phone: (32) 9 264 3242
Voice Phone: (32) 91 626 384
Fax Phone: (32) 9 223 7326
Contact # 1: Pierre Boone
Business Description: Consultancy, education, problem-solving for display holography. Museum applications and (mainly!) non-destructive testing.
****

IF YOUR COMPANY PROVIDES GOODS OR SERVICES TO THE HOLOGRAPHY INDUSTRY, GET YOUR FREE LISTING IN THE NEXT EDITION OF HMP!

Lake Forest College
Center for Photonics Studies
555 N. Sheridan Roadcolo
City: Lake Forest
S tatelProvince: IL
Postal Code: 60045
Country: United States of America
Voice Phone: (1) 8477355160
Fax Phone: (1) 847 615 0835
email: jeong@lfc.edu
Web: http://www.lfc.edu
Contact # 1: T.H. Jeong
Contact #2: Hans Bjelkhagen
Business Description: Lake Forest College offers regularly scheduled classes and workshops about holography. Beginners and advanced student welcome. Well-equipped holography laboratory. T.H. Jeong and College also host the International Symposium on Display Holography.

Hans Bjelkhagen can also be reached in care of the college. He is an acknowledged expert in the field of si Iver-halide recording materials for holography and color reproduction, including Lippman photography. Consulting services.
****

Larry Liebernlan Holography
The Hologram Gallery & The Living Portrait Studio
1642 Euclid Ave.
City: Miami Beach
State/Province: FL
Postal Code: 33139
Country: United States of America
Voice Phone: (1) 305 604 9986
Fax Phone: (1) 305 604 9998
email: lieber741@aol.com
Contact # 1: Larry Lieberman
Business Description: Mastering facility for full color holograms and stereogram portraiture. A collection of full color limited edition artworks available for sale and exhibition. Call for details and consultation.
****

Lasart Ltd.
291 1 San Isidro Ct.
City: Santa Fe
State/Province: NM
Postal Code: 8750 I
Country: United States of America
Voice Phone: (1) 505 438 8224
Fax Phone: (1) 505 438 8224
Contact # 1: August Muth
Business Description: Lasart Ltd. is a manufacturer of production holographic jewelry and gifts as well as one-of-a-kind holographic glass sculpture. We also specialize in unique corporate gifts, taking the project from modeling through completion of the product in our in house facilities.
****

Laser Affiliates
2047 Blucher Valley Road
City: Sebastopol
State/Province: CA
Postal Code: 95472
Country: United States of America
Voice Phone: (1) 707 823 7171
Fax Phone: (1) 707 823 8073
Contact # 1: Nancy Gorglione
Business Description: Laser Affiliates is an award-winning non-profit organization that d signs innovative holographic and laser theatrical productions, installations and exhibitions. Services include curatorial guidance, videotapes and media lectures.
****

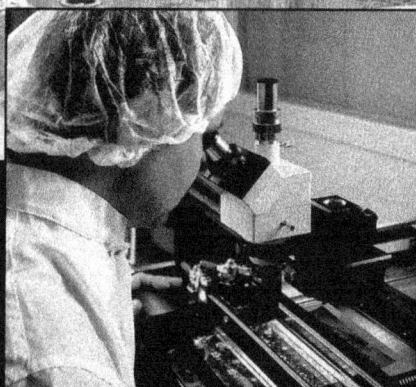

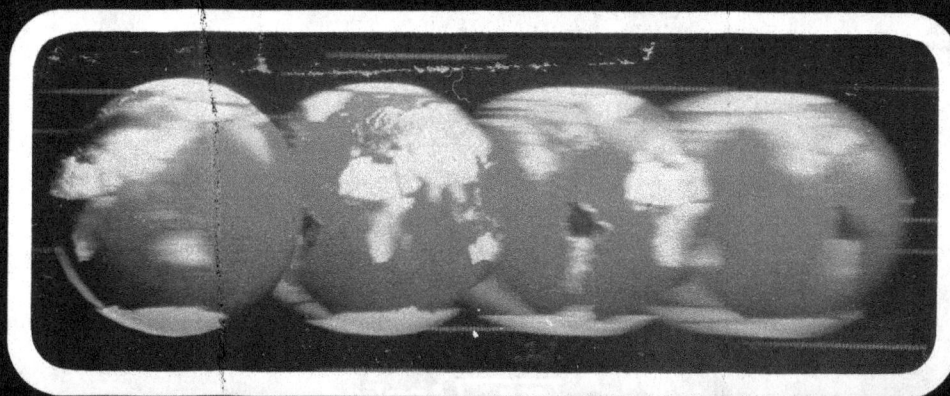

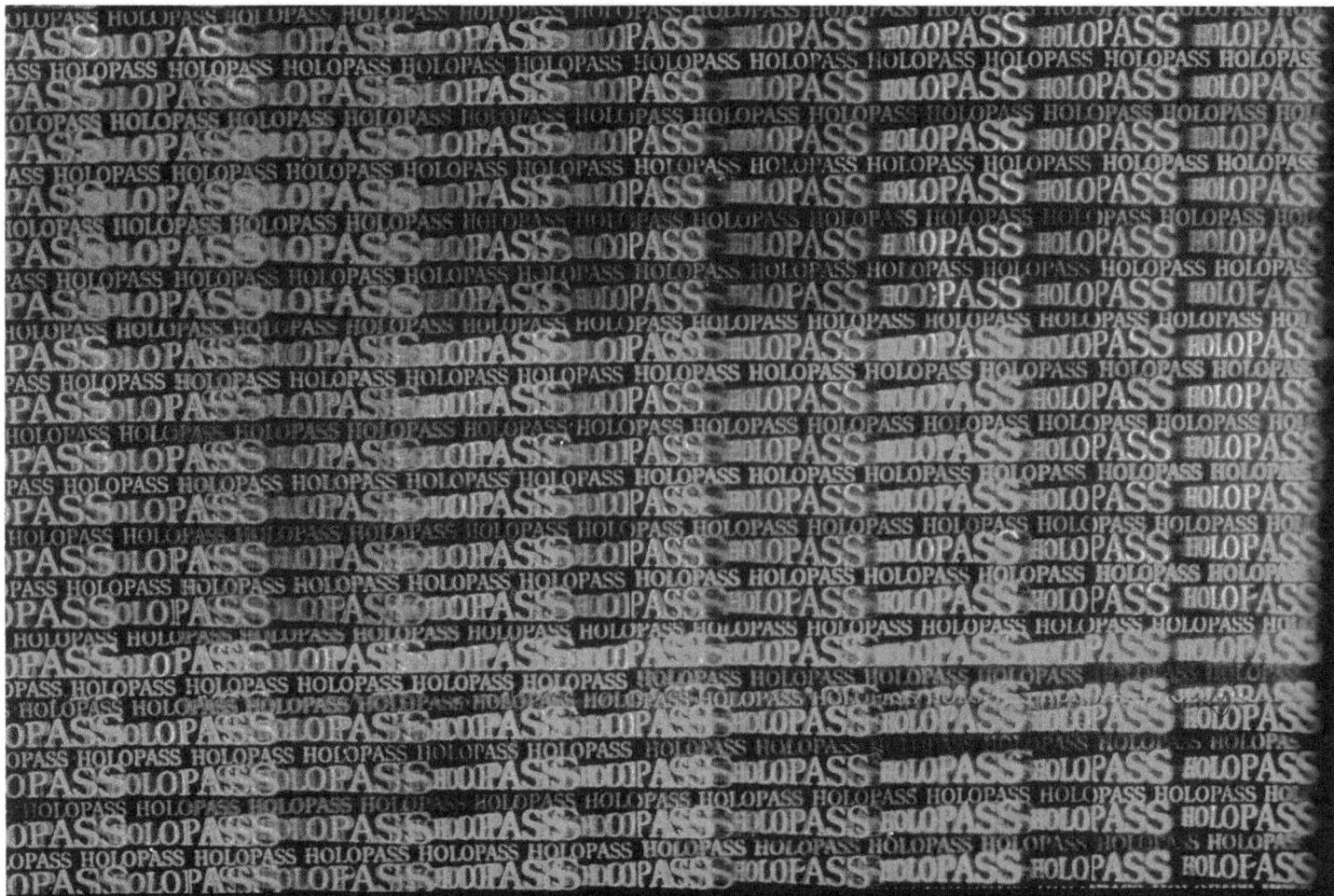

Laser and Motion Development Company
Professional Equipment Exchange
3101 Whipple Road
City: Union City
State/Province: CA
Postal Code: 94587-1216
Country: United States of America
Voice Phone: (1) 5104291060
Fax Phone: (1) 510 429 1065
email: em@mediacity.com
Web: lasern10tion.com
Contact # 1: Ed Monberg
Business Description: LMDC is a buyer and seller of lasers, motion and optical equipment. We also integrate laser processing systems at significant savings to our customers. We offer experience and engineering in galvanometer, X-Y table, and laser based optical systems.
****

Laser Arts Society For Education and Research
PO Box 24-153
City: San Francisco
State/Province: CA
Postal Code: 94 124 - 0153
Country: United States of America
Voice Phone: (1) 4158227123
email: hologram@well.com
Web: http ://www.hmt.com/holography/ laser/
index.html
Contact # 1: Jeffrey Murray
Business Description: Vol unteer staffed nonprofit organization dedicated to holography and laser education and research. Members receive the quarterly L.A.S.E.R. News. One year membership USA $30, outside USA $40.
****

Laser Drive Inc.
5465 WM Flynn Hwy.
City: Gibsonia
State/Province: PA
Postal Code: 15044
Country: United States of America
Voice Phone: (1) 412 443 7688
Fax Phone: (1) 412 444 6430
Contact #1: Carol Smith
Business Description: Manufacturer high voltage power supplies for Helium Neon, Argon Ion and C02 lasers.
****

Laser Focus World
(a division of Penwell Publishing)
10 Tara Boulevard - 5th floor
City: Nashua
State/Province: NH
Postal Code: 03062
Country: United States of America
Voice Phone: (1) 603 891 0123
Voice Phone: (1) 800 331 4463
Fax Phone: (1) 603 891 0574
email: barbaraw@pennwell.com
Web: http://www. lfw.com/WWW/home.htm
Business Description: Trade publication covering the field of optics, lasers, electro-optics, and related imaging research, as well as commercial applications.
****

Laser Holography Workshop
320 South Willard Street
City: New Buffalo
State/Province: MI
Postal Code: 49117
Country: United States of America
Voice Phone: (1) 616 469 4658
Fax Phone: (1) 616 469 4658
email: 102017.1330@compuserve.com
Contact # 1: Joseph A. Farina

Business Description: Specializing in modelmaking for holography. Clients sending us their reference material by facsimile will receive a noobligation proposal within 24 hours. Call for our free brochure.
****

Laser Images
P.O. Box 6873
City: Leawood
State/Province: KS
Postal Code: 66206
Country: United States of America
Voice Phone: (1) 913 648 2525
Voice Phone: (1) 800 346 2430
Fax Phone: (1) 9136486898
Contact # 1: Steve Larson
Business Description: R&D and production capabilities in dichromate, silver halide, photoresist and photopolymer.
****

Laser Innovations
668 Flinn Ave. #22
City: Moorpark
State/Province: CA
Postal Code: 93021
Country: United States of America
Voice Phone: (1) 805 529 5864
Fax Phone: (1) 805 529 6621
Contact # 1: R. Eric King
Business Description: Laser Innovations offers sales, repair, and support of ion laser systems. Specializing in the repair and service of CO HER-ENT lasers; Laser Innovations stocks remanufactured INNOVA plasma tubes for fast and reliable suppOl1 of your ion laser system.
****

Laser Institute Of America
12424 Research Parkway 11125
City: Orlando
State/Province: FL
Postal Code: 32826
Country: United States of Alflt!l'ica
Voice Phone: (1) 407 380 1553
Fax Phone: (1) 407 380 5588
email: lia@mail.creol.ucf.edu
Web: http://www.laserinstitute.org/
Contact #1: Jackie Thomas
Business Description: Laser safety training courses. Publishes ""Journal of Laser Applications"". Hosts annual International Congress on Applications of Lasers and Electro-Optics (ICALEO), including holographic applications and International Laser Safety Conference. Call for membership details and publication catalog.
****

Laser International
19 Nonnanton Rise
Holbeck Hill
City: Scarborough, N Yorks
State: England
Postal Code: YOll 2XE
Country: United Kingdom
Voice Phone: (44) 172 3364452
Contact #1: Keith Dutton
Business Description: Specializing in laser display systems for exhibitions. Auto & manual control avai lable. Full range of prices and features.
****

Laser Las Vegas
5725 N. Fort Apache
City: Las Vegas
State/Province: NV
Postal Code: 89129
Country: United States of America

Voice Phone: (1) 702 645 0477
Fax Phone: (1) 7026450477
email: laserlv@ao1.com
Contact # 1: Bill Aymar
Business Description: Laser sa les, repairs and rentals. Specialists in high power laser systems.
****

Laser Light Designs
2412 Kennedy Way
City: Antioch
State/Province: CA
Postal Code: 94509
Country: United States of America
Voice Phone: (1) 510 754 3144
Contact # 1: Michael Malott
Business Description: New product designs using embossed foil, tinsel and holographic films. I specialize in jeweled novelty designs. Designer of the original Rainbow Flasher, inventor of original Rainbow Sparkler.
****

Laser Light Ltd.
28 Old Fulton SI.
City: Brooklyn Heights
State/Province: NY
Postal Code: 11201
Country: United States of America
Voice Phone: (1) 212 226 7747
email: ddt-laser@ao1.com
Contact #1: Abe Rezny
Business Description: All formats of holography. Designers and producers of 3-D imaging.
****

Laser Media, Inc.
6383 Arizona Circle
City: Los Angeles
State/Province: CA
Postal Code: 90045
Country: United States of America
Voice Phone: (1) 310 338 9200
Fax Phone: (1) 310 338 9221
email: lmilasers@aol.com
Contact # 1: Kevin McCarthy
Business Description: Custom design of entertainment lighting and multimedia installations including la se rs, waterworks, holograms, etc. Specialize in large corporate presentations.
****

Laser Optics, Inc.
III Wooster St.
City: Bethel
State/Province: CT
Postal Code: 0680 I
Country: United States of America
Voice Phone: (1) 2037444160
Fax Phone: (1) 2037987941
Contact # I : Jim Larim
Business Description: A complete line of laser and optical components for ultraviolet, visible and infrared applications from 250 nm to 16 microns, including focusing lenses, windows, cavity components, prisms, beamsplitters, mirrors and coatings.
****

Laser Reflections
589 Howard SI.
City: San Francisco
State/Province: CA
Postal Code: 94105
Country: United States of America
Voice Phone: (1) 4158965958
Fax Phone: (1) 4158965171
email: hologram@access.com.com
Web: http://www.access.com.comi-hologram
Contact # 1: Ron Olson

INTERNATIONAL DIRECTORY

Business Description: Highest quality holographic portraiture. Advanced Nd: YAG laser recording technology producing visibly superior holograms of living subjects up to 14" x 24". Self-standing displays and custom installations. Unique commercial applications for signage and advertisements. Extensive portfolio of limited fine art editions. Stock images available on silver halide and photopolymer.
SEE OUR ADVERTISEMENT
****

Laser Resale Inc.
54 Balcom Road
City: Sudbury
State/Province: MA
Postal Code: 01776
Country: United States of America
Voice Phone: (1) 508 443 8484
Voice Phone: (1) 978 443 8484
Fax Phone: (1) 508 443 7620
email: LaseResale@aol.com
Web: http://www.laserresale.com
Contact #1: Jack Kilpatrick
Business Description: Laser Resale provides a marketplace for buying and selling pre-owned lasers, laser systems, optical tables and associated equipment for holographers.
****

Laser Technical Services
1396 River Road, Box 248
City: Upper Black Eddy
State/Province: PA
Postal Code: 18972
Country: United States of America
Voice Phone: (1) 610 982 0226
Fax Phone: (1) 610 982 0226
email: lasertek@ptd.net
Contact # 1: Dan Morrison
Business Description: Technical consultant and field repair of lasers. Full customer service of laser equipment - Specifically Pulsed Ruby holographic lense"s. Specialize in Lumonics lasers
SEE OUR ADVERTISEMENT
****

Laser Technology, Inc.
1055 West Germantown Pike
City: Norristown
State/Province: PA
Postal Code: 19403
Country: United States of America
Voice Phone: (1) 610 631 5043
Fax Phone: (1) 610 631 0934
Contact # 1: John Newman
Business Description: Manufacture equipment for laser-based NOT; Holography and Shearography equipment and inspection services. Portable and production units available.
****

Laser-Solution-Management
Ingenieur-Biiro fur Laser-Systeme
Barkhorstruecken Sa
City: Essen
Postal Code: 0-45239
Country: Germany
Contact # 1: Vera Hartmann
Business Description: Development of lasers for air traffic control, light-shows and holography
****

Laserfilm Eckard Knuth
Milchstr. 12
City: Munich
Postal Code: 0-81667
Country: Germany
Voice Phone: (49) 89 480 77 14
Fax Phone: (49) 89 48 56 66

Contact # 1: Eckard Knuth
Business Description: 120 / 360 degree Multiplex holograms (stereograms). Moving images. Diameter 45 cm or 65 cm. Most representative work in 1996 - a 360 degree hologram with an imperial crown in original size for the exhibition "Austria 996 -1996".
****

LaserMax, Inc.
3495 Winton Place,Building B
City: Rochester
State/Province: NY
Postal Code: 14623
Country: United States of America
Voice Phone: (1) 716 272 5420
Fax Phone: (1) 716 272 5427
email: blarabee@lasermax-inc.com
Web: http://www.lasermax-inc.com
Business Description: LaserMax is a manufacturer of diode laser systems and is the industry leader for reliability and technical support. Our engineers have drawn upon years of experience in laser design and manufacturing in order to develop the most robust and complete miniature diode drive circu its currently on the market. We offer the most advanced optical systems and circuitry, capable of operation in harsh environments from space flight to mining.
****

Lasermetrics, Inc.
(a division of Fastpulse Technology, Inc.)
220 Midland Ave.
City: Saddle brook
State/Province: NJ
Postal C. 07663
Country: United States of America
Voice Phone: (1) 2014785757
Voice Phone: (1) 800449 FAST
Fax Phone: (1) 2014786115
Contact # 1: Robert Goldstein
Business Description: Laser components and electronic drivers for lasers.
****

Laserworks
PO Box 2408
City: Orange
State/Province: CA
Postal Code: 92859
Country: United States of America
Voice Phone: (1) 714 832 2686
Fax Phone: (1) 714 832 1451
email: 110344.2454@compuserve.com
Web: http:\\www.laser-works.com
Contact # 1: Selwyn Lissack
Business Description: Company manufactures programmable laser scanning equipment for sign age and display applications. Selwyn Lissack has been a holographic artist and researcher since 1969. He has produced numerous holographic exhibitions in addition to the ""Salvador Dali "" holograms.
****

Lasing S.A. ,
Marques De Pico Velasco 64
E-28027
City: Madrid
Country: Spain
Voice Phone: (34) 0 I 268 3643
Business Description: Branch office of Newport Corporation
****

Lasiris Inc.
Main Office
3549 Ashby
City, Prov.: Ville St Laurent, Quebec

Postal Code: H4R 2K3
Country: Canada
Voice Phone: (1) 514 335 1005
Voice Phone: (1) 800 814 9552
Fax Phone: (1) 514 335 4576
email: sales@lasiris.com
Web: http ://www.lasiris.coml
Contact # 1: Alain Beauregard
Business Description: HOE optics in stock gratings. HOE special projects. Beamsplitters. Laser pattern projectors for industrial inspection and machine vision.
****

Lauk & Partner GmbH
(a division of Lauk Kommunikation)
Augustinusstr 9B
City: Frechen
Postal Code: 0-50226
Country: Germany
Voice Phone: (49) 02234 51055-66
Fax Phone: (49) 02234 65019
email: 101624331 @compuserve.com
Contact # 1: Mathias Lauk
Business Description: Holograms, holographic projects, hologram museum.
****

Lawrence Berkeley Laboratory
University Of California
I Cyclotron Road
City: Berkeley
State/Province: CA
Postal Code: 94720
Country: United States of America
Voice Phone: (1) 5 I 0 486 4000
Business Description: Industrial & academic holography research. Will do commercial research projects.
****

LAZA Holograms Ltd
68 Katesgrove Lane
City: Reading
State/Province: England
Postal Code: RG I 2ND
Country: United Kingdom
Voice Phone: (44) 1734391 731
email: sjkyle@cableo1.co.uk
Contact # 1: Stephen Kyle
Business Description: Large range of stock holograms available in silver halide and soon to be photo-polymer New image range being introduced presently. Supplier of both shock spex and holographic fun glasses in large and small quantities.
****

Lazart Holographics
22 Erina Valley Road
City: Erina
State/Province: NSW
Postal Code: 2250
Country: Australia
Voice Phone: (6 1) 2 4367 6245
Fax Phone: (61) 2 4367 2306
email: lazart@ozemail. com.au
Web: http://www.acay.com.aU/-lazartl
Contact # 1: Brett Wilson
Business Description: Artistic holography; buying & selling holograms. Wholesale distribution and retail sales of artist editions and stock images. Production of jewelry items and novelty products from embossed images. Gallery exhibition open 7 days.
****

Lenox Laser
12530 Manor Road
City: Glen Arn1

State/Province: MD
Postal Code: 21057
Country: United States of America
Voice Phone: (1) 410 592 3 106
Fax Phone: (1) 410 592 3362
email: sa les@lenoxlaser.com
Web: http://www.lenoxlaser.com
Contact # 1: Joseph P. D'Entremont
Business Description: Laser-systems laboratory specializing in laser drilling and etching. Manufactures prefabricated aperture kits and pinholes suitable for precision industrial/optical applications. -Also se lls used laser systems and low cost optical kits for holography.
****

Leonhard Kurz GmbH
Schwabacher Strasse 482
City: Fuerth
Postal Code: 0-90763
Country: Germany
Voice Phone: (49) (0)9 11 714 10
Fax Phone: (+49) (0)91 1 7141507
Contact # 1: Werner Reinhart
Business Description: Manufacturer of embossing equipment; broker for hologram embossing.
****

Leseberg, Dr. Detlef
Kamener Str 172
City: Lunen-Beckinghausen
Postal Code: 0-4670
Country: Germany
Voice Phone: (49) (0)230 1794
Fax Phone: (+49) (0)230 1793
Contact #1: Detlef Leseberg
Business Description: Scientific holography research, HOE,computer-generated holography.
****

Letterhead Press, Inc.
W226 N880 Eastmound Drive
City: Waukesha
State/Province: WI
Postal Code: 53 186-1 689
Country: United States of America
Voice Phone: (1) 4145741717
Fax Phone: (1) 414 574 1718
email: let@execpc.com
Contact # 1: Mike Graf
Business Description: Full service trade finisher with 24-hour, 7-days/week manufacturing. Featuring
19 x 25 inch and 40 inch formats for holographic stamping. Complete projects from
print to fina l bindery assuring single-source responsibility.
SEE OUR ADVERTISEMENT
****

Lexel Laser, Inc.
48503 Milmont Drive
City: Fremont
State/Province: CA
Postal Code: 94538
Country: United States of America
Voice Phone: (1) 510 770 0800
Voice Phone: (1) 800 527 3795
Fax Phone: (1) 510 65 16598
email: lexel@aol.com
Contact # 1: Ben Graham
Business Description: Lexel produces the highest quality Argon, Krypton and mixed gas laser systems. In particular, Lexel specializes in production
of single frequency systems which are very stable over a variety of environmental situations.
****

LichtBlicke ClaBen & Voss
Strahlenberger Weg 8
City: Frankfurt
Postal Code: 0-60599
Country: Germany
Voice Phone: (49) 69 628492
Fax Phone: (+49) 69 610176
Contact # 1: Walter ClaBen
Business Description: Custom made holograms, small exhibition
****

LiCONiX
3281 Scott Boulevard
City: Santa Clara
State/Province: CA
Postal Code: 95054
Country: United States of America
Voice Phone: (1) 408 496 0300
Voice Phone: (1) 800 825 2554
Fax Phone: (1) 408 492 1303
email: sales@liconix.com
Web: http: \\www.liconix.com
Contact # 1: Mark Dowley
Business Description: LiCONiX has long been recognized as the leader in Helium Cadmium laser technology. LiCONiX offers the worlds most extensive range of HeCd lasers for commercial, OEM and scientific users. Small, light and robust, LiCONiX HeCd lasers require little maintenance, no water cooling and have lifetimes up to 10,000 hours.
SEE OUR ADVERTISEMENT
****

Light Dimension, Inc.
Sunfami ly Hongo #403
5-10, Hongo 4-Chome, Bunkyo-ku
City: Tokyo
Postal Code: I 13
Country: Japan
Voice Phone: (8 1) 3 3812 9201
Fax Phone: (81) 338129422
Contact #1: Mariko Oishi
Business Description: Handling the whole range of holograms/holographic products from embossed holograms to fine art images; also focusing on exhibitions on holography.
****

Light Fantastic
(See Optical Security Group - UK)
Voice Phone: (44) 1420488000
****

Light Impressions International, Ltd.
430 West Diversey Parkway - Suite 501
City: Chicago
State/Province: IL
Postal Code: 60614
Country: United States of America
Voice Phone: (1) 773 665 1579
Fax Phone: (1) 773 665 1679
email: postbox@lightimpressions.co.uk
Web: http ://www.lightimpressions.co.uk
Contact # 1: Pamela Jamison
Business Description: A full service manufacturer offering the highest quality custom mastering. Embossed product available in hot stamp foil or pressure sensitive labels for security or promotional application. Stock images and holographic equipment available.
****

Light Impressions International, Ltd.
5 Mole Business Park 3
City: Leatherhead, Surrey
State: England
Postal Code: KT22 7BA
Country: United Kingdom

Voice Phone: (44) 1372 386 677
Fax Phone: (44) 1372 386 548
email: postbox@lightimpressions.co.uk
Web: http ://www.lightimpressions.co.ukl
Contact # 1: John Brown
Business Description: A full service manufacturer offering the highest quality custom mastering. Embossed product available in hot stamp foil or pressure sensitive labels for security or promotional application. Stock images and holographic equipment available.
****

Light Wave Gallery
North Pier
435 East JIlinois Street
City: Chicago
State/Province: IL
Postal Code: 60611
Country: United States of America
Voice Phone: (1) 312 321 1123
Voice Phone: (1) 888 873 5473
Fax Phone: (1) 312 3210892
email: Lightwv@aol.com
Web: http://www.hologramsource.com
Contact # 1: Jim Harden
Business Description: Gallery, retail shop. Complete line of holograms and related holographic giftware. Extensive collection of limited edition artworks available for sale or rental. Full color holograms for sale.
****

Lightgate, Ltd
Novy Svet 21
P.O.Box 2,
Post 012 HRAD
City: Prague
Postal Code: 11900
Country: Czech Republic
Voice Phone: (42) 2 352325
Fax Phone: (42) 2 352479
email: light@boohem-net.cz
Contact # 1: Jana Vancurova
Business Description: Description of services:
1) Mastering of si lver halide display holograms a copies on classic materials up to 50 x 50 cm.
2) Mastering for embossed holograms for copies up to 15 x 15 cm oflabel.
3) Stock holograms of Lightgate and European partners.
4) Manufacturing of embossed holograms, label and hot foil s.
****

Lightrix, Inc.
2132 Adams Avenue
City: San Leandro
State/Province: CA
Postal Code: 94577
Country: United States of America
Voice Phone: (1) 510 577 7800
Fax Phone: (1) 510 577 78 16
email: dir@lightrix.com
Web: http://www.lightrix.com
Contact # 1: Deborah Robinson
Business Description: Lightrix manufactures and designs high quality holographic toys, gifts and wall decor. Lightrix offers state of the art holographic product development and graphic design. Lightrix holograms are available to the wholesale trade. We offer a full custom program for embossed and photopolymer holograms.
SEE OUR ADVERTISEMENT
****

LightVision Confections
8 Faneuil Hall
City: Boston
State/Province: MA

INTERNATIONAL DIRECTORY

Postal Code: 02109
Country: United States of America
Voice Phone: (1) 513 469 0330
Fax Phone: (1) 513 489 8222
email: info@lightvis ion.com
Web: http ://www. lightvision.com
Business Description: LightVisions markets a
process that embosses holographic images in
candy. The process is owned by Dimensional
Foods Corp.
****

Lightwave
1161 San Antonio Road
City: Mountain View
State/Province: CA
Postal Code: 94043
Country: United States of America
Voice Phone: (1) 650 526 1249
Contact # 1 : Tom Steele
Business Description: Manufacturers of lasers.
****

Likom
Poslovni Center Ledina
Kotnikova 5
City: Ljubljana
Country: Slovenia
Business Description: Holograms on display and
for sale.
****

Linda Law Holographies
P.O. Box 434
City: Centerport
State/Province: NY
Postal Code: I 1721
Country: United States of America
Voice Phone: (1) 5 16 754 6121
Fax Phone: (1) 516 754 9227
email: llholo@i-2000.com
Contact # 1 : Linda Law
Business Description: State of the art computer
graphics for hologr'aphy. Using Mac and SGI
computers, artwork can be created for mass pro-
duction holograms or large scale disp lay holo-
grams.
SEE OUR ADVERTISEMENT
****

London Holographic Image Studio
9 Warple Mews
Warple Way
City: London
State/Province: England
Postal Code: W3 ORF
Country: United Kingdom
Voice Phone: (44) 181 7405322
Fax Phone: (44) 1817401733
email: easyhologram@easy net
Contact # 1 : Martin Richardson
Business Description: Commissioned holograms
up to 1 x 2 meters, pulse-portraiture, movie ste-
reograms and mass production of silver halide ho-
lograms. Catalogues on request.
****

Lone Star Illusions
2901 Capital Of Texas Highway, # 191
City: Austin
State/Province: TX
Postal Code: 78746
Country: United States of America
Voice Phone: (1) 512 328 3599
Fax Phone: (1) 5 12 328 3599
Contact # 1 : Alan Li fshen
Business Description: Austin's only hologram
gallery and retail shop which features a full range
of holographic giftware and related optical nov-
elties.
****

Loughborough Univ. Of Tech.
Dept Of Physics - Dept of Mechanical Engineering
City: Loughborough, Leicestershire
State: England
Postal Code: LEI I 3TU
Country: United Kingdom
Voice Phone: (44) 1509 263 171
Fax Phone: (44) 1509219702
Web: http://www.lboro.ac.uk
Business Description: Scientific and industrial
research.
****

Louis Paul Jonas Studios. Inco
304 Mi ller Road
City: Hudson
State/Province: NY
Postal Code: 12534
Countly: United States of America
Voice Phone: (1) 518 851 2211
Fax Phone: (1) 518 8512284
email: dmerritt@epix. net
Contact # 1: Dave Merritt
Business Description: Jonas Studios specializes in making models and miniatures for the
museum and film industries . Services include:
sculpting, model making, dioramas, EDM machining, CAD for rapid prototyping, laser cutting, mold making and casting in a wide variety
of materials.
****

Lumenx Technologies, Inc.
PO Box 219
City: New Durham
State/Province: NH
Postal Code: 03855
Country: United States of America
Voice Phone: (1) 603 8593800
Fax Phone: (1) 603 859 250 I
Contact #1: Ed Neister
Business Description: We manufacture and re-
pair a variety of laser equipment, mostly for sci-
entific and medical applications. Call for more
details.
****

Luminer Printing and Converting
1400 Industrial Way
City: Tom's River
State/Province: NJ
Postal Code: 08755
CountlY: United States of America
Voice Phone: (1) 732 341 5727
Fax Phone: (1) 732 341 6175
Busi ness Description: Innovative printer and
converter; labe ls and promotional materials.
Expertise and technology for adhesive coating
imaged holographic materials, including zone
and patterned areas. Overprint, laminate, fold
multiple webs. Complete design and origination
services.
****

Lumonics Ltd.
Cosford Lane
Swift Valley
City: Rugby, Warwickshire
State: England
Postal Code: CV21 I QN
Country: United Kingdom
Voiee Phone: (44) 1788 570 321
Fax Phone: (44) 1788 579 824
Contact # 1 : George Synowiec
Business Description: Due to recent business
changes, see Laser Technical Services in USA
for sa les and service of Lumonics pulsed ruby
lasers. Also see Laser Technical Services ad in
this book,

Lumonics manufactures a variety of lasers and
laser based systems for industrial applications.
These include a range of pulsed ruby lasers specifically designed for holography with output energies spanning from 30 mJ per pulse to greater
than 10 Loules per pulse.
****

Lund Institute Of Tech
Department Of Physics
Box 118
City: Lund
Postal Code: S-221
Country: Sweden
Voice Phone: (46) 046 222 7656
Fax Phone: (46) 046 222 4017
email: seven-goran.pattersson@fysik.lth.se
Business Description: Color H-I ; holography edu-
cation; academic research.
****

M.IT (Massachusetts Institute of Technology)
Spatial Imaging Group
20 Ames Street # E 15-416
City: Cambridge
State/Province: MA
Postal Code: 02139 - 4307
Country: United States of America
Voice Phone: (1) 617 2538145
Fax Phone: (1) 617 2538823
email: sab@media.mit.edu
Web: http://www.media.mit.edu/groups/spi/
Contact # 1: Stephen A. Benton
Business Description: Education & research in-
cluding Computer generated holography research;
Holographic hard copy printer research.
****

M.l.T. Museum
265 Massachusetts Ave.
City: Cambridge
State/Province: MA
Postal Code: 02139-4307
Country: United States of America
Voice Phone: (1) 617 253 4462
Fax Phone: (1) 617 253 8994
Web: http ://web.mit.edu/museum/home/
index.html
Contact # 1 : Diego Garcia
Business Description: Museum has approximately
50 holograms on display from a inventory of ap-
proximately 1,000 holograms. Some of the most
historically-significant holograms ever made are
on display here. Museum shop has holograms for
sale.
****

M.O.M. Inc.
2436 Forest Green Rd.
City: Baltimore
State/Province: MD
Postal Code: 21209
Country: United States of America
Contact #1: Alan Evan
Business Description: Maryland Optical Manufac-
turing. Highest quality. 39 years experience.
****

MacShane Holography
C/O Laser Arts Productions
512 West Braeside Drive
City: Arlington Heights
State/Province: IL
Postal Code: 60004-2060
Country: United States of America
Voice Phone: (1) 847 398 4983
Contact #1: Jim MacShane
Business Description: MacShane Holography/ La-
ser Arts Productions goes to schools and sets up
holography class programs.
****

M

Magic Laser
105 r Moines
City: Paris
Postal Code: F-750 17
Country: France
Voice Phone: (33) I 40 33 17 49
Business Description: Importer and wholesaler
of all holographic products - travelling exhibit.
****

Magick Signs Holografie
Isenburg-Zentrum
Shopteil Wost
City: Neu-Isenburg
Postal Code: D-63263
Country: Germany
Voice Phone: (49) (0)610 328404
Business Description: Hologram retailer.
****

Magick signs Holografie
August-Bebel-Str. 40
City: Egelsbach
Postal Code: D-63329
Country: Germany
Voice Phone: (49) (0)610 45544
Fax Phone: (+49) (0)610 45548
Contact # 1: Andreas Wollenweber
Business Description: PRODUCTION of em-
bossed holograms: Artwork, models, origination,
embossing, many stock images. RETAIL in our
four holographic Magick gift stores (Frankfurt).
****

Man/Environment, Inc.
2251 Federal Avenue
City: Los Angeles
State/Province: CA
Postal Code: 90064
Country: United States of America
Voice Phone: (1) 310477 7922
Voice Phone: (1) 310 477 8960
Fax Phone: (1) 310 477 4910
email: metaphor@ix.netcom.com
Web: http:\\www.armchair.com
Contact #1: Gary Fisher
Business Description: Silver halide and photo-
polymer R&D projects. Design and manufacture
optical printers and holographic systems. Com-
plete website development. Check our website for
additional information.
****

Mario Liedtke Pro Design
PixelerstraBe 34
City: Rheda-Wiedenbrueck
Postal Code: D-33378
Country: Germany
Contact #1: Mario Liedtke
Business Description: modelling for hoiograpil.
****

Marks, Gerald
29 West 26th Street
City: New York
State/Province: NY
Postal Code: 100 I 0
Country: United States of America
Voice Phone: (1) 212 889 5994
Fax Phone: (I) 212 889 5926
email: pulltime3d@aol.com
Web: http: //www.vision3d.com/pulltime3d/
Contact # 1: Gerald Marks
Business Description: Artist specializing in ste-
reoscopic 3D of every type for over twenty years.
He is best known for the 3D music videos he cre-
ated for the Rolling Stones and his 3D museum
exhibits, anaglyph prints and books, lenticulars,
random dot stereograms, computer multimedia
and computer generated holography.
****

Marubun Corporation
8-1 Nihonbashi
Odenmacho, Chuo-ku
Tokyo
Postal Code: 103
Country: Japan
Voice Phone: (81) 3 3639 9872
Fax Phone: (8 1) 336398156
Web: http: //www.newport.com
Business Description: Branch office of Newport
Corp. , Fountain Valley CA, USA.
****

Mazda Motor Corp.
Technical Research Center
POBox 18
City: Hiroshima
Postal Code: 730 91
Country: Japan
Voice Phone: (81) 082 282 11 I I
Fax Phone: (81) 082 252 5343
Web: http://www.mazda.com/
Business Description: Holographic Interferom-
etry
****

McMahan Electro-Optic
2160 Park Avenue
(Orlando Division)
City: Winter Park
State/Province: FL
Postal Code: 32789
Country: United States of America
Voice Phone: (1) 407 645 1000
Fax Phone: (1) 407 644 9000
email : bobmcmahn@aol.com
Contact # 1: Robert McMahan
Business Description: McMahan Electro-Optics
manufactures a laser-based NDT system for test-
ing composite aerospace components and assem-
blies. Mobi le unit.
****

Media Interface, Ltd. _
215 Berkeley Place
City: Brooklyn
State/Province: NY
Postal Code: 11 217
Country: United St~tes of America
Voice Phone: (1) 7 18 3981136
Fax Phone: (1) 718 3981136
email: ronholog@bway.net
Web: bway.net/-ronholog
Contact # 1: Ronald R. Erickson
Business Description: Consulting in holographic
applications and mass produced and commer-
cial holographic image design and production.
Medical holography. Custom holographic optical
configurations. Computer assisted holographic
image design and production - research or com-
mercial.
****

Mefoma Fototechnik GmbH
Ulmer Stral3e I
City: Elchingen
Postal Code: D-73450
Country: Gernlany
Voice Phone: (49) 731 266355
Business Description: Equipment for holographic
development processes
****

Melissa Crenshaw Holography Studio Interna-
tional
Jl RRI
City, Prov.: Bowen Island, B.C.
Postal Code: VON I GO
Country: Canada
Voice Phone: (1) 604 645 2019

Voice Phone: (1) 604 645 20 19
email : mcrensha@capcollege.bc.ca
Web: http://www.capcollege.bc.ca/dept/physics/
Contact # 1: Melissa Crenshaw
Business Description: Holographic fine artist
with experience integrating holographic elements into commercial projects, including architectural and lighting design. Color reflection
hologram mastering and mass production (l2x
16, 4x5 film) services offered. Extensive portfolio.
****

Melles Griot
1770 Kettering Street
City: Irvine
State/Province: CA
Postal Code: 92714
Counl ly: United States of America
Voice Phone: (1) 714261 5600
Voice Phone: (1) 800 645 2737
Fax Phone: (1) 7 14 261 7589
email: mgtech@irvine.mellesgriot.com
Web: http:\\www.mellesgriot.com
Business Description: Melles Griot, worldwide
laser and photonics components manufacturer,
offers a broad spectrum of helium neon, helium
cadmium, argon ion and krypton argon ion lasers covering the blue to infrared range. Ideal
for material analysis, testing and inspection, as
well as interferometric measurement, surface
inspection, scatter measurement, holography,
high resolution metrology, optical disk mastering and more.
SEE OUR ADVERTISEMENT
****

Melles Griot GmbH
Lilienthalstrasse 30-32
City: Bensheim
Postal Code: D-64625
Country: Germany
Voice Phone: (49) (0)625 8406-0
Fax Phone: (+49) (0)625 8406-22
Contact # 1: Daniel Hinz
Business Description: Condensers (optics), fiber
optics construction components, laser diodes,
laser optics,optical filters ,optical lenses,optical
mirrors, optical parts, bulk, optoelectronic com-
ponents, planar optics, planar parallel optics,
prisms.
SEE OUR ADVERTISEMENT
****

Melles Griot Laser Group
2251 Rutherford
City: Carlsbad
State/Province: CA
Postal Code: 92008
Country: United States of America
Voice Phone: (1) 760 438 2131
Voice Phone: (1) 800 645 2737
Fax Phone: (1) 760438 2131
email: sales@carlsbad.mellesgriot.com
Web: http:\\www.mellesgriot.com
Contact # 1: Lisa Tsufura
Business Description: Melles Griot, worldwide
laser and photonics components manufacturer,
offers a broad spectrum of helium neon, helium
cadmium, argon ion and krypton argon ion lasers covering the blue to infrared range. Ideal
for material analysis, testing and inspection, as
well as interferometric measurement, surface
inspection, scatter measurement, holography,
high resolution metrology, optical disk mastering and more.
SEE OUR ADVERTISEMENT
****

INTERNATIONAL DIRECTORY

Menning, Melinda
171 Hopetoun Ave., Vaucluse
City: Sydney
State/Province: NSW
Postal Code: 2030
Country: Australia
Voice Phone: (61) 2 9337 1916
email: mmenning@mpce.mq.edu.au
Web: http ://www.mpce.mq.edu.au
Contact #1: Melinda Menning
Business Description: Practicing Artist. Holographic Consultant. Producing medium format limited edition Art pieces. Producing Laser Transmission masters suitable for mass replication via photopolymer and embossed material. Producer of White light reflection and transmission Holograms for display purposes.
****

Meredith Instruments
5035 North 55th Avenue
Suite 5
City: Glendale
State/Province: AZ
Postal Code: 8530 I
Country: United States of America
Voice Phone: (1) 602 934 9387
Fax Phone: (1) 602 934 9482
email: sales@lasersl.com
Web: www.mi-Iasers.com
Contact #1: Lee Toland
Business Description: Specializing in surplus inventories of HeNe lasers as well as argon and diode lasers, Meredith Instruments is the USA's largest laser discount dealer. Free catalogue. Laser repair.
****

Merrick, Michael G.
1605 Bensington Court
City: Normal
State/Province: IL
Postal Code: 61761-4811
Country: United State~ of America
Voice Phone: (1) 309 452 5228
Contact #1: Michael Merrick
Business Description: Production of master, image plane and reflection holograms using psuedocolor and shadowgram techniques. Experience giving educational programs to elementary and adult groups. Also coordinating holography shows for fundraising . Currently working with children's museum.
****

MesMerized
P.O. Box 295
City: Mount Kisco
State/Province: NY
Country: United States of America
Voice Phone: (1) 914 244 0716
Fax Phone: (1) 914 244 1995
email: sales@mesmerized.com
Contact # 1: Jeffrey Levine
Business Description: Complete design and manufacturing of custom holograms and exclusive holographic products for use in corporate promotions.
MesMerized also specializes in custom hologram P.O.P/P.O.S. displays using exclusive holographic techniques.

****

MetroLaser Inc.
18010 Skypark Ave.
City: Irvine
State/Province: CA
Country: United States of America
Voice Phone: (1) 714 553 0688

Fax Phone: (1) 714 553 0495
email: general@metrolaserinc.com
Web: http ://users.deltanet.com/- metro
Contact #1: Cecil Hess
Business Description: Hi-tech R&D co specializing in optical measurements & diagnostics using lasers. Products include particle sizers, vibrometers, spectroscopic systems, holocameras, holographic NDT, ultra high resolution interferometers.
****

Metrologic Instruments GmbH
Dornierstrasse 2
City: Puchheim
Postal Code: D-82178
Country: Germany
Voice Phone:(49) (0)89 89019 0
Fax Phone (+49) (0)89 89019 200
Contact # 1: Benny Noems
Business Description: see Metrologic Instruments Inc., USA. Please fax us.
****

Metrologic Instruments, Inc.
Coles Road at Route 42
City: Blackwood
State/Province: NJ
Postal Code: 08012
Country: United States of America
Voice Phone: (1) 609 228 8100
Voice Phone: (1) 800 IDMETRO
Fax Phone: (1) 609 228 6673
email: ckendall@metrologic.com
Web: http ://www.metrologic .coml
Contact # 1: Christen Kendall
Business Description: Metrologic Instruments manufacturers holographic laser bar code scanners or HoloTrak (TM). The Holotrak is omnidirectional with a 40 inch depth of field and 26 inch scan width. Applications include pallet scanning, unattended scanning, truck unloading and conveyor belts.
****

Meulien Odile
Mergesst. 16
City: Braunschweig
Postal Code: D-38108
Country: Gern1any
Voice Phone: (49) (0)531 352 816
Fax Phone: (+49) (0)531352816
Contact #1: Odile Meulien
Business Description: Collector and analyst of the holographic trend since 10 years in the US and Europe. Manage a private collection - Conduct studies on future uses of holography - Coordinate holographic and light happenings.
****

MGM Converters Inc.
16604 Edwards Road
City: Cerritos
State/Province: CA
Postal Code: 90703
Country: United States of America
Voice Phone: (I) 562 404 3779
Fax Phone: (1) 562 404 7408
Contact # 1: Steve Meyer
Business Description: Full service converting services for the holography market. Foil hot stamping, including continuous application.
****

Midwest Laser Products
PO Box 262
City: Frankfort
State/Province: IL
Postal Code: 60423
Country: United States of America

Voice Phone: (I) 815 464 0085
Fax Phone: (1) 815 464 0767
email: mlp@nlenx.com
Web: http://www.midwest-Iaser.com
Contact # 1: Steve Garrett
Business Description: We carry HeNe, Argon, HeCd, Nd:YAG and visible diode lasers. We sell complete holography kits, including low-cost HeNe lasers and related materials. Lasers for holography starting under $100!
SEE OUR ADVERTISEMENT
****

Millennium Portraits
(see Spatial Imaging Limited)
Business Description: Spatial Imaging has a separate business, millennium Portraits, that is. devoted to pulsed portraits. Located at same address and phone as Spatial Imaging.
****

Miller, Neal
The Career Center
4203 South Providence Road
City: Columbia
State/Province: MO
Postal Code: 65203
Country: United States of America
Voice Phone: (1) 5738862610
Fax Phone: (1) 573 8862904
Contact # 1: Neal Miller
Business Description: Physics technology instructor. Optics related courses for vocational instuction.
****

Miller, Peter
136 Clinton Road
City: Newfoundland
State/Province: NJ
Postal Code: 07435
Country: United States of America
Voice Phone: (1) 973 697 1773
Contact # 1: Peter Miller
Business Description: Professional holographer - 20 years experience.
****

Ministry Of International Trade
Electrotechnical Laboratory
Optical Information Section
City: Tsukuba Science City
Postal Code: 305
Country: Japan
Voice Phone: (81) 0298 58 5625
Fax Phone: (81) 0298 58 5627
Contact # 1: Dr. Satoshi Ishihara
Business Description: Research using HOEs
****

Minjian Laser Holography
Anticounterfeit Label Factory
Tong-Bei-Lu #540
City: Shanghai
Postal Code: 200082
Country: China
Contact # 1: Wang Rumo
****

Mitsubishi Heavy Industries Ltd.
Nagasaki Technical Institute
1-1 Akunoura-Machi
City: Nagasaki
Postal Code: 850-91
Country: Japan
Voice Phone: (81) 3 3218 2111
email: info@mitsubishi.com
Web: http ://www.mitsubishi .coml
Contact #1: M. Murata

Business Description: Holographic non-destructive testing; industrial research.
****

Mitutoyo Measuring Instruments (MTI Corp.)
965 Corporate Blvd.
City: Aurora
State/Province: IL
Postal Code: 60504
Country: United States of America
Voice Phone: (1) 630 820 9666
Fax lhone: (1) 630 820 1393
Contact #1: Bill Naaman
Business Description: Manufacturers of precision measuring in struments including highly accurate holographic linear tracking systems suitable for precision industrial and research applications.
****

Modem Marketing
Olgastra/3e 2
City: Boennigheim
Postal Code: D-74357
Country: Germany
Voice Phone: (49) 7143 22909
Contact # 1: Wilfried Moedinger
Business Description: Marketing of holography in the area of youth education.
****

Molins PLC
13-13A Westeood Way
Westwood Business Park
City: Coventry
State/Province: England
Postal Code: CV4 8HS
Country: United Kingdom
Voice Phone: (44) 1203421 100
Fax Phone: (44) 1203421 255
Business Description: Makes foil applicator machines to work on currency and other paper substrates.
****

Moonbeamers
1/5 Gibbons Street
City: Telopea
State/Province: NSW
Postal Code: 2 I 17
Country: Australia
Voice Phone: (61) 2 9878 6427
Contact # 1: John Tobin
Business Description: Since 1984, we have been producing commercial holograms and diffractions for security and display applications. We offer a complete service from artwork creation through application & printing within Australia and Asia.
****

Morning Light Holograms
106 Xi Huan Middle Street
Cang Zhou
City: He Bei
Postal Code: 06 I 00 I
Country: China
Voice Phone: (86) 3 I 7 226 164
Fax Phone: (86) 317 226 167
Contact # 1: Chen Guo Tong
Business Description: Hologram di stributor
****

Mu's Laser Works
1328 Dunsterville Avenue
City, Prov.: Victoria, B.C.
Postal Code: V8Z 2X I
Country: Canada
Voice Phone: (1) 250 479 4357
Contact # 1: Ron Meuse

Business Description: Holographic and 3-D photographic services, Can provide lab rental and technical assistance. Laser light show production and rental.
****

Mulhem, Dominique
1, Residense les Camelias,
7, rue du 18 Juin 1940
City: Asnieres
Postal Code: F-92600
Country: france
Voice Phone: (33) 1 47 94 82 42
Fax Phone: (33) I 47948242
email: dwm@mail.dotcom.fr
Web: http://www.alphapix.com
Business Description: Holographic artist.
****

Multiplex Moving Holograms
746 Treat Street
City: San Francisco
State/Province: CA
Postal Code: 94110
Country: United States of America
Voice Phone: (1) 4 I 5 285 9035
Fax Phone: (1) 4152061622
Contact #1: Peter Claudius
Business Description: We are the originators of the Multiplex Hologram. We produce white light viewable moving holograms for trade shows and exhibits. 120, 360 degree and flat fonnat white light viewable holograms made to your specifications. Stock images also available. Ask for our catalogue ! In business since 1973!
****

Mlinchner Volkshochschule e. V.
Am Gasteig, Kellerstra13e 6
City: Muenchen
Postal Code: D-81667
Country: Germany
Voice Phone: (49) 89 41 8060
Business Description: Sf:Il;l4i1ars on holography
****

Museu D' Holografia
Jaume I, I
City: Barcelona
Postal Code: 08002
Country: Spain
Voice Phone: (343) 3 102 172
Fax Phone: (343) 3 319 1676
email: museuholos@mx3.redestb.es
Web: museuholos
Business Description: Holographic Gallery, Itinerant exhibitions, sale of holograms. Teaching. Holographic courses. Holographic laboratory.
****

Museum 3. Dimension
StadtmuhlelNoerdlinger Tor
City: Dinkelbuehl
Postal Code: D-91550
Country: Gennany
Voice Phone: (49) 9851 6336
Fax Phone: (+49) 69 787777
Business Description: Museum for holography and stereography
****

Museum fUr Holographie & neue visuelle Medien
Pletschmuhlenweg 7
City: Pulheim
Postal Code: D-50259
Country: Germany
Voice Phone: (49) 223851053
Fax Phone: (+49) 2238 52158
Contact # 1: Matthias Lauk

Business Description: Museum for holography, one of the worlds largest private collection of holograms
****

Museum Of Holography Chicago
1134 West Washington Blvd.
City: Chicago
State/Province: lL
Postal Code: 60607
Country: United States of America
Voice Phone: (1) 312 226 1007
Fax Phone: (1) 312 829 9636
email: museum@concentric.net
Web: http://www.cris.coml-museumh
Contact # 1: Loren Billings
Business Description: Founded in 1978, the MOHC is now the world's oldest institution d voted to the display, acquisition and maintenance of holography as well as education and research in the field. Permanent collection is now the largest in the world. At least two major exhibitions a year featuring artists from around the world.
SEE OUR ADVERTISEMENT
****

MWK Industries
1269 West Pomona Road # 112
City: Corona
State/Province: CA
Postal Code: 91720
Country: United States of America
Voice Phone: (1) 909 278 0563
Voice Phone: (1) 8003567714
Fax Phone: (1) 909 278 4887
email: mwk@worldnet.att.net
Web: http ://www. mwkindustries.com/
Contact # 1: Mike Ket!.ny
Business Description: Large selection of surplus and used lasers from major manufacturers. Save 30% to 60% on brand name laser purchases. We offer the beginning, intennediate and advanced holographer a large selection of lasers suitable for holography (including HeNe lasers ranging up to 25 mw), as well as other related materials .
SEE OUR ADVERTISEMENT
****

Nakamura, Ikuo
864 President St. #3
City: Brooklyn
State/Province: NY
Postal Code: 11215
Country: United States of America
Voice Phone: (1) 718 636 9112
email: ikuo@spacelab.net
Web: http: //www.spacelab.net/-ikuo/
Contact # 1: Ikuo Nakamura
Business Description: Artist with portfolio.
****

Nanjing Sanle Laser Technology R&D
Nanjing Pukou New Technology Development Zone
PO Box 62
City, Prov.: Nanjing, Jiangsu
Postal Code: 210032
Country: China
Voice Phone: (86) 25 8840357
Fax Phone: (86) 25 3304991
Contact # 1: Lu Zhongming
****

National Physical Laboratory
Queens Road
City: Teddingron, Middlesex
State: England
Postal Code: TW II OLW
Country: United Kingdom
Voice Phone: (44) 181 977 3222

INTERNATIONAL DIRECTORY

Fax Phone: (44) 181 9432155
email: library@newton.npl.co.uk
Web: http://www.npl.co.uk
Contact # 1: David Robinson
Business Description: Scientific and industrial research; holographic non-destructive testing.
****

Navidec Inc. (formerly AC[Systems, Inc.)
14 Inverness Drive East
Suite F-116
City: Englewood
State/Province: CO
Postal Code: 80112
Country: United States of America
Voice Phone: (1) 303 790 7565
Voice Phone: (1) 800 797 7565
Fax Phone: (1) 303 790 8845
email: patt@navidec.com
Contact # 1: Patrick Townsend
Business Description: Exclusive agent (US, Canada, Mexico) for Kimmon Helium Cadmium laser systems used for holography.
****

NEC Electronics (Europe) GmbH
Oberrather Str. 4
City: Duesseldorf
Postal Code: 0-40472
Country: Germany
Voice Phone: (49) 211 65 03-0 I
Fax Phone: (+49) 211 65 03-488
Business Description: Manufacturer of lasers, distribution argon ion laser, he-ne laser
****

NeoVision Productions
PO Box 74277
City: Los Angeles
State/Province: CA
Postal Code: 90004
Country: United States of America
Voice Phone: (1) 21'3 387 0461
Contact #1: Bill Hillard
Business Description: Traveling show. Fine art originals. Produce holograms for home and industry. Consulting.
****

New Dimension Holographics
23 Victoria Ave.
City: Concord West
State/Province: NSW
Postal Code: 2138
Country: Australia
Voice Phone: (61) 2 9743 3767
Voice Phone: (61) 15435076
Fax Phone: (61) 2 9743 3241
Contact #1: Tony Butteriss
Business Description: Retail shop. Wholesale distribution. Origination consultant. Educational consultant. Gallery.
****

New Focus, Inc
2630 Walsh Ave.
City: Santa Clara
State/Province: CA
Postal Code: 95051-0905
Country: United States of America
Voice Phone: (1) 408 980 8088
Fax Phone: (1) 408 980 8883
email: Contact@NewFocus.com
Web: http://www.newfocus.com/
Contact # 1: Milton Chang
Business Description: New Focus is a supplier of photonics tools for laser applications. Products include narrow-linewidth tunable diode lasers, ultrafast photo detectors (DC-60 GHz),

electro-optic modulators, wavelength meters, mechanical positioners, motorized positioners, and high-performance optics.
****

New Light Industries
West 9713 Sunset Hwy.
City: Spokane
State/Province: WA
Postal Code: 99224
Country: United States of America
Voice Phone: (1) 509 456 8321
Fax Phone: (1) 5094568351
email: stevem@compch.iea.com
Web: www.iea.com/-nli
Contact # 1: Steve McGrew
Business Description: Extensive experience with technology transfer, consulting and R & D for em ossed holography. Complete origination and 1 production system installations, worldwide.
****

New York Hall Of Science
47-01 III Th Street Nimbus Manufacturing, Inc.
City: Corona
State/Province: NY
Postal Code: 11368
Country: United States of America
Voice Phone: (1) 718 699 0005
email: mweiss@nyhallsci.org
Web: http://www.nyhallsci.org/
Contact # 1: Beth Weinstein
Business Description: The New York Hall of Science is New York's only hands-on science and technology museum. Lasers and optics demonstrated daily. Color hologram depicting quantum atom is on display.
****

New York Holographic Laboratories
P.O. Box 20391
Thomkins Square Station
City: New York
State/Province: NY
Postal Code: 10009
Country: United States of America
Voice Phone: (I) 212 674 1007
Fax Phone: (1) 212 677 6304
email: dan@waena.edu
Web: http://waena.edul-danl9999a.htm
Contact #1: Dan Schweitzer
Business Description: Fine art editions. Tutorial courses, lectures and consultations.
****

Newport Corporation
1791 Deere Ave.
City: Irvine
State/Province: CA
Postal Code: 92606
Country: United States of America
Voice Phone: (1) 800 222 9980
Voice Phone: (1) 714 863-3144
Fax Phone: (1) 714 253 1800
email: sales@newport.com
Web: http://www.newport.com
Contact # 1: Gary Spiegel
Business Description: Designer and manufacturer of E/O components, optics, spatial filters, optical & beamsteering instruments, magnetic bases, fiber optic components, vibration isolation systems, and holographic recording matrls.
****

Newport GmbH
European Headquarters
Holzhofallee 19
City: Darmstadt
Postal Code: D-64295
Country: Germany

Voice Phone: (49) (0)615 362 10
Fax Phone: (+49) (0)6 15 362152
Business Description: Designer and manufacturer of laserlholographic systems, E/O components, optics , spatial filters , optical & beamsteering instruments, magnetic bases, fiber optic components, vibration isolation systems, and holographic recording materials.
****

Nihon University
Dept Electronic Engineering
7-24-1 Narashinodai
City: Funabashi-Shi
State/Province: Chiba
Postal Code: 274
Country: Japan
Voice Phone: (81) 0474 69 5391
Fax Phone: (81) 0474 67 9683
Contact #1: Dr. Hiroshi Yoshikawa
Business Description: Research NDT
****

Nimbus Manufacturing, Inc.
(a division of Nimbus CD International)
P.O. Box 7427
City: Charlottesville
State/Province: VA
Postal Code: 22906
Country: United States of America
Voice Phone: (1) 800 231 0778 x457
Fax Phone: (1) 804 985 4625
email: Ihaney@nimbuscd.com
Web: http://www.nimbuscd.com/
Contact # 1: Lorri Haney
Business Description: Nimbus manufactures holographic CDs, CD ROMs, CDls, enhanced CDs, and DVD's, as well as providing packaging, prepress, print procurement, and spine labels.
****

Nippon Polaroid K.K.
Business Development Division - Mori Bldg No. 30
3-2-2 Toranomon, Minato-ku
City: Tokyo
Postal Code: 105
Country: Japan
Voice Phone: (81) 03 3438 8883
Fax Phone: (81) 0354738637
Contact #1: Makoto ide
Business Description: Subsidiary of Polaroid Corp., Cambridge, MA USA
****

Nippondenso Co., Ltd.
System Development Engineering
I-I Showa-Cho Kariya-Shi
City: Aichi-Ken
Postal Code: 448
Country: Japan
Voice Phone: (81) 0566 256 924
Contact #1: Hiroshi Ando
Business Description: Manufacture HeadsUp Display. Also Mr. Toru Mizuno, Mr. Tatsuya Fujita, or Mr. Shinji Nanba.
****

Nissan Motor.
Central Research Lab
Natsushima Machi
City: Yokosuka
Postal Code: 237
Country: Japan
Voice Phone: (81) 0468 625 182
Fax Phone: (81) 046 654183
Business Description: Hologram manufacturer head-up display
****

Norland Products, Inc.
PO Box 7145
City: North Brunswick
State/Province: NJ
Postal Code: 08902
Country: United States of America
Voice Phone: (1) 908 545 7828
Fax Phone: (1) 908 545 9542
email: sales@norlandprod.com
Web: http://www.norlandprod.coml
Business Description: Optical adhesives (which cure with UV light). Used to adhere HOEs and for splicing fiber optic cables.
****

North Light Holograms Ltd.
PO Box 40
City, Prov.: Tai-an, Shandong
Postal Code: 271039
Country: China
Voice Phone: (86) 538 651100 I
Fax Phone: (86) 538 651100 I
Contact # 1: Wang Jizhang
Business Description: Holography sales and distribution.
****

Northern Illinois University
Department Of Physics
City: Dekalb
State/Province: IL
Postal Code: 60115
Country: United States of America
Voice Phone: (1) 815 753 1772
Contact # 1: Thomas Rossing
Business Description: Scientific holography research. Projects vary in nature. Holographic interferometry for studying vibration modes such as in musical instruments.
****

NovaVision
419 Gould Street
City: Bowling Green
State/Province: OH
Postal Code: 43402
Country: United States of America
Voice Phone: (1) 419 354 1427
Voice Phone: (1) 800 990 6682
Fax Phone: (1) 4193537908
email: nova@wcnet.org
Web: http://www.wcnet.org/-nova/
Contact # 1: Albert J. Caperna
Business Description: NovaVision (TM) direct holographic embossing technology allows printers to produce embossed holograms and other OVDs in- line, on-press at production speeds on a variety of substrate materials. Whereas conventional holograms aretransferred from preimage foil, the patented NovaVision proceifs allows holograms to be created as an integrafpart of the printing process ... no inventory, lower costs, higher application speed
****

Numazu College Of Technology
Dept Of Mechanical Engineering
360000ka
City: Numazu-City
State/Province: Shizuoka
Postal Code: 410
Country: Japan
Voice Phone: (81) 0559 212 700
Contact # 1: Dr. Koji Lkegami
Business Description: Holographic research.
****

NW Systems Ltd.
6 The Hatches
City: Farnham
State/Province: Surrey

Postal Code: GU9 8UE
Country: United Kingdom
Voice Phone: 44 252 711 060
Fax Phone: 44 252 711 060
Contact # 1: Francis Townsend
Business Description: Holographic consultants specializing in technology transfers. We also manufacture and market embossing-machines, light meters, electro forming equipment, shutter controllers, etc. to our client 's specification. Worldwide contacts.
****

Odhner Holographies
I New South Drive
City: Amherst
State/Province: NH
Postal Code: 03031
Country: United States of America
Voice Phone: (l) 603 673 8651
Fax Phone: (1) 603 673 8685
Contact # 1: Jefferson E. Odhner
Business Description: Exclusive distributor of the Stabilock II inch fringe stabilizer (used to make brighter holograms), manufacture of custom holograms (trans/refl.). Specializing in HOE arrays (to 8"" x 10"") on silver halide.
SEE OUR ADVERTISEMENT
****

Oeserwerk Ernst Oeser & Sohne KG
Rigistrai3e 20
City: Goeppingen-Holzheim
Postal Code: 0-73037
Country: Gernlany
Voice Phone: (49) 7161 8009-0
Fax Phone: (+49) 7161 8009-10
Contact # 1: Ernst Dr. Qeser
Business Description: embossed hologram foils, holographic labels.
****

Ojasmit Holographics
409 Vardhman Market ~tor 17, VASHI
City: New Bombay
Postal Code: 400703
Country: India
Voice Phone: (91) 22 768 3526
Voice Phone: (91') 22 763 0373
Fax Phone: (91) 22 763 2509
Contact # 1: Kailesh Shah
Business Description: Manufacturing and marketing of embossed holograms, photopolymers and dichromates for varied applications.
****

Ontario College of Art/Holography
Art Division
100 McCaul St.
City, Prov.: Toronto, Ontario
Country: Canada
Voice Phone: (1) 416 977 6000 X263
email: mpage@ocad.on.ca/
Web: http: //www.ocad.on.ca/
Contact #1: Michael Page
Business Description: Education & Seminars. Regular holography courses.
****

Ontario Science Centre
770 Don Mills Road
City, prov.: Don Mills, Ontario
Postal Code: M3C I T3
Country: Canada
Voice Phone: (1) 416 429 4100 x 2820
Fax Phone: (1) 416 696 3197
email: alena_kottova@fcgatel.osc.on.ea
Contact # 1: Alena Kottova
Business Description: We have gallery of 15 holograms on permanent display and laser dem- on-

stration area. Holography workshops cover theory and practical uses of holography. Participants make their own reflection hologram .
****

Op-Graphics (Holography) Ltd.
Unit 4 - Technorth
7 Harrogate Road
City: Leeds
State/Province: England
Postal Code: LS7 3NB
Country: United Kingdom
Voice Phone: (44) 1132628687
Fax Phone: (44) 1132374182
email: n.hardy@ukonline.co.uk
Contact # 1: Valerie Love
Business Description: Manufacturer of display holograms. Large selection of stock images in variety of formats and sizes. Commissioned work undertaken. Copying work for holographers undertaken.
****

OpSec - USA
38 Loveton Circle
City: Sparks
State/Province: MD
Postal Code: 21152
Country: United States of America
Voice Phone: (1) 410 666 I 144
Voice Phone: (1) 410 472 2141
Fax Phone: (1) 410 472 4911
email: 103320.340@compuserve.com
Contact #1: Dean Hill
Business Description: Embossed hologram manufacturing facility specializing in security and authentication allplications. Custom work accepted. Stock items available, including 38 patterns of foil (16 colors). Company pioneered mass replication of embossed holograms (formerly Difco.)
****

Optical Coating Laboratory GmbH
MMG Division
Alte Heerstrasse 14
City: Goslar
Postal Code: 0 -38644
Country: Germany
Voice Phone: (49) (0)532 359 0
Fax Phone: (+49) (0)532 359 103
Contact #1: Mr. Koch
Business Description: Flat glass, refined front surface mirrors, glass components, mirrors, optical mirrors, surface-coated mirrors.
****

Optical Research Associates
3280 East Foothill Blvd., Suite 300
City: Pasadena
State/Province: CA
Postal Code: 91107 -3103
Country: United States of America
Voice Phone: (1) 626 795 9101
Fax Phone: (1) 626 795 9102
email: service@opticalres.com
Web: http:\\www.opticalres.com
Contact # 1: Lia Titizian
Business Description: Optical Design Software. We sell the programs that allow you to create Holographic Optical Elements.
****

Optical Security Group
Corporate Headquarters - Suite 920
535 16 St.
City: Denver
State/Province: CO
Postal Code: 80202
Country: United States of America

INTERNATIONAL DIRECTORY

Voice Phone: (l) 303 5344500
Voice Phone: (1) 773 665 8932
Fax Phone: (l) 3035341010
email: staff@opticalsecurity.com
Web: http://www.opticalsecurity.com/
Contact # 1: Richard Bard
Business Description: Produce custom holographic labels as optical security devices for authentication applications. Full range of production services offered. Also produce lenticulars.
****

Optical Security Group - England
4E/F Gelders Hall Road
Shepshed, Leicestershire
State: England
Postal Code: LE12 9NH
Country: United Kingdom
Voice Phone: (44) (1)509 600 220
Voice Phone: (44) 1420488000
Fax Phone: (44) (1)509 508 795
Business Description: A total secure service from concept design artwork to finish ed productspecializing in customer service and delivering quality embossed holographic security and nonsecurity work on time.
****

Optical Society of America (OSA)
2010 Mass Avenue NW
City: Washington
State/Province: DC
Postal Code: 20036-1023
Country: United States of America
Voice Phone: (1) 202 223 8 130
Voice Phone: (1) 800 762 6960
Fax Phone: (1) 202 223 1096
email: osamem@osa.org
Web: http://www.osa.org
Business Description: Organization devoted to promoting optics and photonics research and applications. Publications include: Applied Optics, Optics Letters, Pptics and Photonics News, Journal of Optical Society of America.
****

Optical Test Equipment
(a division of J.D . Moeller Optische Werke GmbH)
Rosengarten 10
City: Wedel
Postal Code: 0-22880
Country: Germany
Voice Phone: (49) 4103 709 345
Fax Phone: (49) 4103 709375
email: mail@moeller-wedel.com
Web: moeller-wedel.com
Contact # 1: Carsten Schlewitt
Business Description: Manufactures custom optical components and optical test equipment including auto collimators, testing telescopes, focometers, goniometers, goniometer-spectrometers, and Fizeau-type interferometers.
****

Optical Works Ltd.
Ealing Science Centre
Treloggan Lane
City: Newquay, Cornwall
State: England
Postal Code: TR7 lHX
Country: United Kingdom
Voice Phone: (44) 1637877222
Fax Phone: (44) 16378772 11
Contact # 1: E.O. Frisk
Business Description: Make optical components, lenses and scientific instruments.
****

Optics Plus Inc.
1369 East Edinger Avenue
City: Santa Ana
State/Province: CA
Postal Code: 92705
Country: United States of America
Voice Phone: (1) 714 972 1948
Fax Phone: (1) 714 835 6510
Contact #1. Allison Valdivia
Business Description: Manufacture optics; precision tool mounts (including lens and mechanical mounts).
****

Optimation
6765 south 400 west
City: Midvelle
State/Province: UT
Postal Code: 84047
Country: United States of America
Voice Phone: (1) 801 263 6575
Fax Phone: (1) 80 I 263 6576
email: optimat@info.net
Contact # 1: Dean Jorgensen
Business Description: Specialize in the' manufacture of excellent quality, burr-free pinholes for holographic and related optical applications. Call for catalog.
****

Optimation Holographi~s
3200 South Haskell , Suite 160
City: Lawrence
State/Province: KS
Postal Code: 66046
Country: United States of America
Voice Phone: (1'j"913 841 1642
Fax Phone: (1) 913841 0439
Web: optiholosa@aol.com
Contact # 1: Terry F add is
Business Description: Large format holographic embossing facility. Complete in-house system including resi st master and shim making (up to 16 inches wide).
****

Optineering
2247 E. La Mirada St.
City: Tucson
State/Province: AZ
Postal Code: 18719
Country: United States of America
Voice Phone: (1) 520 882 2950
Fax Phone: (1) 520 882 6976
email: kcreath@primenet.com
Web: http ://www.primenet.coml-kcreath
Contact # 1: Kathy Creath
Business Description: Optical engineering consulting services specializing in optical testing, metrology, nondestru ctive evaluation, and optical design. Application areas include holography, NOT, speckle interferometry, microscopy, photography, and process control and monitormg.
****

Optitek, Inc.
1330 West Middlefie ld Rd.
City: Mountain View
State/Provi nce: CA
Postal Code: 94043
Coun~ry: United States of America
Voice Phone: (1) 650 938 3300
Fax Phone: (1) 650 938 3896
email: sales@optitek.com
Web: http:\\www.optitek.com
Contact # 1: Burt Hesselink
Business Description: Holographic data storage research and development.
****

Optopol Panoramic Metrology Consulting
Csiksomlyo u. 4.
City: Budapest
Postal Code: H-l025
Country: Hungary
Voice Phone: (36) 1 463 2518
Voice Phone: (36) I 335 5139
Fax Phone: (36) I 463 3178
email: gregyss@next-lb.manuf.bme.hu
Contact # 1: Pal Greguss
Business Description: Nonmultiplexed singleshot 360 degree panoramic holograms, based on the Panoramic Annular Lens (Pal-optic) invented by Dr. Greguss, are produced for metrological and other applications in science, tech. and arts.
****

Oregon Institute of Teclmology
Laser Optical Engineering Technology
3201 Campus Drive
City: Klamath Falls
State/Province: OR
Postal Code: 97601 -8801
Country: United States of America
Voice Phone: (1) 541 885 1698
Fax Phone: (1) 541 885 1666
email: piecer@oit.osshe.edu
Web: http://www.oit.osshe.edu/
Contact # 1: Robert Pierce
Business Description: The LOET program provides state of the art education by combining an applied laboratory approach to optical engineering technology together with theoretical classroom discussion. For program and course information, contact Dr. Robert Pierce.
****

Oregon Laser Consultants
455 Hillside Ave.
City: Klamath Falls
State/Province: OR
Postal Code: 9760 1-2337
Country: United States of America
Voice Phone: (1) 541 8823295
emai l: olcbill@aip.org
Contact # 1: Bill Deutschman
Business Description: Specialists in laser safety consulting and laser safety training. ANSI services for laser users and CDRH services for laser manufacturers. Also laser safety audits, employee training and electronic consulting.
****

Oriel Instruments
250 Long Beach Boulevard
City: Stratford
State/Province: CT
Postal Code: 06497
Country: United States of America
Voice Phone: (1) 203 377 8282
Fax Phone: (1) 203378 2457
email: res_sales@oriel.com
Web: http://www.oriel.com
Contact # 1: Nancy Fernandez
Business Description: A full line of optical components for holographic and related laboratory applications. Call for our catalog.
****

Ose Holografie-Design
Lenneperstralle 13-15
City: Wuppertal
Postal Code: 0-42289
Country: Germany
Voice Phone: (49) 202 825636
Contact # 1: Christian Ose-Wiese
Business Description: Manufacturing of masterand display-holograms
****

OWLS Gmbh
1m Gaisgraben 7
City: Staufen
Postal Code: D-79219
Country: Germany
Voice Phone: (49) 7633 95040
Fax Phone: (49) 7633 950444
Contact # \: Hubert Munzer
Business Description: Optical benches in different sizes. Mirror mounts (gimbal and kenematic), mirrors and lenses.
****

Oxford Holographics
71 High Street
City: Oxford
State/Province: England
Postal Code: OXI 4BA
Country: United Kingdom
Voice Phone: (44) 1865250505
Fax Phone: (44) 1865250505
Web: http://www.oxfordshire.co.uk
Contact #1: Nick Cooper
Business Description: Oxford Holographics has both a very well established retail and an expanding distribution operation, focusing on unusual and unique giftware, including holograms.
****

Pacific Holographics Inc.
503 Caledonia Street
City: Santa Cruz
State/Province: CA
Postal Code: 95062
Country: United States of America
Voice Phone: (1) 408 425 4739
Fax Phone: (1) 408 425 4739
Contact #\: Randy James
Business Description: Photo-resist mastering for embossed holography. Origination, design and consulting services offered. Extensive commercial portfolio.
SEE OUR ADVERTISEMENT
****

Page, Michael
22 Bayview Avenue
City: Toronto, Wards Island
Postal Code: M5J 1ZI
Country: Canada
Voice Phone: (1) 416 203 7243
Fax Phone: (1) 416 203 7243
Contact #1: Michael Page
Business Description: Holographic artist and technician.
****

Panatron Inc.
P.O. Box 2687
City: Pomona
State/Province: CA
Postal Code: 91769-2687
Country: United States of America
Voice Phone: (1) 909 629 0748
Voice Phone: (1) 800 669 7945
Fax Phone: (1) 909 620 0378
email: panatron@aol.com
Business Description: Supplies complete support, parts and service on all lasers. Also manufactures mirrors, lenses, rods and other parts for lasers. Laser repair and used lasers.
****

Pangaea Design
PO Box 2028
City: New York
State/Province: NY
Postal Code: 10009
Country: United States of America
Voice Phone: (1) 888 772 6423

Business Description: Model maker for holography.
****

Parallax Gallery
Shop R-I, Harbor Rocks Hotel
Nurses Walk, The Rocks
City: Sydney
State/Province: NSW
Postal Code: 2000
Country: Australia
Voice Phone: (61) 2 9247 6382
Fax Phone: (61) 29 247 6382
Contact # 1: Tony Butteriss
Business Description: Hologram Gallery. Large variety of holograms for sale to the public.
****

Pasco Scientific
10101 Foothills Blvd.
City: Roseville
State/Province: CA
Postal Code: 95661
Country: United States of America
Voice Phone: (1) 916 786 3800
Voice Phone: (l) 800 772 8700
Fax Phone: (1) 916 786 8905
email: sales@pasco.com
Web: http:\\www.pasco.com
Business Description: Distributor for science supplies and educational materials. Catalog available.
****

Peacock Laboratories, Inc.
1901 S. 54th St.
City: Philadelphia
State/Province: PA
Postal Code: 19143
Country: United States of America
Voice Phone: (1) 215 729 4400
Fax Phone: (1) 215 729 1380
email: plabs@bellatlantic.net
Contact # \: Sagar Venkate~ran
Business Description: Established 1930. Dedicated to developmental research in mirror manufacturing, silver metalizing, and protective coatings. Innovators of silver spray processes and dual-nozzle spray glinS. Consultants and suppliers of silvering solutions and chemicals for electroconductive, decorative and reflective applications.
****

Pennsylvania Pulp & Paper Co.
Prismatic Square
2874 Lime Kiln Pike
City: Glenside
State/Province: PA
Postal Code: 19038
Country: United States of America
Voice Phone: (1) 215 572 8600
Fax Phone:(1) 215 572 8154
email: hologramer@aol.com
Web: http://www.holoprism.com
Contact # 1: Brian J. Monaghan
Business Description: Manufacturer of hologaphic images in paper and paperboard. 4 digital "dot matrix" origination labs, electrofonning and narrow web embossing service. We stock press ready sheets and rolls of holographic paper and board. Masters up to 31" x 41".
****

Pepper, Andrew
46 Crosby Road
City: West Bridgford, Nottingham
State: England
Postal Code: NG2 5GH
Country: United Kingdom

Voice Phone: (44) 7050 133 624
Fax Phone: (44) 7050 133625
email: pepper@monand.demon.co.uk
Web: http://www.holo.comlpeper/search.htrnl
Contact # 1: Andrew Pepper
Business Description: Fine art holography. Limited editions, unique pieces, collaborations. Main work in reflection holography.
****

Phantastica
S uchtener Strasse 4a
City: Amsberg
Postal Code: D-59757
Country: Germany
Voice Phone: (49) (0)293 81917
Fax Phone: (+49) (0)293 29441
Contact # 1: Gerd M. Albrecht
Business Description: Makers and distributors of articles related to embossed holograms and diffraction foil, including earrings and other jewelry, badges, pens, mobiles. Main focus is street, crafts and Christmas markets.
****

Photo Research, Inc.
9330 DeSoto Avenue
City: Chatsworth
State/Province: CA
Postal Code: 91311-4926
Country: United States of America
Voice Phone: (1) 818 3415151
Fax Phone: (l) 818 341 7070
email: sales@photoresearch.com
Web: http://www.photoresearch.com
Contact #1: Manjit Daniel
Business Description: Photo Research is the world leader in manufacturing precision instruments that measure light and color. We serve the following markets: CRT/FPD, automotive, aerospace (commercial and military), motion picture, R&D, and many other related industries. For over 50 years, our leadership has delivered worldclass light measurement solutions.
****

Photo Sciences
2542 W. 237th Street
City: Torrance
State/Province: CA
Postal Code: 90505
Country: United States of America
Voice Phone: (1) 310 784 7460
Fax Phone: (1) 310 539 6740
email: scsales@photo-sciences.com
Web: http://www.photo-sciences.coml
Contact # 1: John Stogsdill
Business Description: PSI was among the earliest producers of photomasks for the computer and electronics industry. PSI 's reputation as a high quality photomask manufacturer has become a well accepted fact within the industry.
****

Photon Cantina Ltd.
PO Box 1098
City: La Canada
State/Province: CA
Postal Code: 91012-1098
Country: United States of America
Voice Phone: (1) 818 790 6735
Fax Phone: (1) 818 790 7081
email: cpax@ix.netcom.com
Contact #1: Roy Chiarot
Business Description: Full service producers of high quality artistic and commercial silver halide reflection holograms. Stock image catalog available. Custom origination, mastering and replication services offered.
****

INTERNATIONAL DIRECTORY

Photon League Of Holographers Ontario
401 Richmond Street West Suite B03
City, Prov .. Toronto, Ontario
Postal Code: M5U 3A8
Country: Canada
Voice Phone: (1) 416 599 9332
Voice Phone: (1) 416 203 7243
Contact # 1: Claudette Abrams
Business Description: Artist run non-profit holography studio. Technical workshops throughout the year. 2 tables 50mw HeNe. Copy Lab. Stereogram LCD HOP. Associate Membership $ 15/year Lab Users Program $60/year.
****

Photonics Direct
Weisestrasse II
City: Berlin
Postal Code: D-12049
Country: Germany
Voice Phone: (49) (0)30 62709372
Voice Phone: (+49) (0)3 62709370
Fax Phone: (+49) (0)30 62709371
email: wappelt@t-online .de
Web: http: //home.t-online.de/home/wappelt/
Contact # 1: Andreas Wappelt
Business Description: Distribution of Holograms (fine art & entertainment), he-ne lasers, laser pointer, laser diodes, diode pumped blue and green solid-state lasers, holographic equipment
SEE OUR ADVERTISEMENT
****

Photonics Spectra
Laurin Publishing Co. Inc.
2 South Street
City: Pittsfield
State/Province: MA
Postal Code: 01202
Country: United States of America
Voice Phone: (1) 413 499 0514
Fax Phone: (1) 413 442 3180
email: photonics@laurin.com
Web: http ://www.laurin.com
Business Descripti~n : Trade publication covering the field of optIcs, lasers, electro-optics, and related imaging research, as well as commercial applications. 676-3290
****

Physik Instrumente (PI) GmbH & Co.
Polytecplatz 5-7
City: Waldbronn
Postal Code: 0-76337
Country: Gelmany
Voice Phone: (49) (0)724 604- 100
Fax Phone: (+49) (0)724 604-1 45
Contact # 1: Karl Dr. Spanner
Business Description: Holography, laser, optical components, miscellaneous, osci llation insulators, vibrating dampers, PZT actuators, sensors, PZT ceramics.
****

Pilkington Optronics
Glascoed Road
St. Asaph
City: Clwyd, North Wales
State: England
Postal Code: LLl7 OLL
Country: United Kingdom
Voice Phone: (44) 1745 583 301
Fax Phone: (44) 1745 584 358
Web: http: //www.thejob.com/pi lkingtonl
Business Description: Manufacturer ofDCG and photopolymer HOEs and related optical components for Heads Up Displays, etc.
****

Planet 3-D
201 Silver Fox Lane

City: Downingtown
State/Province: PA
Postal Code: 19335
Country: United States of America
Voice Phone: (l) 610 873 6192
Fax Phone: (1) 6 108736194
email: rcossa@aol.com
Contact # 1: Rich Cossa
Business Description: Marketing of holographic products.
****

Point Source Productions
14670 Highway 9
P.O. Box 55
City: Boulder Creek
State/Province: CA
Postal Code: 95006
Country: United States of America
Voice Phone: (1) 408 457 1426
Contact # 1: Bob Hess
Business Description: We are an independent recording studio offering product design and techni cal imaging consultations, mastering and ganging services (specializing in silver-halide and photopolymer), and ""short-run"" or limited edition transfer services.
****

Polaris Research Group
24400 Highland Road
City: Richmond Heighh
State/Province: OH
Postal Code: 44143
Country: United States of America
Voice Phone: (1) 216 383 9480
Fax Phone: (~ 16 3839488
email: polarisrg@aol.com
Web: http ://w3.gwis.coml-polaris/whatis.html
Contact # 1: Howard Fein
Business Description: Polaris Research Group offers NDT to the commercial market. We use Holography-based technology to measure vibration, stress, and structural characteristics in a wide range of application s. Modal and vibration analysis; bonded and composite structure test; and flaw, defect, and delamination identification are just a few of the primary applications for these techniques.
****

Polaroid Corporation
2 Osborn Street - 2nd Floor
City: Cambridge
State/Province: MA
Postal Code: 02139
Country: United States of America
Voice Phone: (1) 617 386 8676
Voice Phone: (1) 800 237 5519
Fax Phone: (1) 617 386 8671
email: holography. info@polaroid.com
Web: http://www.holoroid.com
Contact # 1: Diane Martin
Business Description: Fully integrated supplier of highest quality, mass produced photopolymer holograms. Services include design, modeling, origination, manufacturing and converting. Our industrial divi sion provides the highest efficiency, mass produced holographic optical elements available, including reflective and transmissive diffusers, projection screens and depixellators.
SEE OUR ADVERTISEMENT
****

Polytec GmbH
Siemensstra13e 13-1 5
City: Waldbronn
Postal Code: D-76337
Country: Germany

Voice Phone: (49) 7243 604-0
Fax Phone: (+49) 7243 69944
Contact # 1: H. G. Lossau
Business Description: Distributor of holographic laser equipment
****

Potomac Photonics, Inc.
4445 Nicole Drive
City: Lanham
State/Province: MD
Postal Code: 20706
Country: United States of America
Voice Phone: (1) 301459-3033
Fax Phone: (1) 30 I 459-3034
email: gbehrmann@potomac-Iaser.com
Web: http://www.potomac-Iaser.com
Contact # 1: Greg Behrmann
Business Description: Manufactures compact UV lasers and tabletop micro machining workstations. Potomac offers rapid prototyping of computer generated holograms and diffractive optical elements.
****

PPM Promotion Products Miinchen GmbH Werbung
Hohenzollernstra13e 10
Ambassadepassage, I. Stock
City: Muenchen
Postal Code: D-8080 I
Country: Germany
Voice Phone: (49) 89 338616
Contact # 1: Thomas Kubeile
Business Description: Promotion and distribution of holograms and holographic jewelry, toys, gifts Custom made holograms .
****

Print-M-Boss
5/24 Kirti Nagar Indl. Area
City: New Delhi
Postal Code: 110017
Country: India
Voice Phone: (91) 11 530586
Fax Phone: (91) II 544 1144
Contact #1: Ravinder Singh
Business Description: Manufa cturers of embossed holograms with in-house facility for shim making. Would be interested in buying copyrights for various images and patterns. Like to make contacts for mastering.
****

Prismacoat Division
Paper Corporation of United States
161 Avenue of the Americas 4th Fl.
City: New York
State/Province: NY
Country: United States of America
Voice Phone: (1) 212 337 5571
Voice Phone: (1) 888 774 7622
Fax Phone: (1) 212 9240663
email: prismacoat@mindspring.com
Contact # 1: Jason Dennis
Business Description: Prismacoat is a line oflay-flat, non-laminated holographic papers and boards (from 40 lb. paper to 24 point board). 28"" x 40"" sheets in stock for immecliate delivery. Metallized and Non-metallized holograms in range of patterns.
****

Process Technologies
436 West Rawson Ave.
City: Oak Creek
State/Province: WI
Postal Code: 53154
Country: United States of America
Voice Phone: (1) 414 571 9200

Fax Phone: (1) 414571 9202
emai l: pti@execpc.com
Web: http://www.execpc.comJ-ptilindex .html
Contact # 1: Manfred Stelter
Business Description: Provides photoresist coated plates, ronchi rulings. reticles and masks.
****

Pull Time 3-D Laboratories
29 West 26 Street
City: New York
State/Province: NY
Postal Code: 10010
Country: United States of America
Voice Phone: (1) 212 889 5994
Fax Phone: (1) 212 889 5926
email: pulltime3d@aol.com
Web: http://www.vision3d.com/pulltime3d!
Contact # 1: Gerald Marks
Business Description: PullTime 3-D Laboratories is responsible for 3D broadcast television and home video for clients including CBS Records, Fox Television, The Rolling Stones, AT&T, Howard Stem and Atlantic Records. Over 25 million Pull Time 3-D glasses produced.
****

Qingdao Gaoguang Holography Tech. Co
Feng-Xian-Lu #8
City, Prov.: Qingdao, Shangdong
Postal Code: 266071
Country: China
Contact # 1: Yuan Baoqing
Business Description: Holography class
****

Qingdao Qimei Images, Inc.
Qingdao Economic Development Zone
City: Qingdao
Postal Code: 266555
Country: China
Voice Phone: (86) 532 6898751
Fax Phone: (86) 532 6898751
Contact #1: Yang Caizhi
Business Description: Holography sales and distribution.
****

Quan Zhou Pacific Laser Images
Bei-Men Huan-Cheng-Lu Gongsi-Da-Xia
City, Prov.: Quanzhou, Fujian
Postal Code: 362000
Country: China
Voice Phone: (86) 595 2783945
Fax Phone: (86) 595 2784626
Contact # 1: Wu Rongkun
Business Description: Hologram distribution.
****

Quantel
17, av de I' Atlantique 1-
ZA de Courtaboeuf, BP 23 ..
City: Les Ulis Orsay Cedex
Postal Code: F-91941
Country: France
Voice Phone: (33) I 6929 1700
Fax Phone: (33) I 6929 1729
Business Description: Manufacturer oflasers and other light sources, laser accessories.
****

Rainbow Symphony Inc.
6860 Canby Ave. # 120
City: Reseda
State/Province: CA
Postal Code: 91335
Country: United States of America
Voice Phone: (1) 818 708 8400
Fax Phone: (1) 818 708 8470
email: 3dglasses@rainbowsymphony.com

Web: http ://www.rainbowsymphony.com/
Contact #1: Mark Margolis
Business Description: Manufacturers of uniquely designed holographic and diffraction products for the gift, novelty, advertising, specialty, premium incentive, souvenir and museum markets.
SEE OUR ADVERTISEMENT
****

Ralcon
Box 142
850 I South 400 West
City: Paradise
State/Province: UT
Postal Code: 84328
Country: United States of America
Voice Phone: (1) 4352454623
Fax Phone: (1) 435 245 6672
email: rdr@ralcon.com
Web: www.xmission.com/-ralcon
Contact # 1: Richard Rallison
Business Description: Design, development and fabrication of volume holographic optical elements, (HOEs) including gratings, scanners, multi focus devices, heads up and down di sp lays and notch filters fOlmed in dichromated gelatin or photopolymer.
SEE OUR ADVERTISEMENT
****

Real Image
PO Box 566
City: Pacifica
State/Province: CA
Postal Code: 94044
Country: United States of America
Voice Phone: (1) 650 355 8897
Fax Phone: (1) 650 355 5427
Contact # 1: Roy Bradshaw
Business Description: Incorporation of patented holographic designs into fishing tackle and fishing lures.
****

Reconnaissance International Ltd.
3003 Arapahoe St., Suite 213
City: Denver
State/Province: CO
Postal Code: 80205 ,
Country: United States of America
Voice Phone: (1) 303 293 3000
Fax Phone: (1) 303 293 8661
email: ReconnUSA@aol.com
Contact # 1: Lewis Kontnik
Business Description: North American office. We are an international consultancy for market and industry information and analysis. Publisher of Holography News, Holo-Pack/Holoprint Guidebook and Authentication News. All clients studies are fully confidential.
SEE OUR ADVERTISEMENT
****

Reconnaissance International Ltd.
Runnymede Malthouse
Runnymede Road
City: Egham, Surrey
State/Province: England
Postal Code: TW20 9BD
Country: United Kingdom
Voice Phone: (44) 1784497008
Fax Phone: (44) 178449700 I
email: 100142.1164@compuserve.com
Web: http://www.hmt.comJholography/hnews/index. html
Contact # 1: Ian Lancaster
Business Description: We are the leading international consultancy for market and industry information and analysis. Publisher of Holography News, Holo-PackIHolo-print Guidebook and Au-

thentication News. All clients studies are fully confidential.
SEE OUR ADVERTISEMENT
****

Red Beam, Inc.
90 11 Skyline Blvd.
City: Oakland
State/Province: CA
Postal Code: 94611
Country: United States of America
Voice Phone: (1) 510 482 3309
Fax Phone: (1) 510 482 1214
Contact # 1: Lon Moore
Business Description: Mastering facility specializing in the design and production of masters suitab le for high volume corporate applications, especially on photopolymer films and embossed materials. Clients include Activision, AT&T, NFL (Superbowl) and Polaroid. Also produces a line of trademarked giftware holograms distributed by"Lightrix" (See ad and listing in this book).
****

Regal Press Inc., The
Holographics Division
129 Guild Street
City: Norwood
State/Province: MA
Postal Code: 02062
Country: United States of America
Voice Phone: (1) 781 769 3900
Voice Phone: (1) 800 447 3425
Fax Phone: (1) 78 1 551 0466
Contact # 1: William Duffey
Business Description: The Regal Press, Inc. has expertise in all areas of print production, including engraving, lithograpfly, thermography, embossing, foil-stamping, and holography. We hold a worldwide patent for REGAL MARQUE, a simulated plivate watermarking process, and we provide our customers with REGAL EXPRESS guaranteed overnight delivery and 24-hour rush Business Cards.
****

Reva's Holographic Illusions
446 South Main Street
City: Frankenmuth
State/Province: MI
Postal Code: 48734
Country: United States of America
Voice Phone: (1) 517 652 3922
Fax Phone: (1) 517 652 6503
Conta ct # 1: Reva Krick
Business Description: Gallery/Retail store, with over 250 holograms on display. We feature a full line of holographic jewelry, gifts, toys, etc.
****

Reynolds Metals Co.
Flexible Packaging Division
6603 West Broad SI.
City: Richmond
State/Province: VA
Postal Code: 23230
Country: United States of America
Voice Phone: (1) 804 281 2000
Voice Phone: (1) 804 281 3969
Fax Phone: (1) 804 281 2238
email: info@wwwrmc.com
Web: http://wwwm1c.com
Contact # 1: Rich Patterson
Business Description: Holographic specialty cmions and printed paper materials for distilled spilits and wine, phalmaceuticals, confections, personal care, and other consumer goods. Holographic flexible light web materials for pouches, lidding, and overwraps. Full service from design to finishing.
****

INTERNATIONAL DIRECTORY

Rice Systems, Inc.
1150 Main Street, Suite C
City: Irvine
State/Province: CA
Postal Code: 92614
Country: United States of America
Voice Phone: (1) 714 553 8768
Fax Phone: (1) 714 553 0307
email: RiceSys@prodigy.com
Contact # 1: Colleen Fitzpatrick
Business Description: Laser metrology and diagnostic measurements, HNDT fluid measurements. Combustion diagnostics. Integrated optics and non linear optical material (R&D and product development). Very successful SBIR company.
****

Richard Bruck Holography
33 12 West Belle Plaine #2
City: Chicago
State/Province: IL
Postal Code: 60618-2316
Country: United States of America
Voice Phone: (1) 773 267 9288
Fax Phone: (1) 773 267 9288
Contact # 1: Richard Bruck
Business Description: Specialists in large format holography. Extensive experience with live models and commercial work. We are accustomed to the advertising world, and know the importance of quality and service.
SEE OUR ADVERTISEMENT
****

Richardson Grating Laboratory
(a division of Spectronic Instruments)
820 Linden Ave.
City: Rochester
State/Province: NY
Postal Code: 14625
Country: United States of America
Voice Phone: (1) 7_ 6 262 1331
Voice Phone: (1) 800 654 9955
Fax Phone: (1) 716 454 1568
email: gratings@spectronic.com
Web: http://www.spectronic.com/
Contact #1: Susan Willard
Business Description: The Richardson Grating Laboratory has been a world leader in the design and manufacture of ruled and holographic diffraction grating for fifty years.
****

Richmond Development Group
(formerly Gray Scale Studios)
63 South 500 West
City: Richmond
State/Province: UT
Postal Code: 84333
Country: United States of America
Voice Phone: (1) 801 2580709
Fax Phone: (1) 801 2580109
email: george@richmondinc.com
Web: http ://www.richmondinc.coml
Contact # 1: George Sivy
Business Description: Specialists in the design and creation of models and sculptures for holographic imaging, including digital origination services for stereo grams. Consultant services offered, II years experience. Samples available upon request.
****

Richmond Holographic Studios Ltd
6 Yorkton St.
City: London
State/Province: England
Country: United Kingdom
Voice Phone: (44) (0)1717399700

Fax Phone: (44) (0) 171 739 9707
email: rhs@augustin.demon.co.uk
Contact # 1: Edwina Orr
Business Description: Pulsed laser dsplay holography. R&D using HOE's for auto stereoscopic displays
****

Robert Sherwood Holographic Design
1380 Wendover Dr.
City: Charlottesville
State/Province: VA
Postal Code: 22901
Country: United States of America
Voice Phone: (1) 804 971 2910
Fax Phone: (1) 804 971 2998
email: evoke@aol.com
http://www.holographicdesign.com/
Contact # 1: Robert Sherwood
Business Description: RSHD, Inc. provides custom commercial holographic products and serivces. Specialized management of complete material constructions and conversion of holographic materials. Products include: PS labels, Heat applied labels, films, foils, laminates and photopolymer products.
****

Rochester Inst. Of Technology
Center for Imaging Science
One Lomb Memorial Drive
City: Rochester
State/Province: NY
Postal Code: 14623 -5604
Country: United States of America
Voice Phone: (1) 716 4756631
Fax Phone: (1) 716 4755988
email: info@rit.edu
Web: http ://www.rit.edu/
Business Description: Research on HOEs, holo graphic materials, CGHs. Instruction in holography and related topics in the Department of Imaging and Photographic Technology, and the Center for Imaging Science.
****

Rochester Photonics Corporation
330 Clay Road
City: Rochester
State/Province: NY
Postal Code: 14623
Country: United States of America
Voice Phone: (1) 7 16 272 3010
Fax Phone: (1) 716 272 9374
email: sales@rphotonics.com
Web: http ://www.rphotonics.com
Contact # 1: Dan McGarry
Business Description: RPC specializes in the design and manufacturing of diffractive optical components and subsystem s. Precision diffractive mastering, molding, replication, and testing services are provided. Products include: hybrid refractive/diffractive lenses and subassemblies, microlens arrays, diffractive phase plates, engineered diffusers, and holographic gratings.
****

Rofin-Sinar Laser GmbH
Berzeliusstrasse 85
City: Hamburg
Postal Code: D-22113
Coqntry: Germany
Voice Phone: (49) (0)40 733 630
Fax Phone: (+49) (0)40 733 63 100
Business Description: C02 and Nd:YAG lasers for materials processing, Laser components, Laser processing devices and machines.
****

Roll s-Royce Plc
Advanced Research Laboratory
POBox 31
City: Derby
State/Province: England
Postal Code: DE2 48BJ
Country: United Kingdom
Voice Phone: (44) 133 224 2424
Fax Phone: (44) 133 2249936
email: info@rolls-royce.co.uk
Web: http //www.rolls-royce.co.uk!
Business Description : NDT for aircraft engines.
****

Rolyn Optics
706 Arrow Grand Circle
City: Covina
State/Province: CA
Postal Code: 91722-2199
Country: United States of America
Voice Phone: (1) 818 915 5707
Fax Phone: (1) 818 915 1379
email: sales@rolyn.com
Web: http:\\www.rolyn.com
Business Description: General selection of optical items. Catalogue available.
****

Ross Books
P.O. Box 4340
City: Berkeley
State/Province: CA
Postal Code: 94704
Country: United States of America
Voice Phone: (1) 800 367 0930
Voice Phone: (1) 510 8412474
Fax Phone: (\) 510 8412695
email: staff@rossbooks.com
Web: http:\\www.rossbooks.com
Contact # 1: Alan Rhody
Contact # 2: Franz Ross
Business Description: Publisher of the **HOLOGRAPHY MARKETPLACE Ed. 1-7** (a worldwide database and sourcebook), the **HOLOGRAPHY HANDBOOK** - Making Holograms the Easy Way (world's best selling laboratory manual), and other. Printed materials, webpages and software. Educational, research and information services also provided.
**Visit our website:
www.rossbooks.com.**
****

Rottenkolber Holo-System GmbH
Bergweg 47
City: Amerang
Postal Code: D-83 123
Country: Germany
Voice Phone: (49) 89 9030021
Fax Phone: (+49) 89 904 39 83
Contact #1: Hans Dr. Rottenkolber - ,
Business Description: Interferometry
****

Rowland Institute For Science
100 Edwin H. Land Blvd.
City: Cambridge
State/Province: MA
Postal Code: 02142
Country: United States of America
Voice Phone: (1) 617 497 4657
Contact #1: Jean-Marc Fournier

Business Description: Scientific holography research. NOT, Lippman photography
****

Royal Holographic Art Gallery
122 Market Square
560 Johnson Street
City, Prov.: Victoria, B.C.
Postal Code: V8W 3C6
Country: Canada
Voice Phone: (1) 250 384 0123
Fax Phone: (1) 250 384 0123
email: royal@islandnet.com
Web: http://www.islandnet.com/-royal
Contact # 1: Derek Galon
Business Description: Gallery offers full range of holographic art, holograms (including fine art Russian holograms), and holo-gifts. Retail, wholesale and low-cost custom work on our new RED STAR film. Also holography equipment from Russia. We ship worldwide.
SEE OUR ADVERTISEMENT
****

Royal Institute of Technology
Dept. of Materials Processing
Industrial Metrology
City: Stockholm
Postal Code: S-10044
Country: Sweden
Voice Phone: (46) 8 790 7832
Voice Phone: (46) 8 796 6899
email: nilsa@matpr.kth.se
Contact # 1: Nils Abramson
Business Description: Industrial Metrology comprises conventional engineering metrology and laser-based metrology, especially industrial applications of display holography, holographic interferometry and Light-in-Flight recording by holography, which is large ly the result of the research and development at the Department. The principal objective of the group is research and education; to develop new measurement principals for applying lasers in the industry and to disseminate knowledge of known laser-based methods of measurement.
****

Ruey-Tung, Miss. Hung
A 202
Chigasati-Coat Nango 6-7-12
City: Chigasaki-Shi
State/Province: Kanagawa
Postal Code: 253
Country: Japan
Voice Phone: (81) 0467 857 750
Contact # 1: Hung Ruey-Tung
Business Description: Holographic Fine Artist,
****

S.O.P.R.A. ,.
Societe de Production et de Recherche Appliquee
26 & 28 rue Pierre Joigneaux
City: Bois-Colombes
Postal Code: F-92270
Country: France
Voice Phone: (33) I 4781 0949
Fax Phone: (33) I 42 42 29 34
Contact # 1: Robert Stehle
Business Description: Laser equipment, holographic kit and camera for interferometry.
****

Saginaw Valley State University
2250 Pierce Road
City: University Center
State/Province: MI
Postal Code: 48710-000 I
Country: United States of America

Voice Phone: (1) 517 790 4000
Fax Phone: (1) 517 790 2717
Contact #1: Hsuan Chen
Business Description: Course instruction on holography; research includes HOEs, multiplex and rainbow holography.
****

Saint Mary's College
Art Department
City: Notre Dame
State/Province: IN
Postal Code: 46556
Country: United States of America
Voice Phone: (1) 2 19 284 4000
Contact # 1: Doug Tyler
Business Description: Holographic fine artist. Extensive portfolio. Holography Instructor. Call for class schedule.
****

SAM Museum
3-27-3 Isoji, Minato-ku
City: Osaka
Postal Code: 552
Country: Japan
Voice Phone: (81) 6 572 0036
Fax Phone: (81) 6 574 8136
Contact #1: Akinobu Fukuda
Business Description: Museum that exhibits holograms.
****

San Jose State University
Physics Dept. and Inst. for Modem Optics
One Washington Square
City: San Jose
State/Province: CA
Postal Code: 95192-0106
Country: United States of America
Voice Phone: (l) 408 924 5245
Fax Phone: (1) 4089242917
Web: http://fire.sjsu.edu
Contact # 1: Ramen Bahu~a
Business Description: Research and development work on 1) holographic fingerprint sensor, 2) holographic fingerprint verification, 3) display holography on DCG, and 4) holographic optical elements.
****

Sandia National Laboratories
P.O. Box 5800
City: Albuquerque
State/Province: NM
Postal Code: 87 185
Country: United States of America
Voice Phone: (1) 505 845 0011
Voice Phone: (1) 505 843 4123
Web: http ://www.sandia.gov/
Contact # 1: James Otega
Business Description: Sandia National Laboratories is able to do research in all phases of holography.
sandia
****

Saxby, Graham
3 Honor Ave.
Goldthorn Park
City: Wolverhampton, West Midlands
State: England
Postal Code: WV4 5HF
Country: United Kingdom
Voice Phone: (441) 902 341 291
Contact # 1: Graham Saxby
Business Description: Research scientist; author of ''''Practical Holography''''
****

Scharr Industries
40 East Newberry Road
City: Bloomfield
State/Province: CT
Postal Code: 06002
Country: United States of America
Voice Phone: (1) 860 243 0343
Voice Phone: (1) 800 284 7286
Fax Phone: (1) 860 242 7499
Contact # 1: Peg Home
Business Desc ription: Scharr holographic embossing: wide web, on polyester, polypropylene, polyethylene, PVC, and nylon. We coat, laminate, metalize standard and custom patterns. Products include film to paper, board, transfer film, PSA and static cling.
****

School Of Holography
Museum Of Holography/Chicago
1134 W. Washington Blvd.
City: Chicago
State/Province: IL
Postal Code: 60607
Country: United States of America
Voice Phone: (1) 312 2261007
Fax Phone: (1) 312 829 9636
email: museum@concentri c.net
Web: http://www.cris.com/-museumh
Contact # 1: Loren Billings
Business Description: Founded in 1978, the oldestcontinuous school of holographic instruction in the world. Basic courses in holography have been taught to thousands of students. In addition there are special workshops and tutorials for advanced study.
SEE OUR ADVERTISEMENT
****

Science Kit & Boreal Labs
777 East Park Drive
City: Tonawanda
State/Province: NY
Postal Code: 14150-6784
Country: United States of America
Voice Phone: (1) 716 874 6020
Fax Phone: (1) 716 874 9572
Business Description: Suppliers of educational science materials especially suitable for junior and senior high school coursework. Comprehensive mail order catalog includes holography kits, holography books, related optical components and more.
****

Shandong Academy of Sciences
Keyuan Road
City: Jinan Shandong
Postal Code: 250014
Country: China
Voice Phone: (86) 615 615102 316
Contact # 1: Zhu De Shun
Business Description: Laser & holography exhibit.
****

Shanghai Dahua Printing Factory
Pu-Dong-Xin-Qu Wang-Gang Xin-Hong-Cun
Tang-Lu-Gong-Lu #2498
City: Shanghai
Postal Code: 201201
Country: China
Contact #1: Gong Yuanzhong
Business Description: Finishing work for holography.
****

Shanghai Kanlian S & T Development Co. Ltd.
Zhongshan-Xi-LU #1521
City: Shanghai

Postal Code: 200233
Country: China
Voice Phone: (86) 21 64813107
Fax Phone: (86) 21 64647030
Contact #1: Zhou Bingda
Business Description: Hologram distribution.
****

Sharon McCormack Holography
P.O. Box 38
City: White Salmon
State/Province: WA
Postal Code: 98672
Country: United States of America
Voice Phone: (1) 509 493 4850
Voice Phone: (1) 541 3865943
Fax Phone: (1) 509 493 4830
email: sharon@gorge.net
Web: http://www.gorge.netlbusiness/holography/
Contact # 1: Sharon McCormack
Business Description: Holographic fine artist.
Complete stereogram production, including 360
degree viewable. Filming, animation and comput-
er graphics services offered. Also exhibit, design,
and consultation services. Extensive commercial
portfolio for major corporate clients.
****

Sharp Corp.
Tokyo Research Laboratories
Research Dept2 ; 271 , Kashiwa
City: Kashiwa
Postal Code: 227
Country: Japan
Voice Phone: (81) 0471 346 166
Fax Phone: (8 1) 0471 346119
Contact # 1: Shunichi Sato
Business Description: Research
****

Shenzhen Reflective Materials Factory
Shenzhen University
Rm 117-119 Lab ~uilding
City, prov.: Shenzhen, Guangdong
Postal Code: 518060
Country: China
Voice Phone: (86) 755 6660277-2236
Voice Phone: (86) 755 6660970
Fax Phone: (86) 755 755 6660462
Contact # 1: Ye Jingde
Business Description: Holography classes.
****

Shipley Chemical Co.
455 Forrest Street
City: Marlboro
State/Province: MA
Postal Code: 01752
Country: United States of America
Voice Phone: (1) 800 345 3100
Voice Phone: (1) 508 481 7950
Fax Phone: (1) 508 4859113
Contact # 1: Stu Price
Business Description: Primary manufacturer of
photoresist. Sold wholesale by quarts and gal-
lons as liquid. For precoated plates, see listing for
Towne and Process Technology.
****

Shriram Holographics
104, Kirti Deep
Nanagal Raya
City: New Delhi
Postal Code: 110046
Country: India
Voice Phone: (9 1) 55 96697
Fax Phone: (91) II 5552986
Contact # 1: Rajeev Jain
Business Description: Embossed holography
****

Shuttlecart
Avenues Mall Unit K5UL
10300 South Side Blvd.
City: Jacksonville
State/Province: FL
Postal Code: 32256
Country: United States of America
Voice Phone: (1) 904 519 7744
email: shjaxl20@aol.com
Web: http://www.starlog.com/
Business Desc ription: Selling all types of holo-
graphic wall art, jewelry and gift items.
****

Silhouette Technology Inc.
10 Wilmot Street
City: Morristown
State/Province: NJ
Postal Code: 07962-1479
Country: United States of America
Voice Phone: (1) 973 539 2110
Fax Phone: (1) 973 539 5797
Contact # 1 : Toicia Murphay
Business Description: Produces custom HOEs
under contract. HOP maker. Heads-up· display.
DOE & DOE printers.
****

Silicon Graphics
2011 North Shoreline blvd.
City: Mountain View '
State/Province: CA
Postal Code: 94043
Country: United States of America
Voice Phone: (1) 650 960 1940
Voice Phone: .4J,1 800 800 7441
Fax Phone: (1) 650 9601737
email: sales@sgi.com
Web: http:\\www.sgi.com
Business Description: Silicon Graphics pro-
duces high end computer graphics stations ideal
for rendering and modeling holographic stereo-
grams.
****

Sillcocks Plastics International
310 Snyder Avenue
City: Berkeley Heights
State/Province: NJ
Postal Code: 07922
Country: United States of America
Voice Phone: (1) 908 665 0300
Voice Phone: (1) 800 526 4919
Fax Phone: (1) 908 665 9254
email: spisales@sillcocks.com
Web: http ://www.sillcocks.com/
Business Description: Producer of flat plas-
tic products, printed or unprinted, which can
feature hot-stamped holograms and laminated
holograms. Products include credit cards, pro-
motional cards and other custom specialties and
POP products.
****

Silverbridge Group
Box 489
City, prov.: Powasson, Ontario
Postal Code: POH IZO
Country: Canada
Voice Phone: (1) 705 724 6164
Fax Phone: (1) 705 724 6249
Corttact # 1: James Hepburn
Business Description: Limited edition DCG holo-
grams in large format size.
****

Simian Co.
298 Harvey West
City: Santa Cruz
State/Province: CA

Postal Code: 95060
Country: United States of America
Voice Phone: (1) 408 457 9052
Fax Phone: (1) 408 457 9051
Contact #1: Debbie Haines
Business Description: Manufacturer of high qual-
ity masters for embossed holography. Originations
can be 20/30, 3D, animation and motion, or any
combination. High production capacity with quick
turnaround.
****

Sinclair Optics, Inc.
6780 Palmyra Road
City: Fairport
State/Province: NY
PostalCode: 14450
Country: United States of America
Voice Phone: (1) 716 425 4380
Fax Phone: (1) 716 425 4382
email: sales@sinopt.com
Web: http://www.sinopt.com/
Contact # 1: Douglas P. Sinclair
Business Description: Software for Computer
Generated Holograms.
****

Slavich Joint Stock Company
2 Mendeleeva Square
City: Pereslavl-Zalessky
Postal Code: 152140
Country: Russia
Business Description: Makers of ultra fine grain
silver halide emulsion for use in holography. USA
distributors include 3Deep Hologram, Keystone
Scientific and Holicon .
****

Smith & McKay Printing Co. Inc.
96 North Almaden Boulevard
City: San Jose
State/Province: CA
Postal Code: 95110-2490
Country: United States of America
Voice Phone: (1) 408 292 890 I
email: smithmckay@aol.com
Contact # 1 : Dave McKay
Business Description: Expert hot-stampers of-
foil holograms onto paper products. Dimensional
printing and fine lithography. Parent company of
""Holographic Impressions"".
****

Sommers Plastic Products
81 Kuller Road
City: Clifton
State/Province: NJ
Postal Code: 07015
Country: United States of America
Voice Phone: (1) 973 7777888
Voice Phone: (1) 800 225 7677
Fax Phone: (1) 973345 1586
email: sales@sommers.com
Web: www.sommers.com
Business Description: Sommers turns holography
into profitable fashions. Epoxy and polyurethane
domings, PVC and rubber labels, stretch fabrics
and vinyls are a few of the va!ue added ingredi-
ents used to transform holography into profitable
products for Nike, Adidas, Warner Bros, Calvin
Klein and many more. See our ad in the color section.
SEE OUR ADVERTISEMENT
****

Sonoma State University
Physics Dept.
1801 E. Cotati Ave.
City: Rohnert Park
State/Province: CA

Postal Code: 94928
Country: United States of America
Voice Phone: (1) 707 664 2119
Contact # 1: Steve Anderson
Business Description: Holography workshops offered. Call for class schedule.
****

Sophia University
Faculty Of Science & Technology
7-1 , Kioi-Cho Chiyoda-Ku
City: Tokyo
Postal Code: 102
Country: Japan
Fax Phone: (81) 03 32383341
Contact # 1: Kazue Ishikawa
Business Description: Holography research.
****

Southern Indiana Holographics
6841 Newburgh Rd
City: Evansville
State/Province: IN
Postal Code: 47715
Country: United States of America
Voice Phone: (1) 8124740604
Fax Phone: (1) 812 473 0981
email: ljohann@msn.com
Contact #1: Larry Johann
Business Description: Holographic Fine Artist.
****

Southwest Holographics
4525 E.Turney Ave.
City: Phoenix
State/Province: AZ
Postal Code: 85018
Country: United States of America
Voice Phone: (1) 602 808 0429
Fax Phone: (1) 602 808 0429
email: sschauer@amug.org
Contact # 1: Steve Schauer
Business Description: Commercial applications of display holography.
****

Spatial Holodynamics (India) Pvt. Ltd.
104/ I 05 Shah & Nahar Estate
Off. Dr. E. Moses Road
City: Worli, Bombay
Postal Code: 400 018
Country: India
Voice Phone: (91) 22 493 0975
Voice Phone: (91) 22 492 1069
Fax Phone: (91) 22 495 0585
Contact # 1: Yogesh Desai
Business Description: Holographic embossing using the latest DI-HO System. Total service, from designing of holograms up to holographic shims. Alternately, if desired, glass Photo Resist Masters.
****

Spatial Imaging Limited
6 Marlborough Rd.
City: Richmond, Surrey
State: England
Postal Code: TW I 0 6JR
Country: United Kingdom
Voice Phone: (44) 181 332 1948
Voice Phone: (44) 1932 564899
Fax Phone: (44) 1932564899
email: spatial@dircon.co.uk
Contact #1: Jeffrey Robb
Business Description: Holographic Origination for all media including embossed, photopolymer and silver halide. Holographic origination systems including digital stereogram mastering and dot matrix. Pulsed portraiture. Technology transfer.
****

Spectra-Physics GmbH
Siemensstr. 20
City: Darmstadt
Postal Code: D-64289
Country: Germany
Voice Phone: (49) 6151 708-0
Fax Phone: (+49) 6151710795
Business Description: Manufacturer and distributor of lasers, German office.
SEE OUR ADVERTISEMENT
****

Spectra-Physics Lasers Inc.
1330 Terra Bella Ave.
City: Mountain View
State/Province: CA
Postal Code: 94039-7013
Country: United States of America
Voice Phone: (1) 800 775 5273
Voice Phone: (1) 650 966 5596
Fax Phone: (1) 650 964 3584
email: splaser@ix.netcom.com
Web: http://www.splasers.com
Contact # 1: Curt Cavoon
Business Description: World's largest supplier of CW and pulsed gas and solid state laser systems, including a comprehensive optical accessories line and a worldwide customer service network.
SEE OUR ADVERTISEMENT
****

Spectratek Inc.
5405 Jandy Place
City: Los Angeles
State/Province: CA
Postal Code: 90066
Country: United States of America
Voice Phone: (1) 3!O 822 2400
Fax Phone: (1) 310 822 2660
Contact # 1: Randy Bouverat
Business Description: Spectratek manufacturers the highest quality diffractio"iilatterns which are the only ones available without seams or pattern breaks. These patterned films are available in a variety offormats, incl uding adhesive backed for labels, laminated to card stock, or films for packaging and other applications.
****

SPIE
The International Society for Optical Engineering
P.O. Box 10
City: Bellingham
State/Province: WA
Postal Code: 98227
Country: United States of America
Voice Phone: (1) 360 676 3290
Fax Phone: (1) 360 647 1445
email: spie@spie.org
web: http://www.spie.org/
Business Description: SPIE - The international Society for Optical Engineering is a nonprofit educational society dedicated to advancing engineering and scientific applications of optical, electro-optical, and optoelectronic instrumentation, systems, and technologies.
SEE OUR ADVERTISEMENT
****

SPIE's Holography Working Group Newsletter
Society of Photo-Optical Instrumentation Engineers
P.O. Box 10
City: Bellingham
State/Province: WA
Postal Code: 98227-0010
Country: United States of America

Voice Phone: (1) 360 676 3290
Fax Phone: (1) 360 6471445
email: info-holo-request@spie.org
Web: http://www.spie.org/
Business Description: The " Holography Working Group" newsletter is published semiannually by SPIE - The International Society for Optical Engineering, for its International Technical Working Groups on Holography.
****

Spindler & Hoyer GmbH & Co.
Koenigsallee 23
Postfach 33 53
City: Goettingen
Postal Code: D-37070
Country: Germany
Voice Phone: (49) (0)551 6935-0
Voice Phone: (+49) (0)5 6935-971
Fax Phone: (+49) (0)551 6935-166
Contact # 1: Mr. Keilholz
Business Description: Manufacturer of precision optics, mechanics and laser technology.
****

Spot Agentur fur Holographie und Werbung
An St. Katharinen 2
City: Koeln
Postal Code: D-50678
Country: Gernlany
Voice Phone: (49) 221 315500
Fax Phone: (+49) 221 322426
Contact # 1: Walter Trebst
Business Description: Distribution of holograms, exhibition
****

Springer-Verlag New York
175 Fifth Ave.
City: NY
State/Province: NY
Postal Code: 100 I 0
Country: United States of America
Voice Phone: (1) 212 460 1500
Voice Phone: (1) 800 777 4643
Fax Phone: (1) 20 I 348 4505
email: orders@springer-ny.com
Web: http://www.springer-ny.com
Business Description: Publishers of an ""Optical Sciences"" series of books, including ""Silver Halide Recording Materials for Holography'", by H. Bjelkhagen.
****

Stanford University
Mechanical Engineering Dept.
Mail Code 402 I
City: Stanford
State/Province: CA
Postal Code: 94305-4021
Country: United States of America
Voice Phone: (1) 650 723 2123
Fax Phone: (1) 650 723 3521
email: dnelson@leland. stanford.edu
Contact #1: Drew Nelson
Business Description: Use of holographic interferometry (with rapid thermoplastic recording of holograms) for measurements of small deformations and for residual stresses in materials via stress release technique.
****

Star Magic
745 Broadway (below 8th St.)
City: New York
State/Province: NY
Postal Code: 10003
Country: United States of America
Voice Phone: (1) 212 228 7770
email: staff@stannagic.com

Web: http: \\www.stannagic.com
Business Description: Retail store featuring
Space Age gifts, holograms, nove lties, etc.
****

Star Magic
1256 Lexington Ave. (85th St.)
City: New York
State/Province: NY
Postal Code: 10028
Country: United States of America
Voice Phone: (1) 212 988 0300
emai l: staff@stannagic.com
Web: http://www.stannagic.com
Business Description: Retail store featuring Space
Age gifts, holograms, novelties, etc.
****

Star Magic
275 Amsterdam St.
City: NY
State/Province: NY
Postal Code: 10023
Country: United States of America
Voice Phone: (1) 212 7692020
email: sa les@stannagic.com
Web: http:\\www.stannagic.com
Business Description: Retail store featuring Space
Age gifts, holograms, novelties, etc.
****

Star Magic
4026 24th Street
City: San Francisco
State/Province: CA
Postal Code: 94114
Country: United States of America
Voice Phone: (1) 415 641 8626
email: sales@stannagic.com
Web: http:\\www.stannagic.com
Business Description: Retail store featuring Space
Age gifts, holograms, novelties, etc.
****

Starcke, Ky.
Ratastie 6
City: Kokemaki
Postal Code: FIN-32800
Country: Finland
Voice Phone: (358) 39 5460 700
Fax Phone: (358) 39 5467 230
Contact #1: Ari-Veli Starcke
Business Description: Starcke K Y is the leading
company selling holograms in Scandinavia. Holo-
gram Hot Stamping.
****

Starlog
Mall of America
Space # N275
City: Bloomington
State/Province: MN
Postal Code: 55425
Country: United States of America
Voice Phone: (1) 612 853 9988
email: starlog@ero ls.com
Web: http://www.starlog.coml
Business Description: Selling all types of holo-
graphic wall art, jewelry and gift items.
****

Starlog
700 Paramus Park Mall, Space #622
City: Paramus
State/Province: NJ
Postal Code: 07652
Country: United States of America
Voice Phone: (1) 20 I 265 9799
emai l: starlog@erols.com
Web: http: //www.starlog.com/

Business Description: Selling all types of holographic wall a11, jewelry and gift items.
****

Starlog
Arrowhead Towne Center
7700 West Bell
City: Glendale
State/Province: AZ
Postal Code: 85308
Country: United States of America
Voice Phone: (1) 602 412 0202
Business Description: Selling all types of holo-
graphic wall art, jewelry and gift items.
****

Starlog
Fox Valley Center
Space #2256
City: Aurora
State/Province: IL
Postal Code: 60504
Country: United States of America
Voice Phone: (1) 630 585 1414
email: starlog@erols.com
Web: http ://www.starlog.coml
Business Description: Selling all types of holo-
graphic wall art, jewelry and gift items.
****

Starlog
Arden Fair Mall Space 111246
1689 Arden Way
City: Sacramento
State/Province: CA
Postal Code: 93566
Country: United States of America
Voice Phone: (1) 916 927 5669
email: starlog@erols.com
Web: http ://www.starlog.coml
Business Description: Selling all types of holo-
graphic wall art, jewelry and gift items.
****

Starlog
Rockaway Town Square, Space #20 12
Route 80 & Mount Hope Avenue
City: Rockaway
State/Province: NJ
Postal Code: 07866
Country: United States of America
Voice Phone: (1) 210 328 4050
email: starlog@erols.com
Web: http ://www.starlog.coml
Business Description: Selling all types of holo-
graphic wall art. jewelry and gift items.
****

Steinbichler Optotechnik GmbH
Am Bauhof4
City: Neubeuern
Postal Code: D-83115
Country: Germany
Voice Phone: (49) 8035 87040
Fax Phone: (49) 8035 1010
email: sales@steinbichler.com
Web: http://www.steinbichler.com
Contact # 1: H. Steinbichler
Business Description: Development and sales of
optical measuring and test systems, e.g. holo-
graphic interferometer, ESPI (electronic speckle
Pattern Interferometer), contour measurement
systems, non-destructive inspection (shearogra-
phy), image analysis software.
****

Steinbichler U.S.A. inc.
40000 Grand River Ave., Suite 101
City: Novi
State/Province: MI

Postal Code: 48375
Country: United States of America
Voice Phone: (1) 888 349 5641
Fax Phone: (1) 248 426 0643
emai l: sa les@steinbichler.com
Web: http://www.steinbichler.com
Business Desc ription: 3-D Digitizing systems,
Holographic Non-Destructive testing. Parent busi-
ness in Germany.
****

Stensborg Inc.
Center for Advanced TechnoiogyIRIS0
Frederiksborgvej 399
City: Roskilde
Country: Denmark
email: jan.stensborg@cat.risoe.dk
Web : http: //www.catscie nce.dk/compani/
descrip.html#stensborg
Contact # 1: Mr. Jan Stensborg 1
Business Description: We offers advisory and
consultancy services to industries wishing to
use holography as a visual means of communi-
cation.
****

Stephens, Anait
1685 Fernald Point Lane
City: Santa Barbara
State/Province: CA
Postal Code: 93108
Country: United States of America
Voice Phone: (1) 805 969 5666
Fax Phone: (1) 805 969 5666
Contact # 1: Anait Arutunoff Stephens
Business Description: Pioneer Holographic Fine
Artist. Extensive portfolio includes reflection and
transmission holograms, pulsed works and true
color reflection.
****

Steuer KG GmbH & Co.
Ernst-Mey-Strasse 7
City: Leinfelden-Echterdingen
Postal Code: D-70771
CountlY: Germany
Voice Phone: (49) (0)7 11 160680
Fax Phone: (+49) (0)71 1 1606863
Contact #1: Mr. Seitz
Busi ness Description: Manufacturer of holo-
graphic hot-stamping machines.
****

STl (see Bridgestone Technologies)
****

STI- Europe
Huttons Yard, Mapledurwell
City: Basingstoke, Hampshire
State: England
Postal Code: RG25 2LP
Country: United Kingdom
Voice Phone: (44) 1256 346208
Fax Phone: (44) 1256329238
Business Description: STI is a manufacturer of
complex holographic products. We offer a wide
variety of products for security, promotional, col-
lectable, and packaging applications. The com-
pany specializes in the development of security
devices for anti counterfeiting and identification
applications.
****

Stiletto Studios
Freinwalder Str. 13a
City: Berlin
Postal Code: D-1 3359
Country: Germany
Voice Phone: (49) 304936829

Business Description: holography in art, furniture with holographic elements
****

Studio Fuer Holographie
Waldfriedenweg 10
City: Eichenau
Postal Code: D-82223
Country: Germany
Voice Phone: (49) (0)814 70831
Fax Phone: (49) (0)814 70831
Contact # 1: Carlo Schmelzer
Business Description: Products: mastering and copy services (rainbow/reflection), production of mass-run embossed holograms, open stock images, art-pieces.
****

Superbin Co. Ltd
3F-339
Section 2 Ho Ping E Road
City: Taipei
Postal Code: 10662
Country: Taiwan
Voice Phone: (886) 02 70 I 3626
Fax Phone: (886) 02 70 I 3531
Contact #1: Edward Hwang
Business Description: Exclusive Chinese representative of Coherent (Argon, Krypton Laser, Dye Laser); Continuum (ruby Laser, Nd: YAG laser); Newport (optical components).Also supply embossed hologram-manufacturing equipment/material and consulting service.
****

Suzhou University
Laser Research Lab
City, Prov.: Suzhou, Jiangsu
Postal Code: 215006
Country: China
Voice Phone: (86) 512 5215257
Fax Phone: (86) 512 5215257
Contact # 1: Chen Linsen
Business Description: Holography classes.
****

Swift Instruments
I 190 North 4th St.
City: San Jose
State/Province: CA
Postal Code: 95112
Country: United States of America
Voice Phone: (1) 408 293 2380
Business Description: Microscope objectives and related optics.
****

Synchron Pty Ltd.
Chempet
P.O. Box 36921
City: Capetown
Postal Code: 7442
Country: South Africa
Voice Phone: (27) 21 551 1790
Fax Phone: (27) 21 52 5291
Contact # 1: Sean Kritzinger
Business Description: Agent and distributor of holographic foils and labels. Also in house hot stamping of holographic images.
****

Syracuse University
Department of Chemistry
III College Place, 1-014C
City: Syracuse
State/Province: NY
Postal Code: 13244-4100
Country: United States of America
Voice Phone: (1) 315 443 2925
Fax Phone: (1) 315 443 4070

email: ijbarani@mailbox.syr.edu
Web: http ://www-che.syr.edu/Sponsler.html
Contact # 1: Michael B. Sponsler
Business Description: Research in Liquid Crystals as Holographic Recording Media.
****

Tair Hologram Company
Behterevsky, 8
City: Kiev
Postal Code: 252053
Country: Ukraine
Voice Phone: (38) 044 269 13 77
email: feofan@megamed.kiev.ua
Contact # 1: Alexander Monchak
Business Description: Fine art holograms.
****

Taiyuan Shiji Holography Ltd.
Shi-Fan-Jie #23
City, Prov.: Taiyuan, Shanxi
Postal Code: 030006
Country: China
Voice Phone: (86) 351 9001996
Fax Phone: (86) 351 7060457
Contact # 1: An Shouzhong
****

Tama Art Umversity
Department Of Physics
1723 Yarimizu Hachiouji-Shi
City: Tokyo
Country: Japan
Voice Phone: (81) 0426 768 611
Fax Phone: (81) 0426 762 935
Contact # 1: H idetoshi Katsuma
Business Description: Research on Holographic TV, Holography Movie
****

Tamarack Storage Devices
12112 Technology Blvd. , Suite 101
City: Austin
State/Province: TX -
Postal Code: 78727
Country: United States of America
Voice Phone: (1) 512 250 3 100
Contact # 1: John Stockton
Business Description.; R&D on optical memory for computers.
****

TAVEX, Ltd.
45 prospect Nauki
City: Kiev
Country: Ukraine
Voice Phone: (380) 44 265-6178
Voice Phone: (380) 44 265 6178
Fax Phone: (380) 44 265-5871
email: sav@tav.kiev.ua
Web: http://www.tav.kiev.ua
Contact # 1: Alexander Stolyarenko
Business Description: Develop and produce photothermoplastic cameras - reversible photosensitive medium and devices on its basis for hologram registration. Produce optical elements: cube coner prisms, other prisms, lenses etc.
****

Technical Marketing Services
925 Park Avenue
City: Laguna Beach
State/Province: CA
Country: United States of America
Voice Phone: (1) 7144971659
Fax Phone: (1) 714 4975331
Contact # 1: Marcus Noble
Business Description: Combines technical strength with extensive experience in marketing in the electro-optics industry. Offers support

in all areas of marketing, including PR and ad agency services, marketing planning and technical writing.
****

Technical University Zvolen
Faculty of Wood Technology
Dept. of Physics and Applied Mechanics
City: Zvolen
Postal Code: SK-960 53
Country: Slovakia
Voice Phone: (42) 855 635
Fax Phone: (42) 855 321 811
email: stano@tuzvo.sk
Contact # 1: Stanislav Urgela
Business Description: Holographic interferometry. Measurement of temperature fields, deformations and vibrations applied to wood technology, wooden plates, musical instruments and material quality control.
****

Technische Fachhochschule Berlin
FB 2 / Labor Fuer Laseranwendungen
Seestrasse 64
City: Berlin
Postal Code: D-13347
Country: Germany
Voice Phone: (49) (0)30 4504 3917
Voice Phone: (+49) (0)3 4504 3918
Fax Phone: (+49) (0)30 4504 3959
email: eichler@tfh-berlin.de
Contact #1: Juergen Prof. Dr. Eichler
Business Description: Holographic Interferometry, Display Holography, Medical Applications.
****

Technische Hochschule Darmstadt
Institut fUr Angewandte Physik
Hochschulstrafle 2
City: Darmstadt
Postal Code: D-64289
Country: Germany
Contact # 1: Theo Prof. Dr. Tschudi
Business Description: HOEs, computergenerated holograms, holographic interferometry
****

Technische Universitaet Berlin
Optisches Institut, Sekr. P II
Strasse des 17. Juni 135
City: Berlin
Postal Code: D-I0623
Country: Germany
Voice Phone: (49) (0)30 314 22498
Voice Phone: (+49) (0)3 31422097
Fax Phone: (+49) (0)30 314 26888
emai l: eichler@physik.tu-berlin.de
Web: http://www.physik.tu-berlin.de
Contact # 1: Hans Joachim Prof. Dr. Eichler
Business Description: Scientific holography research: Optical holographic data storage, realtime holography, material research, semiconductors, liquid crystals, new lasers.
****

Technoexan Ltd
Polytechnicheskaya, 26
City: St. Petersburg
Postal Code: 194021
Country: Russia
Voice Phone: (7) 812 247 9383
Voice Phone: (7) 812 247 5273
Fax Phone: (7) 812 247 5333
Contact # 1: Igor Lovygin
Business Description: Power semiconductors, Lasers, optoelectronics devices (lR range), many channel 1- and p- diodes, equipment for high format art and picture hologram, holographic registers, school packages for showing optic effects.
****

INTERNATIONAL DIRECTORY

Textile Graphics, Inc.
(see Holography Presses On)
201 North Fruitport Road
City: Spring Lake
State/Province: MI
Postal Code: 49456
Country: United States of America
Voice Phone: (1) 616 842 5626
Fax Phone: (1) 6168425653
Contact #1: Jan Bussard
Business Description: Holographic stock or custom images (all shapes and sizes) applied with heat or pressure for adhesion to all substrates. Specialize in stickers and textile applications. Sealed edges prevent delamination in all weather; washable. Worldwide distributors sought!
SEE OUR ADVERTISEMENT
****

The Foreign Dimension Ltd.
190 I Manley Commercial Bldg.
367-375 Queen's Road, Central
Country: China
Voice Phone: (852) 2 542 0282
Fax Phone: (852) 2 541 6011
email: schvarzy@netvigator.com
Web: http://www.dimension.com.hkI
Contact # 1: Frederic Schvartzman
Business Description: Specialists in manufacturing all kinds of holographic and illusion products (watches, keyrings, etc.). If you are a hologram manufacturer, we can also make top quality products at unbeatable prices using your holograms!
SEE OUR ADVERTISEMENT
****

The Foreign Dimension, Ltd.
The Peak Galleria
Level 2, Shops 29 & 42, The Peak
Country: China
Voice Phone: (852) 2 849 636 I
Fax Phone: (852) 2541 6011
email: schvarzy@~tvigator.com
Web: www.dimension.com.hk
Contact # 1: Frederic Schvartzman
Business Description: Holography shop/showroom offering all varieties of holograms for sale to the public. We also offer holograms for sale wholesale to other businesses.
SEE OUR ADVERTISEMENT
****

The Hologram Company # I
Barefoot Landing Suite 4722A
Highway 17 South
City: North Myrtle Beach
State/Province: SC
Postal Code: 29582
Country: United States of America
Voice Phone: (1) 803 272 3583
email: starlog@erols.com
Web: http://www.starlog.com/
Business Description: Located in the golfing capital of the US, selling all types of holographic wall art, jewelry and gift items.
****

The Hologram Company #2
The Florida Mall Shopping Center Room 344
8001 South Orange Blossom Trail
City: Orlando
State/Province: FL
Postal Code: 32809
Country: United States of America
Voice Phone: (1) 407 856 9072
email: starlog@erols.com
Web: http://www.starlog.com/
Business Description: Central Florida vacation land location, selling all types of holographic wall art, jewelry and gift items.
****

The Hologram Company #3
New Orleans Center - Room 498
City: New Orleans
State/Province: LA
Postal Code: 70112
Country: United States of America
Voice Phone: (1) 504 529 5700
email: starlog@erols.com
Web: http ://www.starlog.com/
Business Description: Located in the mall adjoining the Superdome, selling all types of holographic wall art, jewelry and gift items.
****

The Hologram Store, Ltd.
#2673, 8770 - 170 St.
State/Province: Edmonton, Alberta
Postal Code: T5T 4J2
Country: Canada
Voice Phone: (1) 403 444 3333
Fax Phone: (1) 403 444 4455
Business Description: Independent retail store selling holograms. Specializing in holographic and science products. Wholesale and mail order available.
****

The Holography, Laser & Photonics Resource Center
2018 R Street, N.W.
City: Washington
State/Province: DC
Postal Code: 20009
Country: United States of America
Voice Phone: (1) 202 667-6322
Fax Phone: (~2 265-8563
Contact # 1: Micheal Iordan
Resource
Business Description: Features a permanent international show: """Hologram, Laser and Photonic Image of the Future."" The collection includes 75 masterpieces from: Europe, Asia and North America. It is the most informative and entertaining Center for the technologies of the future. The museum is visited by guided tour only, and reservations are required.
****

Third Dimension Arts Inc.
(see Holocrafts)
****

Thorlabs Inc.
435 Route 206
City: Newton
State/Province: NJ
Postal Code: 07860
Country: United States of America
Voice Phone: (1) 973 579 7227
Fax Phone: (J) 973 383 8406
email: sales@thorlabs.com
Web: http://www.thorlabs.com
Business Description: High quality equipment for optics and photonics research including first surface mirrors, optical component mounts, power meters, etc.
****

Three Deep Hologram Co.
(see 3 Deep Hologram Co.)
609 California SI.
City:: Huntington Beach
Stat~/Province: CA
Postal Code: 92648
Country: United States of America
Voice Phone: (1) 714 969 5354
Fax Phone: (1) 714 969 5354
email: acheimets@aol.com
Web: http: \\www.3deepco.com
Contact # 1: Alex Cheimets

Business Description: Supplies Russian silver halide emulsions on plates.
Green sensitive, red sensitive, and new full color emulsions available. Can also recommend appropriate developing and processing procedures.
SEE OUR ADVERTISEMENT
****

Three Dimensional Imagery
P.O. Box 858
City: Vienna
State/Province: VA
Postal Code: 22183-0858
Country: United States of America
Voice Phone: (1) 703 573-0935
email: smichael@3dimagery.com
Web: http://www.3dimagery.com
Contact # 1: Steve Michael
Business Description: Production of white-light reflection display holograms, animated computer-generated holograms, and custom holograms.
****

Three-D Light Gallery
109-A The Commons
City: Ithaca
State/Province: NY
Postal Code: 14850
Country: United States of America
Voice Phone: (1) 607 273 1187
Fax Phone: (1) 607 347 6454
Contact # 1: Jonathan Pargh
Business Description: Artistic holography; holography gallery.
****

Tianjin Holdor Optics Inc. China
II Tianjin Binguan Nandao
City: Tianjin
Postal Code: 300061
Country: China
Voice Phone: (86) 22 28359338
Voice Phone: (1) 402 466-7468
Fax Phone: (86) 22 28359338
email: yuanwbtj@public.tpuj.cn
Contact #1: Weiben Yuan
Business Description : Manufacturer of silver-halide emulsions; monochromatic and panchromatic; for reflection and transmission holography; resolution 3 -10,000 Ilmm; various sized precoated plates available. Also manufacture optical isolation tables, holographic cameras and true color art reflection holograms. USA distributor - Control Optics.
****

Tianjin Water Laser Holography Image Co.
Bin-Guan-Nan-Dao #5
City: Tianjin
Postal Code: 300061
Country: China
Contact # 1: Song Qihong
Business Description: Holography sales and distribution.
****

Tokai University
Department of Electro Photo Optics -
I 11 7 Kitakaname
City: Hiratsuka City
Postal Code: 259-12
Country: Japan
Voice Phone: (81) 0463 58 1211
Fax Phone: (81) 0463 59 2594
Contact # 1: Hideshi Yokota
Business Description: Holography research - artistic holography.
****

Tokyo Institute Of Technology
Imaging Science And Engineering
4259 Nagatsuda Midori-Ku
City: Yokohama
Postal Code: 227
Country: Japan
Voice Phone: (81) 045 922 I I I I
Fax Phone: (81) 045 921 1492
Contact # 1: Masahiro Yamaguchi
Business Description: Holographic Display, 3-D
Imaging Science. Also Dr. Toshio Honda.
****

Tokyo Institute of Technology
Imaging Science & Engineering Lab.
4529 Nagatsuta Midori-ku
City: Yokohama
Postal Code: 226
Country: Japan
Voice Phone: (8\) 45 921 5183
Fax Phone: (81) 45921 1492
email: guchi@ho.isl.titech.ac.jp
Contact # 1: Mashahiro Yamaguchi
****

topac GmbH, Department Holography
Carl - Miele Str. 202-204
City: Guetersloh
Postal Code: D-33311
Country: Germany
Voice Phone: (49) 05241 803302
Fax Phone: (49) 05241 8060870
email: http ://members.aol.com/tophol/
hhome_e.htm
Web: tophol@aol.com
Contact # 1: Uwe Sarda
Business Description: topac GmbH is a full ser-
vice hologram producer with complete production
line for all hologram processes. Speciality is em-
bossed hologram production for security and au-
thenticity devices.
****

Topcon Inc.
75-1 Hasunuma-Machi Ttabasi-Ku
City: Tokyo
Postal Code: 174
Country: Japan
Voice Phone: (81) 0339663141
Fax Phone: (81) 0339662140
Contact #1: Reiji Hashimoto
Business Description: Hologram manufacturer.
****

Toppan Printing Co. , Ltd.
Tech. Research Inst. Tsukuba Research Lab.
4-2-3 Takanodai - Minami
Saitama, Sugitomachi Kit. - gun
Postal Code: 345
Country: Japan 1
Voice Phone: (81) 480 339079
Fax Phone: (81) 480 339022
email: fiwata@tri.toppan.co.jp
Web: http://www.toppan.com!
Contact # 1: Fujio Iwata
Business Description: Embossed holography
****

Toshihiro Kubota, dept of electronics
Faculty of Engineering and Design,
Kyoto Institute of Technology
Kyoto, Matsugasaki, Saky-ku
Postal Code: 606
Country: Japan
Voice Phone: (81) 75 724 7443
Fax Phone: (81) 757247400
email: kubota@dj.kit.ac.jp
Contact # 1: Toshihiro Kubota
Business Description: Works on holography in-
clude the imaging characteristics, recording

material, color holography for fundamental study,
and their applications to holographic display and
optical devices.
****

Total Register Inc.
71 Commerce Drive (Box 719)
City: Brookfield
State/Province: CT
Postal Code: 06804
Country: United States of America
Voice Phone: (1) 203 740 0199
Fax Phone: (1) 203 740 0177
email: sales@totalregister.com
Web: http://www.totalregister.com/
Contact # 1: John Gallagher
Business Description: Manufacturer of registra-
tion devices for hot-stamping presses. Manufac-
turer of registered rotary die cutting equipment
and hot-stamping machines. Registered hologram
sheeting and die cutting services.
****

Towne Technologies
6-10 Bell Ave.
City: Somerville
State/Province: NJ
Postal Code: 08876-0460
Country: United States of America
Voice Phone: (1) 908 722 9500
Fax Phone: (1) 908 722 8394
email: sales@townetech.com
Web: http ://www.townetech.com
Contact # 1: Sal LoSardo
Business Description: Towne Technologies is a
producer of fine quality holographic photoresist
plates with or without a sub-layer of Iron-Oxide.
These plates are spin-coated with striation free
photoresist in sizes up to 15" x 15".
SEE OUR ADVERTISEMENT
****

Toyama National College Of Marit
1-2 Ebie-Neriya -
City: Shinminato
Postal Code: 933 02
Country: Japan
Voice Phone: (8 1) 0766 860 511
Contact # 1: Dr. Kenji Kinoshita
Business Description: Holographic Stereogram
****

Trace Holographic Art & Design, Inc.
107 Inglewood Court
City: Charlottesville
State/Province: VA
Postal Code: 22901-2619
Country: United States of America
Voice Phone: (1) 804 984 4239
Fax Phone: (1) 804 984 5490
email: bittimann@traceholo.com
Web: http://www.traceholo.com
Contact #1: Fernado Datta-Pretta
Business Description: Trace Holographic Art &
Design is a company dedicated to the art of em-
bossed holography, from the initial design phase
all the way through to its final application. Please
see our ad in this publication. Come and visit our
web site.
SEE OUR ADVERTISEMENT
****

Transfer Print Foils, Inc.
(a division of Holopak Technologies, Inc. Co.)
21B Cotters Lane - P.O. Box 538
City: East Brunswick
State/Province: NJ
Postal Code: 08816
Country: United States of America
Voice Phone: (1) 908 238 1800

Voice Phone: (1) 800 235 3645 x445
Fax Phone: (1) 908651 1660
Web: http ://www.hmt.com/holography/TPF/
index.html
Contact # 1: Marc O. Wootner
Business Description: T.P.F. provides holographic products for a variety of end uses. From security images (as transfer coatings or laminated
patches) which enhance the security of credit
cards, licenses, bank documents, tickets and gate
passes; to value added decorative finishes for
greeting cards, book jackets, trophies and picture
frames.
SEE OUR ADVERTISEMENT
****

Traumlaboratorium
TraubenstraBe 41
City: ApenlAperberg
Postal Code: D-26689
Country: Germany
Contact # 1: Elmar Spreer
Business Description: High quality art holograms
****

Triple-D Laser Imaging
Bergselaan 13-B
City: Rotterdam
Postal Code: NL-3037 BA
Country: Netherlands
Voice Phone: (31) 10 465 6331
Fax Phone: (31) 10 465 6331
email: aca@wirehub.nl
Web: http ://home.wirehub.nll-acal
Contact # 1: A.C. Akveld
Business Description: Specializing in the sale of
holograms and related Ofltical equipment.
****

Turing Institute
Boyd Orr Bldg.
University Ave.
City: Glasgow
State/Province: Scotland
Postal Code: GI2 8NN
Country: United Kingdom
Voice Phone: (44) 141 3376410
Fax Phone: (44) 141 3390796
email: tim@turing.gla.ac.uk
Web: turing.gla.ac.uk
Contact # 1: Tim Niblett
Business Description: Three dimensional imaging
systems based on multiple cameras, digital pro-
cessing and proprietary software. Designed espe-
cially for medical and clinical applications where
precise 3D modeling and accurate measurement
are required. The Turing Institute is the trading
name of Greenagate Ltd. (Scotland).
****

Tyler Group
218 Linden Avenue
City: Moorestown
State/Province: NJ
Postal Code: 08057
Country: United States of America
Voice Phone: (1) 609 234 1800
Fax Phone: (1) 609 866 0351
Contact # 1: Payton Old
Business Description: Holographic image secu-
rity consulting and application service. Extensive
experience with high production document and
packaging authentication. Specia lists in Novavi-
sion system.
****

U.K. Gold Purchasers, Inc.
DBA Holograms Unlimited
110 Central Park Mall
City: San Antonio

INTERNATIONAL DIRECTORY

State/Province: TX
Postal Code: 78216
Country: United States of America
Voice Phone: (1) 210 530 0045
Voice Phone: (1) 800 722 7590
Fax Phone: (1) 210 530 0048
Web: http://www.eden.com/..TIlainlink/tx/satiart/
rai/index.htm
Contact #1: Marvin Uram
Business Description: Full line distributor of ho-
logram and related products for specialty retailers
- representing more than 80 firms. One top shop-
ping at competitive prices.
****

Uk Optical Supplies
84 Wimborne Road West
City: Wimborne, Dorset
State: England
Postal Code: BH21 2DP
Country: United Kingdom
Voice Phone: (44) 1202 886 831
Fax Phone: (44) 1202 886 831
Contact # 1: Ralph Cullen
Business Description: Supplying probably the
world's largest selection of Holographic/Op-
tical components which are best quality, best
value. Designed by experienced holographers.
Component selection and laboratory/studio
set-up advice freely available. Also buys and
se lls a large selection of used and surplus
equipment.
****

Ultra-Res Corporation
1395 Greg St. - Suite 107
City: Reno
State/Province: NV
Postal Code: 89431
Country: United States of America
Voice Phone: (1) 702 355 1177
Fax Phone: (1) 702 359 6273
email: alex@acds.com
Web: http://www.aoos.comlURiultra-res.html
Contact #1: Alex Chaihorsky
Business Description: Manufacturers of the Ul-
tra-Res Instant Holographic camera system. Ca-
pable of making small transmission and reflec-
tion holograms "" instantly"", without darkroom
processing, on slides or film rolls. Especially
useful for interferometry, setup ve rification,
HOEs for phase filters and proofing for embossed runs.
SEE OUR ADVERTISEMENT
****

Unifoil Corporation
217 Brook Avenue
City: Passaic
State/Province: NJ
Postal Code: 07055
Country: United States of America
Voice Phone: (1) 973 365 2000 Ext.243
Fax Phone: (1) 973 365 0924
email: unifoil@unifoil.com
Web: www.unifoil.com
Business Description: Manufacturer of holo-
graphic paper and board, both laminated and non-
laminated products. Stock or custom holography
avai lable. Paper or board can be optically sheeted
or cut.

****

Uniphase Lasers
(a divi sion of Uniphase Corp.)
163 Baypoint Parkway
City: San Jose
State/Province: CA
Postal Code: 95134

Country: United States of America
Voice Phone: (1) 408 434 1800
Voice Phone: (1) 800 644 8674
Fax Phone: (1) 408 433 3838
email: sales@uniphase.com
Web: http: //www.uniphase.com
Contact # 1: Tanis Mofchetti
Business Desc ription: Manufacturer of lasers
suitable for holography.
****

Uniphase Vertriebs-GmbH
Arbeostr. 5
City: Eching
Postal Code: D-85386
Country: Germany
Voice Phone: (49) 89 3 19 60 26
Voice Phone: (+49) 89 3 19 30 02
Business Description: Manufacturer and distribu-
tor of lasers ion-laser, he-ne laser, diode pumped
solid state laser.
****

United Assn. Manufacturer's Representatives
34071 La Plaza, Suite 120
P.O. Box 986
City: Dana Point
State/Province: CA
Postal Code: 92629
Country: United States of America
Voice Phone: (1) 714 240-4966
Fax Phone: (1) 714 240:700 I
Contact # 1: Karen Mazzola
Business Description: Provide valuable services
that help manufacturers and reps come together
for mutual benefit.
****

Univ. de Liege
Sart Tilman
Hololab, Physique B5
City: Liege
Postal Code: B-4000
Country: Belgium
Voice Phone: (32) 4 166 3626
Fax Phone: (32) 4 166 2355
email: lion@gw.unipc.ulg.ac.be
Contact # 1: Yves F. Lion
Business Description: Holography Courses.
****

Universidade Do Porto
Laboratorio De Fisica
Praca Gomes Teixeira
City: Porto
Postal Code: P-4000
Country: Portugal
Voice Phone: (351) 2 557 0700
Business Description: Holographic non-destruc-
tive testing; Academic holography research.
****

Universita Di Roma
La Sapienza Dipt Di Fisica
Piazzale Aldo Moro 2
City: Rome
Postal Code: I-00185
Country: Italy
Voice Phone: (39) 6 559 9776
Business Description: Scientific research.
****

Universite De Neuchatel
Institut De Microtechnique
2, Rue A L Breguet
City: Neuchatel
Postal Code: CH-2000
Country: Switzerland
Voice Phone: (41) 32 720 09 20
Fax Phone: (41) 327200990

email: info@fsrm.ch
Web: http ://www.fsrm.ch
Contact # 1: Rene Dandliker
Business Description: Industrial research.
****

Universite Laval
Dept Physique - C.O.P.L
Pavillon Vachon
City, Prov.: University City, Quebec
Postal Code: GIK 7P4
Country: Canada
Voice Phone: (1) 418 656 2131
Voice Phone: (1) 418 656 3436
Web: http://www.fsg.ulaval.cal
Contact #1: Roger A. Lessard
Business Description: Holography education. Re-
search in holographic recording materials (photo-
polymer) for optical data storage, CGH and dif-
fractive optics.
****

University Of Tsukuba
Institute Of Art & Design
City: I-I , Tennodai
State/Province: Tsukuba
Postal Code: 305
Country: Japan
Voice Phone: (81) 298 53 2883
Fax Phone: (8 1) 298 53 6508
Contact # 1: Shunsuke Mitamura
Business Description: Artistic holography, holog-
raphy education.
****

Center For Applied Optics
City: Huntsville
State/Province: AL
Postal Code: 35899
Country: United States of America
Voice Phone: (1) 205 890 6030
Fax Phone: (1) 205 895-6618
Contact #1: Chandra Vikram
Business Description: Scientific holography re-
search, NDT.
****

University Of Alicante
Applied Phys ics/Cent De Holograf
Facultad De Ciencias
City: Alicante Apdo
Postal Code: 99
Country: Spain
Voice Phone: (34) 566 1200
Contact # 1: A. Fimia.
Business Description: Artistic holography; HOEs;
workshops.
****

University Of Arizona
Optical Sciences Center
City: Tucson
State/Province: AZ
Postal Code: 85721
Country: United States of America
Voice Phone: (1) 520 621 6997
Fax Phone: (1) 520 62196 13
Contact # 1: Dick Powell
Business Description: Industrial and scientific
holography research; Holographic interferom-
etry; Holographic non-destructive testing. Pri-
marily graduate. Some undergraduate courses
available.
****

University Of Dayton
Research Institute
300 College Park
City: Dayton

State/Province: OH
Postal Code: 45469-0 I 02
Country: United States of America
Voice Phone: (1) 937 229 3515
Fax Phone: (1) 937 229 3433
email: info@udri .udayton.edu
Web: http://www.udri.udayton.edul
Business Description: Scientific research, industrial research; courses.
****

University of Erlanger
Physics Institute, Dept. of Optics
Staudtstrasse 7
City: Erlangen
Postal Code: 0-9 1058
Country: Germany
Voice Phone: (49) 9131858395
Fax Phone: (49) 9131 13508
email: schwider@move.physik. uni-erlangen.de
Contact # 1: 1. Schwider
Business Description: We offer design and manufacturing of microlenses and diffractive optical elements (lenses, beamsplitters, beam shaping elements). We can test micro optical elements by means of interferometers.
****

University of Latvia
Institute of Solid State Physics
8 Kengaraga Str.
City: Riga
Postal Code: LV-I063
Country: Latvia
email: teteris@acad.latnet.lv
Web: htrp :l/www.cfi.lu.lv/
Business Description: R&D and production of photoresists for holography suitable up to 650 nm. Design & production of embossed 20/30 holograms. Ni shim electroforming.
****

University Of Michigan
Dept. of Electrical Engineering
Room 1108 EECS Building
City: Ann Arbor
State/Province: MI
Postal Code: 48109-2122
Country: United States of America
Voice Phone: (1) 313 764 9545
Fax Phone: (1) 3 13 763 1503
Contact # 1: Emmet Leith
Business Description: Scientific holography research. Design HOEs. Courses on holography.
****

University of Muenster
Laboratory of Biophysics
Robert-Koch-Str. 45
City: Muenster
Postal Code: 0-48129
Country: Germany
Voice Phone: (49) (0)251 836888
Fax Phone: (+49) (0)251 838536
email: biophys@gabor.uni-muenster.de
Contact # 1: Gert von Bally
Business Description: holography and Interferometry in medicine. Environmental research and cultural heritage protection.
****

University of Munich
Institute Of Medical Optics
Barbarastrasse 16
City: Munich
Postal Code: 0-80797
Country: Germany
Voice Phone: (49) (0)89 2 105 3000
Fax Phone: (+49) (0)89 1240 630 I
Contact # 1: Mr. Zurek

Business Description: Medical Holography, Scientific holographic research.
****

University Of Rochester
The Institute Of Optics
City: Rochester
State/Province: NY
Postal Code: 14627
Country: United States of America
Voice Phone: (1) 716 275 2322
Fax Phone: (1) 716 273 1072
email: info@optics.rochester.edu
Web: http://www.optics. rochester.edu
Business Description: Scientific and industrial holography research; interferometry; particle testing & measurement. Primarily graduate. Some undergrad courses include holography related studies.
****

University Of Southern California
Department Of Physics
University Park
City: Los Angeles
State/Province: CA
Postal Code: 90089-0484
Country: United States of America
Voice Phone: (1) 213740 11 34
Contact # 1: Jack Feinberg
Business Description: R&D program. Utilizes holographic gratings for for passive and active optical switches, primari ly for fiber opitc communications.
****

University of Stuttgart
Institute Of Applied Optics
Pfaffenwaldring 9
City: Stuttgati
Postal Code: D-70569
Country: Germany
Voice Phone: (49) (0)711 68 ~75
- Fax Phone: (+49) (0)7 11 685 6586
Contact #1: Hans Tiziani
Business Description: Scientific holography research; interferometry, NOT, HOE's
****

University Of Tokyo
Faculty Of Engineering
Hongo 7-3-1 Bunkyo-Ku
City: Toyko
Country: Japan
Voice ·Phone: (8 1) 3 3812 2111
Fax Phone: (81) 3 381 8 5706
email: info@t.u-tokyo.ac.jp
Web: http://www.t.u-tokyo.ac.jp/
Contact # 1: T. Uyemura
Business Description: Scientific and Medical holography research; Interferometry.
****

University Of Wisconsin/Madison
Dept. Of Engineeri ng - Professional Development
432 North Lake Street
Ci ty: Madison
State/Province: WI
Postal Code: 53706
Country: United States of America
Voice Phone: (1) 608 262 8708
Fax Phone: (1) 608 263 3160
email: custserv@epd.engr.wisc.edu
Web: www.engr.wisc.edu/
Business Description: Continuing Education courses on laser system design and application. Call or write for catalog.
****

Unterseher & Associates
709 112 West Glen Oaks Blvd.
City: Glendale
State/Province: CA
Postal Code: 9 1202
Country: United States of America
Voice Phone: (1) 8 I 8 549 0534
Contact # 1: Fred Unterseher
Business Description: Artistic holography and holography education. Originates masters for mass production holograms in embossed, DCG and photopolymer materials. Both pulsed and CW lasers available.
****

Uvex Safety Inc.
10 Thurber Blvd.
City: Smithfield
State/Province: RI
Postal Code: 02917
Country: United States of America
Voice Phone: (1) 40 I 232 1200
Voice Phone: (1) 800 343 34 I I
Fax Phone: (1) 40 I 232 1830
Web: www.uvex.com
Contact # 1: Mark McLear
Business Description: Manufacturer of industrial, medical and laser safety eyewear, as well as disposable and reusable respirators.
****

Van Leer Metallized Products
24 Forge Park
City: Franklin
State/Province: MA
Postal Code: 02038
Country: United States of~erica
Voice Phone: (1) 508 541 7700
Voice Phone: (1) 800 343 6977
Fax Phone: (1) 508 541 7777
email: vlcommdept@vanleer.nl
Web: http://www.vanleer.com
Contact # 1: Harry Mann
Business Description: Producer of unique metallized papers: HoloPRISM holographic papers, Valvac metallized papers and HoloSECURE security papers. Increase product awareness in a variety of applications including labels, advertisements, laminated constructions and specialty promotions. Can be easily converted to meet your needs.
****

Vincennes University
1002 North First Street
City: Vincennes
State/Province: IN
Postal Code: 47591
Country: United States of America
Voice Phone: (1) 812 888 8888
Contact # 1: Richard Duesterberg
Business Description: Offering holography workshops for high school teachers, & college level courses in holography. We have 4 research grade optical tables, as well as argon and krypton lasers. Call for more details.
****

VinTeq, Ltd,
6 1 I November Lane / Autumn Woods
City: Willow Springs
State/Province: NC
Postal Code: 27592-7738
Country: United States of America
Voice Phone: (1) 919 639 9424
Fax Phone: (1) 919 639 7523
email: vinson@vinteq.com
Web: http ://www.vinteq.com
Contact # 1: Joachim Vinson
Business Description: Distributor for HRT sil-

INTERNATIONAL DIRECTORY

ver halide plates. Red, Blue, Green and panchromatic silver halide emulsions. (HRT GmbH in Steinau/Germany).
SEE OUR ADVERTISEMENT
****

Virtual Image (a division of Printpack, Inc.)
PO box 1198
City: Litchfield
State/Province: CT
Postal Code: 06759
Country: United States of America
Voice Phone: (1) 860 567 2022
Fax Phone: (1) 860 567 8699
email: tvinstadt@printpack.com
Web: http://www.virtimage.com
Contact # 1: Tom Vinstadt
Business Description: Manufacturer of high quality, embossed BOPP holographic film, primarily for the packaging industry. Specialty is large runs in wide web.
SEE OUR ADVERTISEMENT
****

Visual Visionaries
2011 Clement St., Suite 4
City: San Francisco
State/Province: CA
Postal Code: 94 121
Country: United States of America
Voice Phone: (1) 415 666 0779
Business Description: Consulting and marketing firm. Exhibitions and educational services. Over 13 years experience with holographic production, display and sales. Professional and reliable.
****

Volkswagen AG
Forschung und Entwicklung
Brieffach 1785
City: Wolfsburg
Postal Code: D-38~36
Country: Germany
Voice Phone: (49) (0)536 925 824
Fax Phone: (+49) (0)536 972 444
Contact #1: M.-A. Dr. Beeck
Business Description: Industrial research, Interferometry; Holographic non-destructive testing, Laser combustion diagnostics.
****

Volvo-Flygmotor
S-461
City: Trollhattan
Postal Code: S-81
Country: Sweden
Voice Phone: (46) 0520 94471
Contact # 1: Robert Frankmark.
Business Description: Holographic non-destructive testing.
****

Voxel
26081 Merit Circle - Suite 117
City: Laguna Hills
State/Province: CA
Postal Code: 92653
Country: United States of America
Voice Phone: (1) 714 348 3200
Fax Phone: (1) 714 348 8665
email: sales@voxel.com
Web: http://www.voxel.com
Business Description: Medical imaging research. Company is developing a 3D visual display for non-invasive imaging techniques to be used as a diagnostic tool.
****

W. Cordes GmbH + Co.
Offsetdruckerei und Kartonagenfabrik
Wasserbreite 71
City: Bunde
Postal Code: D-32257
Country: Germany
Voice Phone: (49) 5223 1590 I
Fax Phone: (+49) 5223 15900
Contact #1: H. E. Dipl.-Ing.Cordes
Business Description: Hotstamping holograms, sticker-holograms, packaging using paper and carton for advertising, exhibitions
****

Waseda University
Dept Of Applied Physics
School Of Science & Engineering
City: Tokyo
Postal Code: 160
Country: Japan
Voice Phone: (81) 03 209 321
Business Description: Medical holography research.
****

Wave Mechanics
450 North Leavitt
City: Chicago
State/Province: IL
Postal Code: 60612
Country: United States' of America
Voice Phone: (1) 312 829 WAVE
Fax Phone: (1) 312 829 8557
Contact # 1: Deni Drinkwater
Business Description: Artistic holographer; silver halide transmission and reflection; consultant.
****

Wavefront Research, Inc.
616 West Broad Street
City: Bethlehem
State/Province: PA
Postal Code: 180 18-5221
Country: United States of America
Voice Phone: (1) 610 974 8977
Fax Phone: (1) 610 974 9896
email: tws@wavres.com
Contact # 1: Thomas Stone
Business Description: Basic and applied research and development in holographic optical elements, applications of holography, non linear optics, and novel optical systems.
****

Wavefront Technology
15149 Garfield Ave.
City: Paramount
State/Province: CA
Postal Code: 90723
Country: United States of America
Voice Phone: (1) 562 634 0434
Fax Phone: (1) 562 634 0434
Contact # 1: Joel Petersen
Business Description: Specialists primarily serving the embossing industry. Embossed hologram mastering, recombining and ganging. Prototype, short run embossing. Rigid sheet embossing up to 4 x 8 foot in transmission or reflection.
SEE OUR ADVERTISEMENT
****

Wesley, Ed
2124 West Irving Park Road
City: Chicago
State/Province: IL
Postal Code: 60618-3924
Country: United States of America
Voice Phone: (1) 312 539 3672
Contact # 1: Ed Wesly

Business Description: Holographic fine artist, author, and researcher. Consulting services offered. Pulsed laser specialist.
****

Whiley Foils Limited
Firth Road
Houston Industrial Estate
City: Livingston, West Lothian
State: Scotland
Postal Code: EH54 5DJ
Country: United Kingdom
Voice Phone: (44) 1506438611
Fax Phone: (44) 150643 8262.
email:
  graeme_milloy@whileyfoi ls@compuserve.com
Contact #1: David Hedley
Business Description: Whiley Foils Limited is a long-established manufacturer of stamping foils. We have developed special base materials for Holographic embossing and market these and other Holographic foils worldwide.
****

Wild Style Entertainment
1201 Park Ave. Suite 203A
City: Emeryville
State/Province: CA
Postal Code: 94608
Country: United States of America
Voice Phone: (1) 510 654 8395
Fax Phone: (1) 510 654 8396
email: STaylorWS@AOL.com
Contact # 1: Steven Taylor
Business Description: Computer imaging and animation for holography.
****

Witchcraft Tape Products, Inc.
P.O. Box 937
City: Coloma
State/Province: MI
Postal Code: 49038
Country: United States of America
Voice Phone: (1) 616 468 3399
Voice Phone: (1) 800 521 0731
Fax Phone: (1) 616 468 3391
Contact # 1: Ronald Warczynski
Business Description: Full service manufacturer of quality embossed holograms, precision registered die cutting hot stamping, specialty lamination, product assembly and packaging. This is our 25th year of furnishing quality products to the industry.
SEE OUR ADVERTISEMENT
****

Wonders of Holography Gallery
PO Box 1244
City: Jeddah
Postal Code: 21431
Country: Saudi Arabia
Voice Phone: (966) 2 652 0052
Voice Phone: (966) 2 653 4004
Fax Phone: (966) 2 651 1325
Contact #1: A.M. Baghdadi
Business Description: Retail holograms of all kinds. We are the first and only holography gallery in the Gulf Countries. We resell all types of holograms and we produce laser shows.
****

Worcester Polytechnic Institute
Mechanical Engineering Department
100 Institute Road
City: Worcester
State/Province: MA
Postal Code: 01609-2280
Country: United States of America
Voice Phone: (1) 508 831 5000

Zero Gravity

Fax Phone: (1) 508 8315713
email: rjp@wpi.edu
Web: https:llwww.wpi.edu
Contact #1: Ryszard Pryputniewicz
Business Description: Center for Holographic Studies and Laser Micro-Mechatronics - Photonics studies for undergrad, graduate and post graduate students. Scientific, medical & industrial holography research; Interferometry; Holographic non-destructive testing.
****

World Holographics
2934 Beverly Glen Circle #400
City: Los Angeles
State/Province: CA
Postal Code: 90077
Country: United States of America
Voice Phone: (1) 310 4741935
Fax Phone: (1) 310 446 9 194
email: world3D@ao1. com
Contact #1: Greg Schuman
Business Description: We are a full service producer of both holographic and lenticular product. We specialize in the production of promotional premiums for our clients who include many of the nations largest marketers.
****

Wuhan Packaging and Printing United Co.
Han-Kou Da-Shuang-Jie #335
City, Prov.: Wuhan, Hubei
Postal Code: 430022
Country: China
Contact # 1. Luo Wenjun
Business Description: Hologram finisher
****

Wuxi Light Impressions Inc.
Mei Cun
City, Prov.: Wuxi, Jiangsu
Postal Code: 214000
Country: China
Voice Phone: (86) 510 8150292
Fax Phone: (86) 5 I 0 8150292
Contact # 1: Wang Guoping
Business Description: Holography distribution
****

WYKO-VEECO
2650 East Elvira Road
City: Tucson
State/Province: AZ
Postal Code: 85706-7123
Country: United States of America
Voice Phone: (1) 520 741 1044
Fax Phone: (1) 520 294 1799
email: sales@wyko.com
Web: http://www.wyko.com
Contact #1: Don McNeil
Business Description: Scientific holography research; Interferometry and analysis.
****

Xiamen Grand World Laser Label Products
Si-Ming-Nan-Lu 412-# 12
2nd Floor
City, Prov.: Xiamen, Fujian
Postal Code: 361005
Country: China
Voice Phone: (86) 592 2083178
Fax Phone: (86) 592 2083179
Contact # 1: Huang Lishan
****

Yu Feng Laser Images Co.
Shantou Economic Zone
Long-Hu-Qu Zhu-Chi-Lu Zhong-Duan
City: Shantou, Guangdong
Postal Code: 515041

Country: China
Voice Phone: (86) 754 8893697
Voice Phone: (86) 754 8891015
Fax Phone: (86) 754 8893677
Contact # 1: Xue Minqin
Business Description: Holography distribution.
****

Zanders Feinpapiere AG
Veldener Str. 121-131
City: Dueren
Postal Code: D-52349
Country: Germany
Voice Phone: (49) 2202 15-0
Fax Phone: (+49) 2202 15-2806
Business Description: Manufacturer of paper, carrier material for hot stamped holograms
****

Zec, Peter.
Lerchenstr. 142A
City: Osnabrueck
Postal Code: D-49088
Country: Germany
Voice Phone: (49) 541 186059
Contact # 1: Peter Dr. Zec
Business Description: consulting, organization of hologram exhibitions, publications on holography
****

Zentrum fur Kunst und Medientechnologie
Kaiserstral3e 64
City: Karlsruhe
Postal Code: D-76133
Country: Germany
Voice Phone: (49) 721 1333855
Business Description: Education for artists and other interested parties on holography and computergenerated holograms
****

Zero Gravity
3A River Valley Road -
#01 -04 Clark Quay
Postal Code: 0617
Country: Singapore
Voice Phone: (65) 334 3648
Business Description: , Retail gift store offering holograms and other unique products.
****

Zero Gravity
Changi Airport - Terminal 2
Departure Transit North 026-11 2/026-113
Country: Singapore
Voice Phone: (65) 542 3426
Business Description: Retail gift store offering holograms and other unique products.
****

Zero Gravity
Streets of Mayfair
3390 Mary SI. # 182
City: Coconut Grove
State/Province: FL
Postal Code: 33133
Country: United States of America
Voice Phone: (1) 305 569 0077
email: info@spectore.com
Web: http://www.spectore.com
Business Description: Retail gift store offering holograms and other unique products.
****

Zero Gravity
#F3 Forum Shops at Ceasar's
3500 Las Vegas Blvd. S.
City: Las Vegas
State/Province: NV

Postal Code: 89109
Country: United States of America
Voice Phone: (1) 702 731 3565
email: info@spectore.com
Web: http://www.spectore.com
Business Description: Retail gift store offering holograms and other unique products.
****

Zero Gravity
Aloha TowerlSpace 155
101 Ali Moana Blvd.
City: Honolulu
State/Province: HW
Postal Code: 96813
Country: United States of America
Voice Phone: (1) 808 545 2355
email: info@spectore.com
Web: http://www.spectore.com
Business Description: Retail gift store offering holograms and other unique products.
****

Zero Gravity
Pointe Orlando Suite 1120
9101 International Drive
City: Orlando
State/Province: FL
Postal Code: 32819
Country: United States of America
Voice Phone: (1) 954429 1017
email: info@spectore.com
Web: http: //www.spectore.com
Business Description: Retail gift store offering holograms and other unique products.
****

Zero Gravity / Galaxies lfnlimited, Inc.
637 N.W. 12th Ave.
City: Deerfield Beach
State/Province: FL
Postal Code: 33442
Country: United States of America
Voice Phone: (1) 954 429 10 17
Fax Phone: (1) 954 4212391
email: info@spectore.com
Web: http ://www.spectore.com
Contact # 1: Larry DeBerry
Business Description: Parent company for Zero Gravity retail stores.
****

Zhuhai Xiangzhou Great Wall Laser Anticounterfeiting Label Co.
Zi-Jing-Lu #40
City, prov.: Zhuhai, Guang-Dong
Postal Code: 519000
Country: China
Contact # 1: Wu Ziping
****

Zone Holografix
5338 B Vineland Ave.
City: North Hollywood
State/Province: CA
Postal Code: 91601
Country: United States of America
Voice Phone: (1) 8 I 8 985 8477
Fax Phone: (1) 818 549 0534
email: 100142.1543@compuserve.com
Contact # 1: Fred Unterseher
Business Description: Originates masters for mass production holograms in embossed, DCG and photopolymer materials. Both pulsed and CW lasers available.
SEE OUR ADVERTISEMENT
****

IF YOUR COMPANY PROVIDES GOODS OR SERVICES TO THE HOLOGRAPHY INDUSTRY, GET YOUR FREE LISTING!

A List of Holography-Related Businesses, Sorted by Country

Australia - 3D Optical Illusions
Australia - 3DIMAGE
Australia - Australian Holographics
Australia - Holograms Fantastic & Optical..
Australia - HopSec/dii
Australia - Lazart Holographics
Australia - Menning, Melinda
Australia - Moonbeamers
Australia - New Dimension Holographics
Australia - Parallax Gallery

Austria - Holography Center of Austria

Belgium - Agfa - Gevaert N.V
Belgium - Center for Applied Research
Belgium - Free University Of Bmssels
Belgium - Laboratory Vinckiner
Belgium - Univ. de Liege

Canada - Abrams, Claudette
Canada - Capilano College
Canada - Centre d'Art Holographique et. ..
Canada - Concordia University
Canada - Deep Space Holographics
Canada - Dimension 3
Canada - Fringe Research Holographics
Canada - General Holographics, Inc.
Canada - Harvard Apparatus, Canada
Canada - Holocrafts
Canada - Hologram Development Corp.
Canada - Holo. & Media Inst of Quebec
Canada - Holostar
Canada - Inspeck, inc
Canada - Island Graphix
Canada - Lasiri~ Inc.
Canada - M Crenshaw Holo.Studio Internnl
Canada - Mu's Laser Works
Canada - Ontario College of Art/Holography
Canada - Ontario Science Centre
Canada - Page, Michael
Canada - Photon League Of Holographers
Canada - Royal Holographic Art Gallery
Canada - Silverbridge Group
Canada - The Hologram Store, Ltd.
Canada - Universite Laval

China - Beijing Dongfang Laser Printing ..
China - Beijing Fantastic Hologram Products
China - Beijing Hologram Printing Tech. Co
China - Beijing Sanyou Laser Images Co.
China - Beijing University of Posts & Tele.
China - Beijing Xi Ji Wo Computer Graphic.
China - Chengdu Xinxing lnts Developmt..
China - China Ann Arbor Holographical In st.
China - Chongqing Yinhe Laser Products Ltd
China - Far East Holographics
China - Foreign Dimension
China - Foshan Holosun Packaging Co. Ltd
China - Guangdong Dongguan South Holo
China - Guangzhou Chuntian Industrial Tech
China - Guangzhou Inst.of Electronics
China - Holography Development Co.Ltd.
China - Image Technical Development Co.
China - Institute for Holographic Tech.
China - Jiangsu Sida Images, Inc.
China - Minjian Laser Holography
China - Morning Light Holograms

China - Nanjing Sanle Laser Technology.
China - North Light Holograms Ltd.
China - Qingdao Gaoguang Holography ..
China - Qingdao Qimei Images, Inc.
China - Quan Zhou Pacific Laser Images
China - Shan dong Academy of Sciences
China - Shanghai Dahua Printing Factory
China - Shanghai Kanlian S & T Develmt. ..
China - Shenzhen Reflective Materials Fact ..
China - Suzhou University
China - Taiyuan Shiji Holography Ltd.
China - riapjin.H,zldpr.Qgtics.Jns.. ~hi!!a .
China - Tianjin Water Laser'l-iolographllmgf
China - Wuhan Packaging and Printing Untd :r
China - Wuxi Light Impressions In c. i.
China - Xiamen Grand World Laser Label..
China - Yu Feng Laser Images Co.
China - Zhuhai Xiangzhou Great Wall Laser

Columbia - Imagenes Holo De Columbia

Czech Rep. - Lightgate, Ltd

Denmark - Ibsen Micro Stmctures NS
Denmark - Stensborg Inc

Finland - Heptagon Oy
Finland - Starcke, Ky.

France - Aerospatiale
France - Atelier Holographique De Paris
France - GEHOL sari
France - Holo-Laser
France - Hologram Industries
France - Holomedia France
France - Magic Laser
France - Mulhem, Dominique
France - Quantel
France - S.O.P.R.A.

Germany - 3D Vision
Germany - AB Rueck Holoart
Germany - Academy of Media Alt s Cologne
Germany - AD HOC Public Relations
Germany - Adlas G.M.B.H. & Co Kg.
Germany - AHT 3D-Medien
Germany - AKS Holographie-Galerie GmbH
Germany - AKS Holograp hie-Gallerie
Germany - AlDirksen + Solm ..
Germany - Photonics Direct
Germany - Arbeitskreis Holografie B.V
Germany - Arn1in Klix Holographie
Germany - Art Agentur Kaln
Germany - Artbridge Light Studios
Germany - Baier Praegepressen
Germany - Bauer, Josef
Germany - Bernhard Halle Nachf. GmbH
Germany - BIAS
Germany - Burgmer, Brigitte
Germany - CHIRON Technolas GmbH
Germany - Coherent Luebeck GmbH
Germany - Curt Abramzik
Germany - Daimler Benz Aerospace
Germany - Decolux GmbH
Germany - Deutsche Gesellschaft fur Holo.
Germany - D.Dmcker Verlagsgese llschaft
Germany - Die Dritte Dimension

Germany - Dietmar Oehlmann
Germany - Directa GmbH
Germany - Dornier Medizintechnik GmbH
Germany - Dr. Steeg & Reuter GmbH
Germany - EPA - Elektro-Physik Aschen
Germany - ETA-Optik Gmbh
Germany - Fielmann-Verwaltung KG
Germany - F.M. Schenuker's Aquarius...
Germany - Galerie 3D
Germany - Galerie 7000
Germany - Galerie Westerland/Sylt
Germany - G. Winopal Forschungsbedarf
Germany - Gigahertz-Optik
Germany - Grau, G., Prof. Dr. techn.
Germany - Gresser, E., KG
Germany - H & W, Holo. & Werbung...
Germany - H. Kallenbach - H.M.V.
Germany - Haus der Technik e. V
Germany - HeiB, Peter, Dr., Priv.-Doz.
Germany - Highlite
Germany - HMS-Elektronik
Germany - Hofmann-Lange, Brigitte, Dr.
Germany - HOL 3, Galerie fur Holographie
Germany - Holar Seele KG
Germany - Holarium
Germany - Holo GmbH
Germany - Holo Service/Service-Druck
Germany - Holo Time Gericke
Germany - Holo-Idee Reiner Kleinherne
Germany - Holodesign
Germany - Holografie-Hofmann
Germany - Hologram Company RAKO
Germany - Hologramm
Germany - Holographic Laserdesign
Germany - Holographic Studios Rosowski
Germany - Holographic Systems Muenchen
Germany - Holographie & Design
Germany - Holographie Anubis
Germany - Holographie Fachstudio.
Germany - Holographie Konzept GmbH
Germany - Holographie Labor
Germany - Holographie Roth
Germany - Holographie-Labor Mielke
Germany - Holoptics
Germany - Holopublic Unbehaun
Germany - Holosta Holographie-Galerie
Germany - Holostudio Beate Krengel
Germany - Holotec
Germany - HoloVision
Germany - Holtronic GmbH
Germany - HRT Holo. Recording Tech.
Germany - Ingenieurbi.iro Geiger
Germany - Institut fur Angewandte Physik
Germany - K. Thielker & T. Rost
Germany - Kolbe-Dmck mit Tochtergese ..
Germany - Krystal Holographics Vertriebs
Germany - Laser-Solution-Manag::ment
Germany - Laserfilm Eckard Knuth
Germany - Lauk & Partner GmbH
Germany - Leonhard Kurz GmbH
Germany - Leseberg, Dr. Detlef
Germany - LichtBlicke Clallen & Voss
Germany - Magick Signs Holografie
Germany - Magick signs Holografie
Germany - Mario Liedtke pro design
Germany - Mefoma Fototechnik GmbH
Germany - Melles Griot GmbH

Germany - Metrologic Instruments GmbH
Germany - Meulien Odile
Germany - Modem Marketing
Germany - Miinchner Volkshochschule e. V.
Germany - Museum 3. Dimension
Germany - Museum flir Holographie ..
Germany - NEC Electronics (Europe)
Germany - Newport GmbH
Germany - Oeserwerk Ernst Oeser & S6hne
Germany - Optical Coating Laboratory
Germany - Optical Test Equipment
Germany - Ose Holografie-Design
Germany - OWIS Gmbh
Germany - Phantastica
Germany - Physik Instrumente (PI) GmbH
Germany - Polytec GmbH
Germany - Promotion Products Miinchen
Germany - Rofin-Sinar Laser GmbH
Germany - Rottenkolber Holo-System
Germany - Spectra-Physics GmbH
Germany - Spindler & Hoyer GmbH & Co.
Germany - Spot Agentur flir Holographie ..
Germany - Steinbichler Optotechnik GmbH
Germany - Steuer KG GmbH & Co.
Germany - Stiletto Studios
Germany - Studio Fuer Holographie
Germany - Technische Fachhochschule
Germany - Technische Hochschule Darn1st ...
Germany - Technische Universitaet Berlin
Germany - Technolas Laser Technik Gmbh
Germany - topac GmbH, Dept Holography
Germany - Traumlaboratorium
Germany - Uniphase Vertriebs-GmbH
Germany - University of Erlanger
Germany - University of Muenster
Germany - University of Munich
Germany - University of Stuttgart
Germany - Volkswagen AG
Germany - W. Cordes GmbH + Co.
Germany - Zanders Feinpapiere AG -
Germany - Zec, Peter, Dr. phil.
Germany - Zentrum flir Kunst & Medien.

Greece - Cavomit
Greece - Hellenic Institute Of Holography

Hungary - Artplay Holographika Studio
Hungary - Hologram Varga Miklos
Hungary - Optopol Panoramic Metrology ...

India - Holostik India Pvl. Ltd.
India - I.S. Gill
India - Ojasmit Holographics
India - Print-M-Boss
India - Shriram Holographics
India - Spatial Holodynamics (India) Pvt.

Israel - Glaser - Technical Consulting
Israel - Holo-Or Ltd
Israel - Holography Israel

Italy - Diavy srl
Italy - Holo 3D S.pA
Italy - Universita Di Roma

Japan - Asahi Glass Co.
Japan - Brainet Corporation - International
Japan - Canon Inc. R&D Headquarters
Japan - Central Glass Co., Ltd.
Japan - Chiba University
Japan - Dai Nippon Printing Co., Ltd.

Japan - Fuji Electric Co. Ltd
Japan - Fujitsu Laboratories Ltd.
Japan - Holographic Display Artists & ..
Japan - Hyogo Prefectual Museum of Art
Japan - Ishii, Ms. Setsuko
Japan - Japan Communication Arts Co.
Japan - Keio University
Japan - Kimmon Electric Co., Ltd.
Japan - Laboratories of Image Information
Japan - Light Dimension, Inc.
Japan - Marubun Corporation
Japan - Mazda Motor Corp.
Japan - Ministry Of International Trade
Japan - Mitsubishi Heavy Industries Ltd.
Japan - Nihon University
Japan - Nippon Polaroid K.K.
Japan - Nippondenso Co., Ltd.
Japan - Nissan Motor.
Japan - Numazu College Of Technology
Japan - Ruey-Tung, Miss. Hung
Japan - SAM Museum
Japan - Sharp Corp.
Japan - Sophia University
Japan - Tama Art Umversity
Japan - Tokai University
Japan - Tokyo Institute Of Technology
Japan - Topcon Inc.
Japan - Toppan Printing Co., Ltd.
Japan - Toshihiro Kubota
Japan - Toyama National College Of Marit
Japan - University Of Tsukuba
Japan - University Of Tokyo
Japan - Waseda University

Korea - Dan Han Optics

Latvia - University of Latvia

Lithuania - Geola -

Mexico - Centro de Investigaciones en Opti ..
Mexico - Corp Mexicana De Impresion
Mexico - HOLOGRAMAS, SA DE c.v.

Netherlands - 3-D Hologrammen
Netherlands - Dutch Holographic Lab. BV
Netherlands - HOLOTECH -Texel
Netherlands - Triple-D Laser Imaging

Norway - Interferens Holografi D.A

Pakistan - Dimensions

Panama - HDIPanama

Poland - Holografia Polska
Poland - Holographic Dimensions, Poland
Poland - Hololand S.c.

Portugal - INETI - Institute of Info. Tech.
Portugal - Universidade Do Porto

Russia - Denisyuk, Yuri N.
Russia - Siavich Joint Stock Company
Russia - Technoexan Ltd

Saudi Arabia - Wonders of Holo. Gallery

Singapore - Zero Gravity
Singapore - Zero Gravity

Slovakia - Technical University Zvolen

Slovenia - 3D Technologies & Arts
Slovenia - Likom

South Africa - Synchron Pty Ltd.

Spain - Holosco, Ernest Barnes
Spain - Karas Studios Holografia
Spain - Lasing SA,
Spain - Museu D' Holografia
Spain - University Of Alicante

Sweden - Hologram Center Holmby
Sweden - Holography Group TEMOO
Sweden - HoloMedia Ab/Hologram Museum
Sweden - Holovision AB
Sweden - Lund Institute Of Tech
Sweden - Royal Institute of Technology
Sweden - Volvo-Flygmotor

Switzerland - Galerie Illusoria
Switzerland - Holos Art Galerie
Switzerland - Universite De Neuchatel

Taiwan - Ahead Optoelectronics, Inc.
Taiwan - Fong Teng Technology
Taiwan - Hiat Image Technology Group, Inc .
Taiwan - Holo Images Tech Co., Ltd.
Taiwan - Holo Impressions Inc
Taiwan - Holoart Studio
Taiwan - Industrial Tech. Research Inst.
Taiwan - Infox Catporation
Taiwan - Institute Of Optical Science
Taiwan - Superbin Co. Ltd

Thailand - Electro-Optics Lab, NECTEC

Ukraine - Feofaniya Ltd.
Ukraine - Institute of Applied Optics
Ukraine - Tair Hologram Company
Ukraine - TAVEX, Ltd.

United Arab Em. - Hololaser Gallery

U.K. - 3D Holographics
U.K. - 3D Images Ltd.
U.K. - 3D-4D Holographics
U.K. - A.H. Prismatic, Ltd.
U.K. - Action Tapes
U.K. - Advanced Holographic Laboratories
U.K. - Ag Electro-Optics Ltd.
U.K. - Amazing World Of Holograms
U.K. - Applied Holographics, Pic.
U.K. - Astor Universal Ltd.
U.K. - Beddis Kenley (Machinery) Ltd.
U.K. - Bemrose Security and Promotional..
U.K. - Benyon, Margaret
U.K. - Boyd, Patrick
U.K. - British Aerospace Pic.
U.K. - Checkpoint Security Services Lim ited
U.K. - Colour Holographics
U.K. - Courtauld Chemicals
U.K. - Creative Holography Index. The
U.K. - Customer Service Instrumentation
U.K. - Datasights Ltd.
U.K. - De La Rue Holographies
U.K. - Dimuken (GB)
U.K. - Electro Optics Developments Ltd.
U.K. - Embossing Technology Ltd
U.K. - Expanded Optics Limited

BUSINESSES BY COUNTRY

U.K. - Focal [mage Ltd
U.K. - Galvoptics Ltd.
U.K. - Global Images
U.K. - Holocrafts Europe Limited.
U.K. - Holograms 3D
U.K. - Holographic Consulting Agency
U.K. - Holographics (Uk) Ltd.
U.K. - Holomex
U.K. - iC Holographics
U.K. - [mperial College Of Science
U.K. - INNOLAS (UK) Ltd.
U.K. - International Data Ltd.
U.K. - Intml Hologram Manufacturers Assn.
U.K. - layco Holographics
U.K. - John, Pearl
U.K. - K.C. Brown Holographics
U.K. - Kendall Hyde Ltd.
U.K. - Laser International
U.K. - LAZA Holograms Ltd
U.K. - Light Fantastic
U.K. - Light Impressions Intemational, Ltd.
U.K. - London Holographic Image Studio
U.K. - Loughborough Univ. Of Tech.
U.K. - Lumonics Ltd.
U.K. - Millennium Portraits
U.K. - Molins PLC
U.K. - National Physical Laboratory
U.K. - Op-Graphics (Holography) Ltd.
U.K. - Optical Security Group - England
U.K. - Optical Works Ltd.
U.K. - Oxford Holographics
U.K. - Pepper, Andrew
U.K. - Pilkington Optronics
U.K. - Reconnaissance International Ltd.
U.K. - Richmond Holographic Studios Ltd
U.K. - Rolls-Royce Plc
U.K. - Saxby, Graham
U.K. - Spatial Imaging Limited
U.K. - STI - Europe
U.K. - Turing Institute
U.K. - Uk Optical Supplies
U.K. - Whiley Foils Limited

U.S.A. - 21st Century Finishing Inc.
U.S.A. - 3 Deep Hologram Company
U.S.A. - 3-D Systems
U.S.A. - 3M - Safety and Security Systems
U.S.A. - A.D. Tech (Advanced Deposition ..)
U.S.A. - A.H. Prismatic, Inc.
U.S.A. - Accuwave Corp.
U.S.A. - Acme Holography
U.S.A. - AD 2000, [nco
U.S.A. - Advanced Holographic Laboratories
U.S.A. - Advanced Optics, Inc.
U.S.A. - Advanced Prec ision Technology, Inc.
U.S.A. - Advanced Technology Program
U.S.A. - Aerotech [nco
U.S.A. - Alabama A&M University
U.S.A. - Amagic Technologies Inc.
U.S.A. - American Bank Note Holographics
U.S.A. - American Holographic Inc.
U.S.A. - American Laser Corporation
U.S.A. - American Paper Optics Inc.
U.S.A. - American Propylaea Corporation
U.S.A. - Am. Society Nondestructive Testing
U.S.A. - Ana MacArthur
U.S.A. - Another Dimension Inc. (Spectore)
U.S.A. - Applied Optics
U.S.A. - Art Institute Of Chicago
U.S.A. - Art Lab
U.S.A. - Art, Science & Technology Institute

U.S.A. - Automated Holographic Systems
U.S.A. - Avant-Garde Studio
U.S.A. - Barilleaux, Rene Paul
U.S.A. - Batelle Pacific NW National Lab
U.S.A. - Bellini, Victor
U.S.A. - Berkhout, Rudie
U.S.A. - Blue Ridge Holographics, Inc.
U.S.A. - Bobst Group Inc.
U.S.A. - Booth, Roberta
U.S.A. - Brandtjen & Kluge, Inc. ,
U.S.A. - Bridgestone Technologies, Inc.
U.S.A. - Broadbent Consulting
U.S.A. - Burleigh Instruments, Inc.
U.S.A. - California Institute of the Arts
U.S.A. - Cambridge Laser Labs
U.S.A. - Capitol Converting Equipment, Inc.
U.S.A. - Carl M. Rodia And Associates
U.S.A. - Casdin-Silver Holography
U.S.A. - CFC Applied Holographics
U.S.A. - Cherry Optical Holography
U.S.A. - Chromagem Inc.
U.S.A. - Chronomotion
U.S.A. - Cifelli, Dan
U.S.A. - City Chemical
U.S.A. - Coburn Corporation
U.S.A. - Coherent, Inc. - Laser Group
U.S.A. - Continental Optical
U.S.A. - Control Module Inc.
U.S.A. - Control Optics
U.S.A. - Corion Corp.
U.S.A. - Coming Incorporated
U.S.A. - Creative Label
U.S.A. - Cro~ Roll Leaf, Inc.,
U.S.A. - CVI Laser Corporation
U.S.A. - D. Brooker & Associates
U.S.A. - Datacard Corporation
U.S.A. - David Dann Modelmaking Studios
U.S.A. - Deem, Rebecca
U.S.A. - Dell Optics Company, Inc.
U.S.A. - Diamond Images, Inc.
U.S.A. - Diffraction Ltd.
U.S.A. - Digital Matrix Co.
U.S.A. - Dimensional Arts
U.S.A. - Dimensional Foods Co.
U.S.A. - Direct Holographics
U.S.A. - Doris Vila Holographics
U.S.A. - Dri-Print Foils
U.S.A. - DuPont (E.I. DuPont De Nemours)
U.S.A. - E.C.S.C.
U.S.A. - Ealing Electro-Optics Inc.
U.S.A. - Eastman Kodak Company
U.S.A. - Edmund Scientific Company
U.S.A. - EI Don Engineering
U.S.A. - Electro Optical Industries, [nco
U.S.A. - Elusive Image
U.S.A. - Engineering Animation, Inc.
U.S.A. - Evolution Design, Inc.
U.S.A. - Excitek Inc.
U.S.A. - Fantastic Holograms
U.S.A. - Feroe Holographic Consulting
U.S.A. - Fisher Scientific
U.S.A. - FLEX con
U.S.A. - Flight Dynamics
U.s.A. - Floating Images, Inc.
U.S.A. - Foil Stamping and Embossing Assn.
U.S.A. - FoilMark Holographic Images
U.S.A. - Fornari, Arthur David
U.S.A. - Forth Dimension Holographics
U.S.A. - Frank DeFreitas Holography Studio
U.S.A. - Frank J. Deutsch Inc.
U.S.A. - Fresnel Technologies Inc.

U.S.A. - G.M. Vacuum Coating Lab, Inc.
U.S.A. - General Design
U.S.A. - Glass Mountain Optics
U.S.A. - Gorglione, Nancy
U.S.A. - Hallmark Capital Corp.
U.S.A. - Holicon Corporation.
U.S.A. - Holo Art
U.S.A. - Holo Sciences, LLC
U.S.A. - Holo-Spectra
U.S.A. - Holo/Source Corporation
U.S.A. - HoloCom
U.S.A. - Holografica
U.S.A. - Hologram Fantastic
U.S.A. - Hologram Land
U.S.A. - Hologram Research, Inc.
U.S.A. - Hologram World, Inc.
U.S.A. - Hologram, etc .
U.S.A. - Holograms and Lasers International
U.S.A. - Holograms International
U.S.A. - Holograms Unlimited
U.S.A. - Holographic Applications
U.S.A. - Holographic Design Systems
U.S.A. - Holographic Dimensions
U.S.A. - Holographic Finishing
U.S.A. - Holographic Images Inc.
U.S.A. - Holographic Impressions
U.S.A. - Holographic Industries, [nco
U.S.A. - Holographic Label Converting HLC
U.S.A. - Holographic Optics Inc.
U.S.A. - Holographic Products
U.S.A. - Holographic Studios
U.S.A. - Holographics Inc.
U.S.A. - Holographics North Inc.
U.S.A. - Holography Institute of San Fran ..
U.S.A. - Holography Marketplace
U.S.A. - Holography Presses On (HPO)
U.S.A. - Holographyx inc
U.S.A. - Holographyx Inc.
U.S.A. - Holonix
U.S.A. - Holophile, Inc.
U.S.A. - Holotek
U.S.A. - Holovision Systems Inc.
U.S.A. - Holoworld.com
U.S.A. - Honeywell Technology Center
U.S.A. - IBM Almaden Research Center
U.S.A. - ICI Polyester
U.S.A. - Illinois Institute Of Technology
U.S.A. - Illuminations
U.S.A. - Imagen Holography. Inc.
U.S.A. - Images Company
U.S.A. - Imagination Plantation
U.S.A. - ImEdge Technology
U.S.A. - Industrial Technology Institute
U.S.A. - Infinity Laser Laboratories
U.S.A. - Infrared Optical Products, Inc.
U.S.A. - Innovative Technology Associates
U.S.A. - Inrad, Inc.
U.S.A. - Inside Finishing Magazine
U.S.A. - Integraf
U.S.A. - Interactive Industries, Inc.
U.S.A. - Intrepid World Communications
U.S.A. - Ion Laser Technology, Inc.
U.S.A. - James River Products
U.S.A. - Jeffery Murray Custom Holography
U.S.A. - lodon Inc.
U.S.A. - Jurewicz, Arlene
U.S.A. - Kaiser Optical Systems, Inc.
U.S.A. - Kauffman, John
U.S.A. - Keystone Scientific Co.
U.S.A. - Kinetic Systems, Inc.
U.S.A. - Krystal Holographics International

U.S.A. - Krystal Holographics International
U.S.A. - Kurz Foils
U.S.A. - Lab. for Optical Data Processing
U.S.A. - Lake Forest College
U.S.A. - Larry Liebennan Holography
U.S.A. - Lasa1l Ltd .
U.S.A. - Laser Affiliates
U.S.A. - Laser and Motion Development Co.
U.S.A. - Laser Arts Society For Edu. and Rsch
U.S.A. - Laser Drive Inc.
U.S.A. - Laser Focus World
U.S.A. - Laser Holography Workshop
U.S.A. - Laser Images
U.S.A. - Laser Innovations
U.S.A. - Laser Institute Of America
U.S.A. - Laser Las Vegas
U.S.A. - Laser Light Designs
U.S.A. - Laser Light Ltd.
U.S.A. - Laser Media, Inc.
U.S.A. - Laser Optics, Inc.
U.S.A. - Laser Reflections
U.S.A. - Laser Resale Inc.
U.S.A. - Laser Technical Services
U.S.A. - Laser Technology, Inc.
U.S.A. - LaserMax, Inc.
U.S.A. - Lasennetrics, Inc.
U.S.A. - Laserworks
U.S.A. - Lawrence Berkeley Laboratory
U.S.A. - Lenox Laser
U.S.A. - Letterhead Press, Inc.
U.S.A. - Lexel Laser, Inc.
U.S.A. - LiCONiX
U.S.A. - Light Impressions International, Ltd.
U.S.A. - Light Wave Gallery
U.S.A. - Lightrix, Inc.
U.S.A. - LightVision Confections
U.S.A. - Lightwave
U.S.A. - Linda Law Holographics
U.S.A. - Lone Star Illusions
U.S.A. - Louis Paul Jonas Studios, Inc.
U.S.A. - Lumenx Teclmologies, Inc.
U.S.A. - Luminer Printing and Converting
U.S.A. - M.I.T. (Massachusetts Inst. of Tech.)
U.S.A. - M.I.T. Museum
U.S.A. - M.O.M. Inc.
U.S.A. - MacShane Holography
U.S.A. - ManiEnvironment, Inc.
U.S.A. - Marks, Gerald
U.S.A. - McMahan Electro-Optic
U.S.A. - Media Interface, Ltd.
U.S.A. - Melles Griot
U.S.A. - Melles Group Laser Group
U.S.A. - Meredith Instruments
U.S.A. - Merrick, Michael G.
U.S.A. - MesMerized
U.S.A. - MetroLaser Inc.
U.S.A. - Metrologic Instruments, Inc.
U.S.A. - MGM Converters Inc.
U.S.A. - Midwest Laser Products
U.S.A. - Miller, Neal
U.S.A. - Miller, Peter
U.S.A. - Mitutoyo Measuring Instruments
U.S.A. - Multiplex Moving Holograms
U.S.A. - Museum Of Holography/Chicago
U.S.A. - MWK Industries
U.S.A. - Nakamura, lkuo
U.S.A. - Navidec Inc.
U.S.A. - NeoVision Productions
U.S.A. - New Focus, Inc
U.S.A. - New Light Industries
U.S.A. - New York Hall Of Science

U.S.A. - New York Holographic Laboratories
U.S.A. - Newport Corporation
U.S.A. - Nimbus Manufacturing, Inc .
U.S.A. - Norland Products, Inc.
U.S.A. - Northern Illinois University
U.S.A. - NovaVision
U.S.A. - Odhner Holographics
U.S.A. - OpSec - USA
U.S.A. - Optical Research Associates
U.S.A. - Optical Security Group
U.S.A. - Optical Society of America (OSA)
U.S.A. - Optics Plus Inc.
U.S.A. - Optimation
U.S.A. - Optimation Holographics
U.S.A. - Optineering
U.S.A. - Optitek, Inc.
U.S.A. - Oregon Institute of Technology
U.S.A. - Oregon Laser Consultants
U.S.A. - Oriel Instruments
U.S.A. - Pacific Holographics Inc.
U.S.A. - Panatron Inc.
U.S.A. - Pangaea Design
U.S.A. - Pasco Scientific
U.S.A. - Peacock Laboratories, Inc.
U.S.A. - Pennsylvania Pulp & Paper Co.
U.S.A. - Photo Research, Inc.
U.S.A. - Photo Sciences
U.S.A. - Photon Cantina Ltd.
U.S.A. - Photonics Spectra
U.S.A. - Pink, Patty
U.S.A. - Planet 3-D
U.S.A. - Point Source Productions
U.S.A. - Polaris Research Group
U.S.A. - Polaroid Corporation
U.S.A. - Potomac Photonics, Inc.
U.S.A. - Prismacoat Division
U.S.A. - Process Technologies
U.S.A. - PullTime 3-D Laboratories
U.S.A. - Rainbow Symphony Inc.
U.S.A. - Ralcon
U.S.A. - Real Image
U.S.A. - Reconnaissance International Ltd.
U.S.A. - Red Beam, Inc.
U.S.A. - Regal Press In'c.
U.S.A. - Reva 's Holographic Illusions
U.S.A. - Reynolds Metals Co.
U.S.A. - Rice Systems, Inc.
U.S.A. - Richard Bruck Holography
U.S.A. - Richardson Grating Laboratory
U.S.A. - Richmond Development Group
U.S.A. - Robert Sherwood Holo. Design
U.S.A. - Rochester Inst. Of Technology
U.S.A. - Rochester Photonics Corporation
U.S.A. - Rolyn Optics
U.S.A. - Ross Books
U.S.A. - Rowland Institute For Science
U.S.A. - Saginaw Valley State University
U.S.A. - Saint Mary's College
U.S.A. - San Jose State University
U.S.A. - Sandia National Laboratories
U.S.A. - Scharr Industries
U.S.A. - School Of Holography
U.S.A. - Science Kit & Boreal Labs
U.S.A. - Sharon McConnack Holography
U.S.A. - Shipley Chemical Co.
U.S.A. - Shuttlecart
U.S.A. - Silhouette Technology Inc.
U.S.A. - Silicon Graphics
U.S.A. - Sillcocks Plastics International
U.S.A. - Simian Co.
U.S.A. - Sinclair Optics, Inc.

U.S.A. - Smith & McKay Printing Co. Inc.
U.S.A. - Sommers Plastic Products
U.S.A. - Sonoma State University
U.S.A. - Southern Indiana Holographics
U.S.A. - Southwest Holographics
U.S.A. - Spectra-Physics Lasers Inc.
U.S.A. - Spectratek Inc.
U.S.A. - SPIE
U.S.A. - SPIE's Holo. Working Group News.
U.S.A. - Springer-Verlag New York
U.S.A. - Stanford University
U.S.A. - Star Magic Stores
U.S.A. - Starlog Stores
U.S.A. - Steinbichler U.S.A. inc.
U.S.A. - Stephens, Anait
U.S.A. - STI
U.S.A. - Swift Instruments
U.S.A. - Syracuse University
U.S.A. - Tamarack Storage Devices
U.S.A. - Technical Marketing Services
U.S.A. - Textile Graphics, Inc.
U.S.A. - The Hologram Company Stores
U.S.A. - The Halo., Laser & Photonics ... Ctr.
U.S.A. - Thorlabs Inc.
U.S.A. - Three Deep Hologram Co.
U.S.A. - Three Dimensional Imagery
U.S.A. - Three-D Light Gallery
U.S.A. - Total Register Inc.
U.S.A. - Towne Technologies
U.S.A. - Trace Holographic Art & Design, Inc
U.S.A. - Transfer Print Foils, Inc.
U.S.A. - Tyler Group
U.S.A. - U.K. Gold Purchaeers, Inc.
U.S.A. - Ultra-Res Corporation
U.S.A. - Unifoil Corporation
U.S.A. - Uniphase Lasers
U.S.A. - United Assn Manufacturer's Reps.
U.S.A. - University Of Alabama at Huntsville
U.S.A. - University Of Arizona
U.S.A. - University Of Dayton
U.S.A. - University Of Michigan
U.S.A. - University Of Rochester
U.S.A. - University Of Southern California
U.S.A. - University OfWisconsiniMadison
U.S.A. - Unterseher & Associates
U.S.A. - Uvex Safety Inc.
U.S.A. - Van Leer Metallized Products
U.S.A. - Vincennes University
U.S.A. - VinTeq, Ltd.
U.S.A. - Virtual Image (of Printpack, Inc.)
U.S.A. - Visual Visionaries
U.S.A. - Voxel
U.S.A. - Wave Mechanics
U.S.A. - Wavefront Research, Inc.
U.S.A. - Wavefront Technology
U.S.A. - Wesley, Ed
U.S.A. - Wild Style Entertainment
U.S.A. - Witchcraft Tape Products, Inc.
U.S.A. - Worcester Polytechnic Institute
U.S.A. - World Holographics
U.S.A. - WYKO-VEECO
U.S.A. - Zero Gravity Stores
U.S.A. - Zero Gravity/Galaxies Unlimited
U.S.A. - Zone Holografix

A Listing of Individuals (A-Z), Their Affiliation and Country

Last Name,   First Name,   Affiliation,   Country

A

Abendroth, Detlev, AKS Holographie-Galerie GmbH, Germany
Abouchar, Natalalie, Foreign Dimension, China
Abrams, Claudette, Abrams, Claudette, Canada
Abrams, Claudette, Photon League Ontario, Canada
Abramson, Nils, Royal Institute of Technology, Sweden
Abramzik, Curt, Curt Abramzik, Germany
Abramzik, Petra, H & W, Holographie & Werbung, Germany
Ackermann, G., Tech. Fachhochschule Berlin, Germany
Akveld, A.C., Triple-D Laser Imaging, Netherlands
Albrecht, Gerd M., Phantastica, Germany
Albright, Steve, Optimation Holographics, USA
Alten, Susanne, Spindler & Hoyer GmbH & Co. , Germany
Anders, Ulrich, Holographie Konzept GmbH, Germany
Andersen, Chad, Meredith Instruments, USA
Anderson, Mike, Holomex, UK
Anderson, Steve, Sonoma State University, USA
Ando, Hiroshi, Nippondenso Co., Ltd., Japan
Andrade, Ana Alexandra, INETT - Instiof Info Tech, Portugal
Andrews, Mathew, 30-40 Holographics, UK
Anoff, Mark, Another Dimension Inc. (Spectorel ADI), USA
Anton, Theodor 1., D. Drucker Verlagsgesellschaft, Germany
Arkin, Bill, Holo-Spectra, USA
Amott, Bob, Holographyx inc, USA
Amott, Bob, Holographyx Inc. , USA
Aymar, Bill, Laser Las Vegas, USA

B

Baghdadi, A.M., Wonders of Holography Gallery, Saudi Arabia
Baghdadi, Abdul Wahab, Hololaser Gallery, Utd Arab Emerates
Bagley, Sheila, A.H. Prismatic, Inc., USA
Bahuguna, Ramen, San Jose State University, USA
Baker, Marion, Krystal Holographics Intemational Inc. , USA
Balogh, Tibor, Artplay Holographika Studio, Hungary
Baoqing, Yuan, Qingdao Gaoguang Holography Tech. Co, China
Bard, Richard, Optical Security Group, USA
Barefoot, Paul D., Holophile, Inc., USA
Barre, Pascal, Holos Art Galerie, Switzerland
Bassin, Barry, Infrared Optical Products, Inc., USA
Battin, Dave, Holo Alt, USA
Bauer, Josef, Bauer, Josef, Germany
Bazargan, Kaveh, Focal Image Ltd, UK
Bear, Sol, Hologram World, Inc., USA
Beauregard, Alain, Lasiris Inc., Canada
Beeching, Dave, CFC Applied Holographics, USA
Beeck, M.-A., Volkswagen AG, Germany
Begleiter, Erich, Dimensional Foods Co., USA
Behrmann, Greg, Potomac Photonics, Inc., USA
Bellini, Victor, Bellini, Victor, USA
Benito, Ramon, Karas Studios Holografia, Spain
Bentley, John, Dimuken (GB), UK
Benton, Stephen A., M.LT. (Mass. Insti. of Technology), USA
Benyon, Margaret, Benyon, Margaret - Holography Studio, UK
Berkhout, Rudie, Berkhout, Rudie, USA
Bianchi, Herman-Josef, Arbeitskreis Holografie B.Y., Germany
Billeri, Ralph, Control Module Inc., USA
Billings, Loren, Museum Of Holography/Chicago, USA
Billings, Robert, Holographic Design Systems, USA
Billo, Andreas, Institut fur Angewandte Physik, Germany
Bingda, Zhou, Shanghai Kanlian S & T Development, China
Birch, Clive, Kendall Hyde Ltd., UK
Birenheide, Richard, HRT, Germany
Bjelkhagen, Hans, Lake Forest College, USA
Blosvem, Moss, Kinetic Systems, Inc. , USA
Bobeck, Paula, DuPont (E.I. DuPont De Nemours & Co.), USA

Bohan, Brian, Cambridge Laser Labs, USA
Boone, Pierre, Laboratory Vinckiner, Belgium
Boone, Prof. Pierre, Ctr. for Applied Research in Art, Belgium
Booth, Roberta, Booth, Roberta, USA
Bosco, Eric, Centre d'Art Holographique et Photonique, Canada
Bosco, Eric, Holostar, Canada
Bose, Hans J., Highlite, Germany
Botos, Steve A., Aerotech Inc., USA
Bourque, Nina, Hologram Fantastic, USA
Bouverat, Randy, Spectratek Inc. , USA
Boyd, Patrick, BOY~ Patrick, UK
Bradshaw, Roy, RealUmage, USA
Brill, Louis, Illuminations, USA
Broadbent, Donald, Broadbent Consulting, USA
Brooker, Dennis, D. Brooker & Associates, USA
Brown, David, Optical Research Associates, USA
Brown, John, Light Impressions Intemational, Ltd. , UK
Brown, Kevin, Holographic Dimensions, USA
Brown, Kevin, K.C. Brown Holographics, UK
Brown, William, INNOLAS (UK) Ltd., UK
Bruck, Richard, Holicon Corporation. , USA
Bruck, Richard, Richard Bruck Holography, USA
Bruegmalm, Machteld, Hologram Co. RAKO GmbH, Germany
Buell , Richard, James River Products, USA
Bunkenburg, Jo, Rochester Photonics Corporation, USA
Burder, David, 3D Images Ltd. , UK
Burgmer, Brigitte, Burgmer, Brigitte, Germany
Burke, Ed, Hologram Development Corp. , Canada
Bumey, Michael, Chronomotion, USA
Burns, Joseph, Hologram Research, Inc., USA
Bussard, Jan, Holography Presses On (HPO), USA
Bussard, Jan, Textile Graphics, Inc., USA
Bussaut, Laurent, ASTI, USA
Bussaut, Loren, The Holography, Laser & Photonics, USA
Butteriss, Tony, New Dimension Holographics, Australia
Butteriss, Tony, Parallax Gallery, Australia

C

Caizhi, Yang, Qingdao Qimei Images, Inc., China
Capema, Albert 1., NovaVision, USA
Carlsson, Torgny, Royal Institute of Technology, Sweden
Carmichael, Lisa, Checkpoint Security Services Limited, UK
Casasent, David, Laboratory for Optical Data Processing, USA
Casdin-Silver, Harriet, Casdin-Silver Holography, USA
Cason, Thad, Infinity Laser Laboratories, USA
Cavoon, Curt, Spectra-Physics Lasers Inc., USA
Chaihorsky, Alex, Ultra-Res Corporation, USA
Chang, Long, Amagic Technologies Inc., USA
Chang, Milton, New Focus, Inc, USA
Chantler, Sylvia, National Physical Laboratory, UK
Cheimets, Alex, 3 Deep Hologram Company, USA
Chen, Alex c.T., Infox Corporation, Taiwan
Chen, Hsuan, Saginaw Valley State University, USA
Cheng, Fan, Guangdong Dongguan South Holoprint Co., China
Cherry, Greg, Cherry Optical Holography, USA
Chiang, Mark, Fong Teng Technology, Taiwan
Chiarot, Roy, Photon Cantina Ltd., USA
Chiou, Craig, Holo Images Tech Co., Ltd., Taiwan
Chmielewski, Gabriele, Holographie Konzept GmbH, Germany
Chou, Billy, Hiat Image Technology Group, Inc., Taiwan
Chuntian, Xu, Guangzhou Chuntian Industrial Techniques, China
Cifelli, Dan, Cifelli, Dan, USA
Clarke, Walter, Global Images, UK
Clal3en, Walter, LichtBlicke Clal3en & Voss, Germany
Claudius, Peter, Multiplex Moving Holograms, USA

Last Name,	First Name,	Affiliation,	Country	Last Name,	First Name,	Affiliation,	Country

Claytor, Linda H., Fresnel Technologies Inc., USA

Conklin, Don, Glass Mountain Optics, USA

Connors, Betsy, Acme Holography, USA

Cooper, Nick, Oxford Holographics, UK

Cordes, H. E., W. Cordes GmbH + Co. , Germany

Cordner-Guled, Valeska, HOL 3, Galerie fur Holo .. , Germany

Corwin, Jason, Accuwave Corp., USA

Cossa, Rich, Planet 3-D, USA

Cossette, Marie-Andree, Holography & ... of Quebec, Canada

Cote, -Paul, FoilMark Holographic Images, USA

Coursen, Dan, G.M. Vacuum Coating Lab, Inc., USA

Cox, Dr. 1 Allen, Honeywell Technology Center, USA

Creath, Kathy, Optineering, USA

Crenshaw, Melissa, Holography Studio International, Canada

Crenshaw, Milessa, Capilano College, Canada

Cross, Lloyd, 3-D Systems, USA

Cubberly, George, Excitek Inc., USA

Cullen, Karoline, Holocrafts, Canada

Cullen, Ralph, Uk Optical Supplies, UK

Cvetkovich, Thomas l, Chromagem Inc., USA

D

Diihn, Wolfgang G. 0., Gigaheltz-Optik, Germany

Da-hsiung, Hsu, Beijing Univ. of Posts & Telec .. , China

Darner, Cynthia, AD 2000, Inc. , USA

Dandliker, Rene, Universite De Neuchatel, Switzerland

Daniel, Manjit, Photo Research, Inc., USA

Dann, David, David Dann Modelmaking Studios, USA

Datta-Pretta, Fernando, Trace Holographic Art & Design, USA

Dausmann, G., Holtronic GmbH, Germany

Dausmann, Gunther, Holographic Systems Muenchen , Germany

Davis, Ernie, MWK Industries, USA

Dayus, Ian, A.H. Prismatic, Ltd. , UK

de Roos, Marc, Deep Space Holographics, Canada

DeBerry, Larry, Zero Gravity/Galaxies Unlimited, Inc., USA

Deem, Rebecca, Deem, Rebecca, USA

Deem, Rebecca, Zone Holografix, USA

DeFreitas, Frank, Frank DeFreitas Holography Studio, USA

DeFreitas, Frank, Holoworld.com, USA

del-Prete, Sandro, Galerie lliusoria, Switzerland

Denisyuk, Yuri N., Denisyuk, Yuri N., Russia

Dennis, Jason, Prismacoat Division, USA

D'Entremont, Joseph P., Lenox Laser, USA

Desai, Yogesh, Spatial Holodynamics (India) PVI. Ltd., India

Detrich, Ed, OpSec - USA, USA

Deutschman, Bill, Oregon Laser Consultants, USA

Deutschmann, Gunter, AHT 3D-Medien, Germany

Deutschmann, Gunter, Holo. Fachstudio Bad Germany

Devens, Scott, Holographyx inc, USA

Diamond, Mark, Diamond Images, Inc. , USA

Dicker, Mark, Holographic Consulting Agency, UK

Dietrich, Edward, Optical Security Group, USA

Dion, David, FoilMark Holographic Images, lUSA

Dirksen, Alfred, Alfred Dirksen + Sohn, Germany

Doleschel, Brigitte, Haus der Technik e. V, Germany

Dominic, Francis, Photo Research, Inc., USA

Dowley, Mark, LiCONiX, USA

Dreelaw, Dawn, International Data Ltd., UK

Drinkwater, Deni, Wave Mechanics, USA

Duesterberg, Richard, Vincennes University, USA

Duffey, William, Regal Press Inc. , USA

Duignan, Michael, Potomac Photonics, Inc. , USA

Dutton, Keith, Laser International , UK

E

Easterlang, Lund, Imagen Holography, Inc. , USA

Edgar, John, Brandtjen & Kluge, Inc., USA

Edhouse, Simon, 3DIMAGE, Australia

Edhouse, Simon, Australian Holographics, Australia

Eichler, Hans Joachim, Tech. Univ. Berlin, Germany

Eichler, Juergen, Tech. Fachhochschule Berlin, Germany

Eimers, Tilmann, Holographic Laserdesign, Germany

Engelmann, Heiko, Modern Marketing, Germany

Erickson, Ronald R., Media Interface, Ltd. , USA

F

Faddis, Terry, Optimation Holographics, USA

Farina, Joseph A., Laser Holography Workshop, USA

Fattal, Isaac, . Krystal Holographics International Inc., USA

Fee, Renee, Holografica, USA

Fein, Howard, Polaris Research Group, USA

Feinberg, Jack, University Of Southern California, USA

Felix, Patricia, Holograms and Lasers International, USA

Felix, Perry, Holograms and Lasers International, USA

Fernandez, Nancy, Oriel Instruments, USA

Feroe, James, Feroe Holographic Consulting, USA

Fimia., A., University Of Alicante, Spain

Fischer, Julian, HoloVision, Germany

Fischler, Ben, Imagination Plantation, USA

Fisher, Gary, Man/Environment, Inc., USA

Fitzpatrick, Colleen, Rice Systems, Inc., USA

Florence, Mike, James River Products, USA

Foerster, Thomas, Holostudio Beate Krengel, Germany

Ford, Richard, Courtauld Chemicals, UK

Fonnosa, Joe, Van Leer Metallized Products, USA

Fomari, Arthur David, Fornari, Arth ur David, USA

Forsberg, Mona, HoloMedia Ab/Hologram Museum, Sweden

Foryst, Carole, Floating Images, Inc., USA

Fournier, Jean-Marc, Rowland Institute For Science, USA

Frank, A., Bernhard Halle Nachf. GmbH & Co., Germany

Frankmark., Robert, Volvo-Flygmotor, Sweden

Freude, Priv.-Doz., Dr.-Ing., W., Grau, G., Germany

Frieb, M.T., Holographie Anubis, Germany

Frisk, E.O., Optical Works Ltd., UK

Fuechtenbusch, Annette, Holodesign, Germany.

Fukuda, Akinobu, SAM Museum, Japan

Fukuda, Shinichi, Kimmon Electric Co., Ltd., Japan

Fukuma, Mineko, Japan Communication Arts Co., Japan

Fuller, Mary, Foil Stamping and Embossing Association, USA

G

Gabrielson, Dan, Pennsylvania Pulp & Paper Co., USA

Gallagher, Dan, Total Register Inc., USA

Gallagher, John, Total Register Inc., USA

Galon, Derek, Royal Holographic Art Gallery, Canada

Garcia, Diego, M.LT. Museum, USA

Garrett, Steve, Midwest Laser Products, USA

Gaynor, Joseph, Innovative Technology Associates, USA

Geiger, Thomas, Ingenieurbiiro Geiger, Germany

Gericke, H. G., Holo Time Gericke, Germany

Gericke, W. K., Holo Time Gericke, Germany

Gibb, Don, Ion Laser Technology, Inc. , USA

Gibb, Jim, Automated Holographic Systems, USA

Gibson, 1 A. , Ag Electro-Optics Ltd., UK

Gillespie, Don, EI Don Engineering, USA

Gillespie, Mike, Jodon Inc., USA

Ginouves, Paul, Coherent, Inc. - Laser Group, USA

Glaser, Shelly, Glaser - Technical Consulting, Israel

Glazer, Stewart, Crown Roll Leaf, Inc." USA

Goldstein, Robert, Lasermetrics, Inc., USA

Golen, VP, Grace, Holographic Dimensions, Poland S.A. , Poland

Gorglione, Nancy, Cherry Optical Holography, USA

Gorglione, Nancy, Gorglione, Nancy, USA

Gorglione, Nancy, Laser Affiliates, USA

Gott, Barry, E.C.S.C., USA

Gougeon, Pierre, Dimension 3, Canada

Graf, Mike, Letterhead Press, Inc., USA

Graffeo, Gus, Digital Matrix Co., USA

Graham, Ben, Lexel Laser, Inc. , USA

Grau, Prof. Dr. techn., G. , Grau, G., Prof. Dr. techn., Germany

Greenspan, Alex, Digital Matrix Co., USA

Greguss, Pal, Optopol Panoramic Metrology Consulting, Hungary

Groote, Manfred, HMS-Elektronik, Germany

Grueneberg, Joachim, Decolux GmbH, Germany

INDIVIDUALS

Last Name,	First Name,	Affiliation,	Country

Gugg-Helminger, Anton E. A. , Gigahertz-Optik, Germany
Guoping, Wang, Wuxi Light Impressions Inc., China
Gustafsson., Jonny. Holovision AB, Sweden

H

Haidong, Liu, Beijing Dongfang Laser Printing Tech. Co., China
Haines, Debbie, Simian Co., USA
Halkes, Adrian J., Far East Holographics, China
Hall, Patricia M., Hallmark Capital Corp., USA
Haney, Lorri, Nimbus Manufacturing, Inc., USA
Hangjia, Sun, Jiangsu Sida Images, Inc., China
Hankin, Alan, Bridgestone Technologies, Inc. , USA
Harden, Jim, Light Wave Gallery, USA
Harding, Kevin, Industrial Technology Institute, USA
Hardy, Nick, Op-Graphics (Holography) Ltd. , UK
Harms, Bruno, Holarium, Germany
Harris, Ken, Dimensional Arts, USA
Harris, Ken, HoloCom, USA
Harrison, Ann Marie, Intrepid World Communications, USA
Hartman, John, Batelle Pacific Northwest National Lab. , USA
Hartmann, Vera, Laser-Solution-Management, Germany
Hashimoto, Chikara, Central Glass Co., Ltd. , Japan
Hashimoto, Reiji, Topcon Inc., Japan
Hassen, Chuck, Holo Sciences, LLC, USA
Heck, David, Spectra-Physics Lasers Inc. , USA
Hedley, David, Whiley Foils Limited, UK
Heil, Wendy, Advanced Optics, Inc., USA
Hein, Elke, Die Dritte Dimension, Germany
Hepburn, James, Silverbridge Group, Canada
Herr, Doug, Bobst Group Inc., USA
Hess, Bob, Point Source Productions, USA
Hess, Cecil, MetroLaser Inc., USA
Hesselink, Burt, Optitek, Inc., USA
Hill, Dean, OpSec - USA, USA
Hill, Zohra, HOL 3, Galerie fur Holographie GmbH, Germany
Hillard, Bill, NeoVision Productions, USA
Hinz, Daniel, Melles Griot GmbH, Germany
Hocke, Wolfpeterr AD HOC Public Relations GmbH, Germany
Hoefer, Dan, American Laser Corporation, USA
Hoffstadt-Braeutigam, Irmhild. topac GmbH, Germany
Hofmann, Christina, Galerie 7000, Germany
Hofmann, Christina, Galerie WesteriandiSylt, Germany
Hofmann, Christina, Holografie-Hofmann, Germany
Hofmann, Martin, Galerie 7000, Germany
Hofmann, Martin, Galerie WesterlandiSylt, Germany
Hofmann, Martin, Holografie-Hofmann, Germany
Hofmann-Lange, Germany
Holden, Laurence, Astor Universal Ltd., UK
Hollinsworth, TR., Expanded Optics Limited, UK
Hollmann-Langecker, Liesel, Art Agentur Kaln, Germany
Hoose, John, Richardson Grating Laboratory, USA
Hom, Rolf, HoloMedia Ab/Hologram Museum, Sweden
Horne, Peg, Scharr Industries, USA
Hsu, Jonathan, Holo Impressions Inc, Taiwan
Huajian, Gu, Suzhou University, China
Huff, Marilyn, Amagic Technologies Inc. , USA
Hufmann, Michael, Holodesign, Germany
Hui, Chen, Beijing Xi Ji Wo Computer Graphic Workg Co., China
Hurwitz, Noah, Imagination Plantation, USA
Hwang, Edward, Superbin Co. Ltd, Taiwan

I

Infantes, Victoria E. , Karas Studios Holografia, Spain
Inoue, Yutaka, Brainet Corporation - International Division, Japan
Iordan, Micheal, The Holography, Laser & Photonics Ctr .. USA
Iovine, John, Art Lab, USA
Ishihara, Dr. Satoshi, Ministry Of International Trade, Japan
Ishii, Setsuko, Ishii, Ms. Setsuko, Japan
Ishikawa, Kazue, Sophia University, Japan
Iverson, Mark, Datacard Corporation, USA
Iwata, Fujio, Toppan Printing Co., Ltd., Japan

Last Name,	First Name,	Affiliation,	Country

J

Jain, Rajeev, Shriram Holographics, India
James, Randy, Pacific Holographics Inc., USA
Jamison, Pamela, Light Impressions International, Ltd., USA
Jelic, Nikola, 3D Technologies & Arts, Slovenia
Jeong, TH., Integraf, USA
Jeong, TH., Lake Forest College, USA
Jerit, John, American Paper Optics Inc., USA
Jiang, Yaguang, China Ann Arbor Holographical Institute, China
Jingde, Ye, Shenzhen Reflective Materials Factory, China
Jizhang, Wang, N0l1h Light Holograms Ltd., China
Johann, Larry, Southern Indiana Holographics, USA
John, Pearl , John Pearl, UK
Jorgensen, Dean, Optimation, USA
Jung, Dieter, Acadamy of Media Arts Cologne, Germany
Junger, Mr. , CHIRON Technolas GmbH, Germany
Juptner, Werner, B14S, Germany
Jurewicz, Arlene, Jtli.ewicz, Arlene, USA

K

Kalka, Jutta, PPM GmbH Werbung, Germany
Kallenbach, Heinz, H. Kallenbach - H.M.V., Germany
Kane, Brian, General Design, USA
Karaganova, Svetlana, Australian Holographics, Australia
Kassover, Kathy, Crown Roll Leaf, Inc." USA
Katsuma, Hidetoshi, Tama Art Umversity, Japan
Kauffman, John, Kauffman, John, USA
Keilholz, Mr., Spindler & Hoyer GmbH & Co., Germany
Keller, Manuela, Steuer KG GmbH & Co., Germany
Kendall , Christen, Metrologic Instruments, Inc. , USA
Kenny, Mike, MWK Industries, USA
Kerpen, Uta, Fielmllolln-Verwaltung KG, Gernlany
Kettel, Klaus, Krystal Holographics Vel1riebs-GmbH, Germany
Kiely, Annette, De La Rue Holographies, UK
Kilpatrick, Jack, Laser Resale Inc., USA
King, R. Eric, Laser Innovations, USA
Kinoshita, Dr. Kenji, Toyama National College Of Marit, Japan
Kirk, Ronald 1., Holovision Systems Inc., USA
Kleinhenz, G., Holotec, Germany
Kleinherne, Klaus, Holo Service/Service-Dmck, Germany
Kleinherne, Reiner, Holo Service/Service-Dmck, Germany
Klimasewski, Tim, Burleigh Instruments, Inc. , USA
Klix, Almin, Armin Klix Holographie, Germany
Knuth, Eckard, Laserfilm Eckard Knuth, Germany
Koch, Mr., Optical Coating Laboratory GmbH, Gernlany
Koerner, Juergen, Hologramm, Germany
Koizumi , Fumihiko, Asahi Glass Co., Japan
Kollin, Joel, Holonix, USA
Kontnik, Lewis, Reconnaissance International Ltd. , USA
Koril, Gary, Creative Label, USA
Koril, Jerry, Creative Label, USA
Kornienko, Sergey, feofaniya Ltd. , Ukraine
Kottova, Alena, Ontario Science Centre, Canada
Kraak, lO., 3-D Hologrammen, Netherlands
Krengel, Beate, Holostudio Beate Krengel , Germany
Krick, Reva, Reva 's Holographic Illusions, USA
Kritzinger, Sean, Synchron Pty Ltd. , South Africa
Kmeger, Dave, Holograms International, USA
Krueger, Jean, Holograms International, USA
Kubeile, Thomas, Miinchen GmbH Werbung, Germany
Kubota, Toshihiro, Toshihiro Kubota, dept of electronics, - Japan
Kukhtarev, Nicholai, Alabama A&M University, USA
Kuwayama, Tetsuro, Canon Inc. R&D Headquarters, Japan
Kyle, Stephen, LAZA Holograms Ltd, UK

L

Labelle, Scott, Holographic Label Converting (HLC), USA
Lacey, Lee, Holo/Source Corporation, USA
Laeri, F. , Technische Hochschule Darmstadt, Germany
Lancaster, lan, Reconnaissance International Ltd. , UK
Langer, w., Dornier Medizintechnik GmbH, Germany
Lansing, Joseph, Electro Optical Industries. Inc., USA

Holography Presses On

Holographic transfers applied with heat or stickers
with pressure, in stock or custom shapes and
sizes for permanent adhesion to all surfaces.
Sealed edges prevent delamination in
all weather...washable...dry cleanable.

Patented product and process.
Distributors sought. Licensing available.

Jan Bussard
Box 193 Spring lake, MI 49456
Phone 616/842-5626 Fax 616/842-5653

Rainbow reflectivity at its best in these stock transfer designs, much more in catalog. Custom designs and 3D holograms available, too. Display alone or as a focal point with screen printing or embroidery. Sealed edges ... washing / drying 150 times ... no delamination.

Industry's leader for decorative rainbow transfers for any substrate.

**Holography
Presses
On**

Box 193 Spring Lake, MI 49456
Phone 616/842-5626 Fax 616/842-5653

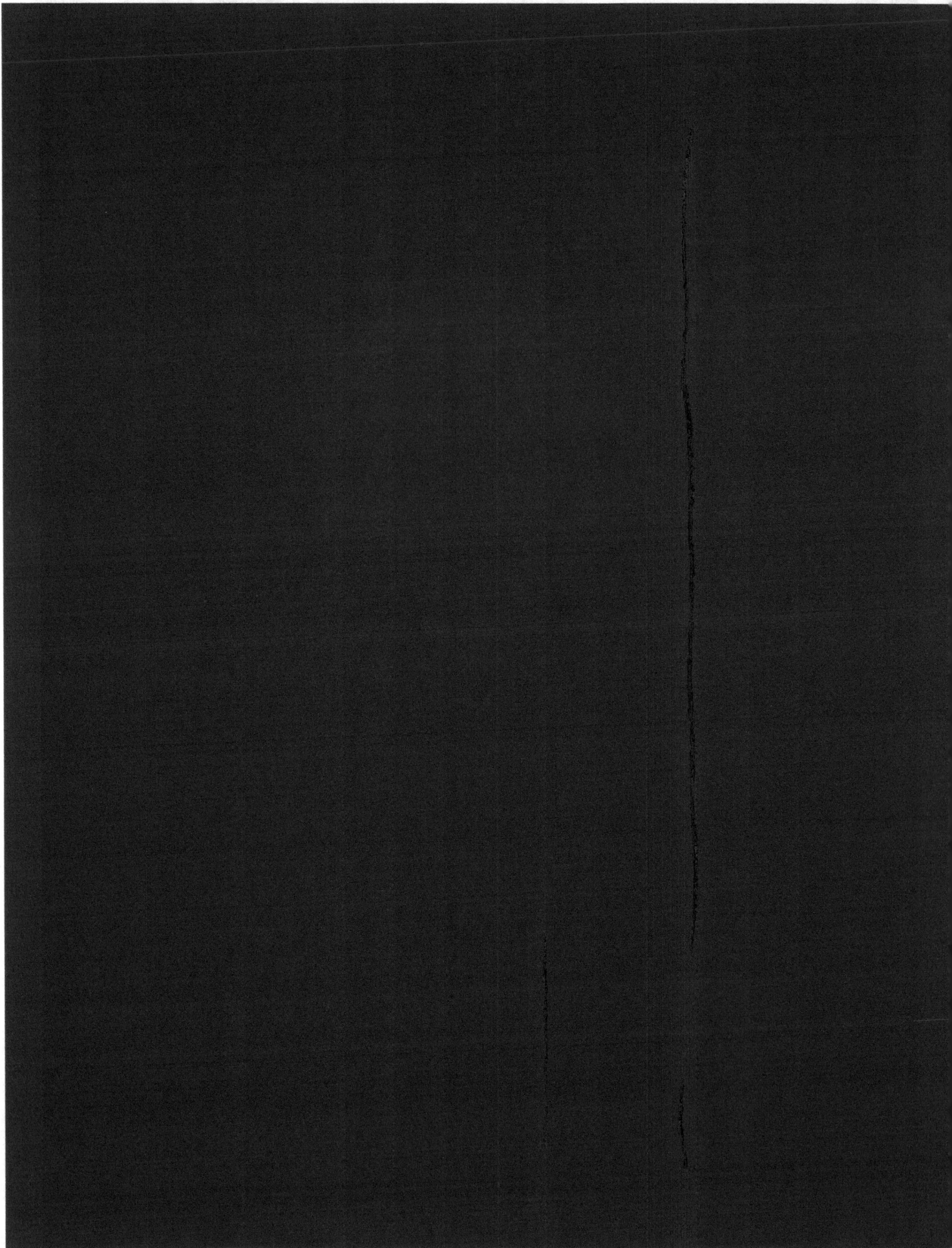

INDIVIDUALS

Last Name, First Name, Affiliation, Country	Last Name, First Name, Affiliation, Country
Larim, Jim, Laser Optics, Inc. , USA	McNeil, Don, WYKO-VEECO, USA
Larson, Ann, Laser Images, USA	Medford, Amy E., Avant-Garde Studio, USA
Larson, Steve, Laser Images, USA	Medora, Michael, Colour Holographics, UK
Lauder, Dea, American Propylaea Corporation, USA	Mendoza, Ph.D., Fernando, Ctr. de Investiga ... en Optica, Mexico
Lauk, Mathias, Lauk & Partner GmbH, Germany	Menning, Melinda, Menning, Melinda, Australia
Lauk, Matthias, Museum fur Holo. & neue visuelle ... , Germany	Merget, Stefa~ , Galerie 3D, Germany
Law, Linda, Linda Law Holographics, USA	Merrick, Michael, Merrick, Michael G. , USA
Lee, Julie, Ahead Optoelectronics, Inc. , Taiwan	Merritt, Dave, Louis Paul Jonas Studios, Inc. , USA
Leith, Emmet, University Of Michigan, USA	Metz, Michael, ImEdge Technology, USA
Lembessis, Alkis, Hellenic Institute Of Holography, Greece	Meulien, Odile, Artbridge Light Studios, Germany
Lembessus, Alkis, Cavomit, Greece	Meulien, Odile, Meulien Odile, Germany
Leseberg, Detlef, Leseberg, Dr. Detlef, Germany	Meuse, Ron, Mu's Laser Works, Canada
Lessard, Roger A. , Universite Laval, Canada	Meyer, Steve, MGM Converters Inc., USA
Lev, Steven, Chromagem Inc. , USA	Michael, Steve, Three Dimensional Imagery, USA
Levine, Chris, iC Holographics, UK	Mielke, H. M., Holographie-Labor Mielke, Germany
Levine, Jeffrey, MesMerized, USA	Mike, Spectratek Inc., USA
Levy, Rob, Holo/Source Corporation, USA	Mikes, Thomas, American Holographic Inc. , USA
Levy, Uri, Holo-Or Ltd, Israel	Miller, Doug, Krystal Holographics International Inc., USA
Liberato, Pablo, Evolution Design, Inc., USA	Miller, Neal, Miller, Neal, USA
Lieberman, Dan, HOLOGRAMAS, S.A. DE C.Y., Mexico	Miller, Peter, Miller, Peter, USA
Lieberman, Larry, Larry Lieberman Holography, USA	Minqin, Xue, Yu Feng Laser Images Co., China
Liedlbauer, Eric, Highlite, Germany	Mistry, Rohit, Jayco Holographics, UK
Liedtke, Mario, mario liedtke pro design, Germany	Mitamura, Shunsuke, University Of Tsukuba, Japan
Lifshen, Alan, Lone Star Illusions, USA	Mitchell, Astrid, Applied Holographics, Pic., UK
Lin, Yow-Snin, Holoart Studio, Taiwan	Moedinger, Wilfried, Modem Marketing, Germany
Lind, Michael, Batelle Pacific Northwest National Lab, USA	Moeller, NikJas, Deutsche Gesellschaft fur Holo, Germany
Linsen, Chen, Suzhou University, China	Mofchetti, Tanis, Uniphase Lasers, USA
Lion, Yves F. , Univ. de Liege, Belgium	Monaghan, Brian, Pennsylvania Pulp & Paper Co., USA
Lishan, Huang, Xi amen Grand World Laser Label Products, China	Monberg, Ed, Laser and Motion Development Company, USA
Lissack, Selwyn, Laserworks. USA	Monchak, Alexander, Tair Hologram Company, Ukraine
Liu, Wai-Min, Control Optics, USA	Moore, Lon, Red Beam, Inc. , USA
Lkegami, Dr. Koji, Numazu College Of Technology, Japan	Moree, Sam, New York Holographic Laboratorie, USA
Ing. Daams, Hans-Jiirgen, EPA, Germany	Morrison, Dan, Laser Technical Services, USA
Long, Mike, Pacific Holographics Inc., USA	Morterud, Alan P. , Imagen Holography, Inc., USA
LoSardo, Sal, Towne Technologies, USA	Mortier, Frank, Agfa - Gevaert N.Y., Belgium
Lossau, H. G. , Polytec GmbH, Germany	Mueller, Joachim, Gresser, E. , KG, Germany
Love, Valerie, Op-Graphics (Holography) Ltd. , UK	Munday, Rob, Spatial Imaging Limited, UK
Lovygin, Igor, Technoexan Ltd, Russia	Munzer, Hubert, OWIS Gmbh, Germany
Lucy, Thomas, Holo GmbH, Germany '"	Murata, M., Mitsubishi Heavy Industries Ltd., Japan
Ludwig, Peter, Holographie & Design, Germany	Murphay, Toicia, Silhouette Technology Inc., USA
Luton, Chris, Holocrafts Europe Limited., UK	Murray, Jeffery, Jeffery Murray Custom Holography, USA

M

MacArthur, Ana, Ana MacArthur, USA	Murray, Jeffrey, Holography Institute of San Francisco, USA
MacShane, Jim, MacShane Holography, USA	Murray, Jeffrey, Laser Arts Society For Education and ... USA
Magarinos, Jose R., Holographic Optics Inc., USA	Murray, Maria, Inrad, Inc., USA
Mairiedl, Horst, Decolux GmbH, Germany	Muth, August, Lasart Ltd. , USA
Malott, Michael, Laser Light Designs, USA	

N

Mann, Harry, Van Leer Metallized Products, USA	Naaman, Bill, Mitutoyo Measuring Instruments, USA
Margolis, Mark, Rainbow Symphony Inc ., USA	Nakajima, Dr. Masato, Keio University, Japan
Mario, Allibrante, Holo 3D S.p.A., Italy	Nakamura, Ikuo, Nakamura, Ikuo, USA
Markov, Vladimir B. , Institute of Applied Opfics, Ukraine	Nakashima, Masato, Fujitsu Laboratories Ltd., Japan
Marks, Gerald, Marks, Gerald, USA	Nava-Calvillo, Salvador, Corp. Mexicana De Impresion, Mexico
Marks, Gerald, PullTime 3-D Laboratories, USA	Neister, Ed, Lumenx Technologies, Inc., USA
Martin, Diane, Polaroid Corporation, USA	Nelson, Drew, Stanford University, USA
Martinez, Guillern10, Evolution Design, Inc.; USA	Neu, Martha, Polaroid Corporation, USA
Maslenkov, Michael , Technoexan Ltd, Russia	Newman, John, L aser Technology, Inc., USA
Massuda, EIIe, Harvard Apparatus, Canada, Canada	Nguyen, Michel, Hologram, etc., USA
Mazzola, Karen, United Association Manufacturer's Reps, USA	Niblett, Tim, Turing Institute, UK
McCarthy, Kevin, Laser Media, Inc. , USA	Nicholas, Fred, Holographics Inc. , USA
McCormack, Sharon, Sharon McCormack Holography, USA	Noble, Marcus, Technical Marketing Services, USA
McGarry, Dan, Rochester Photonics Corporation, USA	Noems, Benny, Metrologic Instruments GmbH, Germany
McGaw, Trevor, 3D Optical Illusions, Australia	Norman, Kenneth, Control Module Inc., USA

O

McGaw, Trevor, Holograms Fantastic and Optical .. , Australia	Odhner, Jefferson E., Odhner Holographics, USA
McGaw, Trevor, HopSec/dii, Australia	Oehlmann, Dietmar, Dietmar Oehlmann, Germany
McGrew, Steve, New Light Industries, USA	Oishi, Mariko, Light Dimension, Inc., Japan
McKay, Dave, Holographic Impressions, USA	Okada, Dr. Katsuyaki, HODIC, Japan
McKay, Dave, Smith & McKay Printing Co. Inc. , USA	Old, Payton, Tyler Group, USA
McLear, Mark, Uvex Safety Inc., USA	Olmo, Anthony, 21 st Century Finishing Inc., USA
McLeod, Don, Corion Corp., USA	Olson, Bernadette, Laser Reflections, USA
McMahan, Robert, McMahan Electro-Optic, USA	Olson, Ron, Laser Reflections, USA

INDIVIDUALS

Orr, Edwina, Richmond Holographic Studios Ltd, UK
Os ada, RB, Fantastic Holograms, USA
Ose-Wiese, Christian, Ose Holografie-Design, Germany
Otega, James, Sandia National Laboratories, USA
Oteri, Lance, Holographic Label Converting (HLC), USA
Ott, Hans-Peter, Museum flir Holographie & ... , Germany

P

Page, Michael, Island Graphix, Canada
Page, Michael, Ontario College of Art/Holography, Canada
Page, Michael, Page, Michael, Canada
Pahnke, Roland, Kolbe-Druck mit Tochtergesell, Germany
Paletz, Jim, Hologram World, Inc. , USA
Pargh, Jonathan, Three-D Light Gallery, USA
Parker, Bill, Diffraction Ltd. , USA
Parker, Dr. Steve, British Aerospace Pic. , UK
Pastorius, Bruce, Advanced Precision Technology, Inc. , USA
Patterson, Rich, Reynolds Metals Co. , USA
Paxton, Chuck, Photon Cantina Ltd., USA
Pepper, Andrew, Creative Holography Index, The, UK
Pepper, Andrew, Pepper, Andrew, UK
Perreau, Jean-Luc, GEHOL sari, france
Perry, Iohn, Holographics North Inc., USA
Petersen, Ioel, Wavefront Technology, USA
Peterson, Ieff, Inside Finishing Magazine, USA
Phillips, Alan I., Action Tapes, UK
Phillips, Jacque, Direct Holographics, USA
Phillips, Ron, Interactive Industries, Inc., USA
Pierce, Robert, Oregon Institute of Technology, USA
Platts, Dave, HOLOTECH -Texel, Netherlands
Plotnick, Harvey, Laser Media, Inc., USA
Poe, Nelson, Blue Ridge Holographics, Inc., USA
Powell, Dick, University Of Arizona, USA
Price, Stu, Shipley Chemical Co., USA
Pricone, Robert, Holographic Industries, Inc. , USA
Provence, Steve, Blue Ridge Holographics, Inc., USA
Pryputniewicz, Ryszard, Worcester Polytechnic Institute, USA

Q

Qeser, Ernst, Oeserwerk Ernst Oeser & Sohne KG, Germany
Qiang, Li, Beijing Sanyou Laser Images Co., China
Qihong, Song, Tianjin Water Laser Holography Image Co., China

R

Rahe, Rolf, W. Cordes GmbH + Co., Germany
Rallison, Richard, Rakon, USA
Randazzo, Dean, Holographic Images Inc. , USA
Rayfield, Dave, Krystal Holographics International Inc., USA
Reichert, Uwe, 3D Vision, Germany
Reindl, Richard, Letterhead Press, Inc., USA
Reinert, R., Holtronic GmbH, Germany
Reinhart, Werner, Leonhard Kurz GmbH, Germany
Rezny, Abe, Laser Light Ltd., USA
Rhody, Alan, Holography Marketplace, USA
Rhody, Alan, Ross Books, USA
Rich, Chris, Wavefront Technology, USA
Richardson, Martin, London Holographic Image Studio, UK
Rickert, Sue, Hologram Land, USA
Robb, Jeffrey, Spatial Imaging Limited, UK
Robiette, Nigel, Colour Holographics, UK
Robinson, David, National Physical Laboratory, UK
Robinson, Deborah, Lightrix, Inc., USA
Robinson, George, Hologram Land, USA
Robur, Lubomir, Feofaniya Ltd., Ukraine
Rodia, Carl M. , Carl M. Rodia And Associates, USA
Rongkun, Wu, Quan Zhou Pacific Laser Images, China
Rosowski, Ralf, Holographic Studios Rosowski, Germany
Ross, Franz, Holography Marketplace, USA
Ross, Franz, Ross Books, USA
Ross, Jonathan, Holograms 3D, UK
Ross, Michael, IBM Almaden Research Center, USA
Rossing, Thomas, Northern Illinois University, USA
Rost, Thomas, K. Thielker & T. Rost, Germany

Roth, Ulrich G., Holographie Roth, Germany
Rottenkolber, Hans, Rottenkolber Holo-System, Germany
Roule, Richard, ST!, USA
Rueck, A. B., AB Rueck Holoart, Germany
Ruey-Tung, Hung, Ruey-Tung, Miss. Hung, Japan
Ruiz-Rosales, Jorge, Corp Mexicana De Impresion, Mexico
Rumo, Wang, Minjian Laser Holography, China

S

Saarinen, Jyrk i, Heptagon Oy, Finland
Sakai, Miss Tomoko, HODIC, Japan
Sander, Ingolf, Optitek, Inc. , USA
Sarda, Uwe, topac GmbH, Department Holography, Germany
Sato, Shunichi, Sharp Corp., Japan
Saxby, Graham, Saxby, Graham, UK
Schaper, Stefan, Directa GmbH, Germany
Schauer, Steve, SouthW.:st Holographics, USA
Scheir, Peter, AD 2000'\ Inc., USA
Schenker, Frank M., A~arius-Vertrieb, Germany
Schipper, Wilfried, Holar Seele KG, Germany
Schipper, Wilfried, Hologram Company RAKO GmbH, Germany
Schlewitt, Carsten, Optical Test Equipment, Germany
Sclunelzer, Carlo, Studio Fuer Holographie, Germany
Schmidt, Helmut, Hologramm, Germany
Schomer, Jurgen, Rottenkolber Holo-System, Germany
Schrieber, Matthew, Holographic Images Inc., USA
Schulze Brockhausen, Eva, EPA - Elektro-Physik ... , Germany
Schulze, Ute, Directa GmbH, Germany
Schuman, Greg, World Holographics, USA
Schvartzman, Frederic, Foreign Dimension, China
Schvartzman, Frederic, Foreign Dimension, China
Schweer, Joerg, Holop~ s, Germany
Schweitzer, Dan, New York Holographic Laboratories, USA
Schwider, J. , University of Erlanger, Germany
Sciammarella, Cesar, Illinois Institute Of Technology, USA
Scott, G.H. , Customer Service Instrumentation, UK
Seele, Gerd, Holar Seele KG, Germany
Seitz, Mr., Steuer KG GmbH & Co., Gern1any
Selbach, H., Polytec GmbH, Germany
Shafer, Brad, Engineering Animation, Inc., USA
Shah, Kailesh, Ojasmit Holographics, India
Shahjahan, Mr., Dimensions, Pakistan
Sharma, Govind, Holostik India Pvt. Ltd. , India
Sharpe, Frank, Datasights Ltd., UK
Sherwood, Robert, Robert Sherwood Holographic Design, USA
Shimon, Hameiri, Holography Israel, Israel
Sholizhong, An, Taiyuan Shiji Holography Ltd. , China
Shun, Zhu De, Shandong Academy of Sciences, China
Simson, Bernd, Capilano College, Canada
Simson, Bernd, General Holographics, Inc., Canada
Simson, Paula, General Holographics, Inc., Canada
Sinclair, Douglas P., Sinclair Optics, Inc., USA
Singh, Ravinder, Print-M-Boss, India
Siveriver, Leonid, Avant-Garde Studio, USA
Sivy, George, Richmond Development Group, USA
Skipnes, Olav, Interferens Holografi D.A., Norway
Smith, Carol, Laser Drive Inc., USA
Smith, S.D., Beddis Kenley (Machinery) Ltd. , UK
Soales, Bob, CVI Laser Corporation, USA
Song, Chung, Dan Han Optics, Korea
Song, Li, Inspeck, inc, Canada - ,
Sott, Gudrun, AKS Holographie-Galerie GmbH, Germany
Souparis, Hughes, Hologram Industries, France
Sowdon, Michael, Fringe Research Holographics, Canada
Spanner, Karl , Physik Instrumente (PI) GmbH, Gelmany
Spiegel, Gary, Newport Corporation, USA
Spierings, Walter, Dutch Holographic Laboratory BV, Netherlands
Sponsler, Michael B. , Syracuse University, USA
Spreer, Elmar, Traumlaboratorium, Germany
St. Cyr, Suzanne, Holographic Applications, USA
Staiger, Brigitta, Holosta Holographie-Galerie, Germany

Last Name,	First Name,	Affiliation,	Country

Starcke, Ali-Veli, Starcke, Ky., Finland
Steele, Tom, Lightwave, USA
Stehle, Robert, S.O.P.R.A., France
Steinbichler, H., Steinbichler Optotechnik GmbH, Germany
Steinfeld, Belle, Dell Optics Company, Inc., USA
Steinmetz, E., Haus der Technik e. V, Germany
Stelter, Manfred, Process Technologies, USA
Stensborg, Mr. Jan, Stensborg Inc. , Denmark
Stephens, Anait Arutunoff, Stephens, Anait, USA
Stepien,- Pawel, Hololand S.c., Poland
Stich, Boguslaw, Holografia Polska, Poland
Stockton, John, Tamarack Storage Devices, USA
Stogsdill, John, Photo Sciences, USA
Stolyarenko, Alexander, TAVEX, Ltd. , Ukraine
Stone, Thomas, Wavefront Research, Inc., USA
Stooss, Richard, Krystal Holographics Vertriebs-GmbH, Germany
Strassner, Hans M., HMS-Elektronik, Germany
Styns, Erik, Free University Of Brussels., Belgium
Su, Dr. 1.1., Industrial Technology Research Ins!., Taiwan
Sugarman, Stephen, Holographic Products, USA
Summar, Carol, Art Lab, USA
Surana, Rajendra, Ojasmit Holographics, India
Swetter, Erik, 3-D Hologrammen, Netherlands
Swinehart, Patricia, CFC Applied Holographics, USA
Synowiec, George, Lumonics Ltd., UK
Sziedat, Hardo, Holarium. Germany

T

Taylor, Rob, Forth Dimension Holographics, USA
Taylor, Steven, Wild Style Entertainment, USA
Taylor, Tom, Direct Holographics, USA
Thielker, Klaus, K. Thielker & TRost, Germany
Thiemon, Ms., Daimler Benz Aerospace, Germany
Tholen, Maureen, 3M - Safety and Security Systems, USA
Thoma, John, Advanced Holographic Laboratories, USA
Thomas, Jackie, Laser Institute Of America, USA
Thompson, Bridget, iC Holographics, UK
Thompson, Naisha, Diffraction Ltd., USA
Thuston - Lighty, Cathy, M.LT Museum, USA
Tianji, Wang, Guangzhou Inst.of Electronics:- China
Tidmarsh, David, Applied Holographics, Pic., UK
Tiemon, M. , Domier Medizintechnik GmbH, Germany
Titizian, Lia, Optical Research Associates, USA
Tiziani, Hans, University of Stuttgart, Germany
Tobin, John, Moonbeamers, Australia
Toland, Lee, Meredith In struments, USA
Tolia, Dr. Arun, Spatial Holodynamics (India) Pvt. Ltd" India
Tomking, Donald, Kurz Foil s, USA
Tong, Chen Guo, Morning Light Holograms, China
Townsend, Patrick, Navidec Inc. USA
Trask, Lany, American Society for Nondestructive Testing, USA
Trayner, David, Richmond Holographic StudiO'S Ltd, UK
Trebst, Walter, Spot Agentur fur Holo. und bung, Germany
Tribillon, Dr.Jean Louis, Holo-Laser, France
Tschudi, Theo, Tech.Hochschule Darmstadt, Germany
Tsufura, Lisa, Melles Group Laser Group, USA
Tsujiuchi., Jumpei, Chiba University, Japan
Tuffy, Francis, Advanced Holographic Laboratories, UK
Tunnadine, Graham, 3D-4D Holographics, UK
Tyler, Doug, Saint Mary's College, USA
Tzong, Tang Yaw, Institute Of Optical Science, Taiwan

U

Unbehaun, Klaus, Deutscher Drucker Veri." Gemlany
Unbehaun, Klaus, Holopublic Unbehaun, Germany
Unterseher, Fred, Unterseher & Associates, USA
Unterseher, Fred, Zone Holografix, USA
Upatnieks, Juris, Applied Optics, USA
Uram, Marvin, Holograms Unlimited, USA
Uram, Marvin, U.K. Gold Purchasers, Inc., USA
Urgela, Stanislav, Technical University Zvolen, Slovakia
Uwe, Saurda, Holographie Labor, Gelmany

Last Name,	First Name,	Affiliation,	Country

Uyemura, T, University Of Tokyo, Japan

V

Valdivia, Allison, Optics Plus Inc., USA
Vancurova, Jana, Lightgate, Ltd, Czech Republic
Varga, Miklos, Hologram Varga Miklos, Hungary
Vamey, Chris, Electro Optics Developments Ltd., UK
Venkateswaran, ' Sagar, Peacock Laboratories, Inc., USA
Vikram, Chandra, University Of Alabama at Huntsville, USA
Vila, Doris, Doris Vila Holographics, USA
Vila, Doris, Doris Vila Holographics, USA
Vinson, Joachim, VinTeq, Ltd., USA
Vinstadt, Tom, Virtual Image (a division of Printpack, Inc.), USA
Vogel, Jon, Holographics (Uk) Ltd., UK
von Bally, Gert, University of Muenster, Germany
Voss, Katharina, LichtBlicke Cla13en & Voss, Germany
Vulcano, Charlie, Holographic Finishing, USA

W

Wada, Takashi, Dai Nippon Printing Co., Ltd., Japan
Wagensonner,M., Holotec, Germany
Wale, R. D., Galvoptics Ltd., UK
Walters, Glenn J., A.D. Tech., USA
Wang, Hunter, Holography Development Co.Ltd., China
Wanlass, Mike
Wappelt, Andreas, Andreas Wappelt - Photonics Direct, Germany
Wappelt, Andreas, Technische Universitaet Berlin, Germany
Warczynski, Ronald, Witchcraft Tape Products, Inc., USA
Wassel, Manfred, AD HOC Public Relations GmbH, Germany
Wegeler, Marc, Lauk & Partner GmbH, Germany
Weil, Jeffrey, Holographic Dimensions, USA
Weili, Zhu, Institute for Holographic Tech., China
Weinstein, Beth, New York Hall Of Science, USA
Wen, Pei, Beijing Sanyou Laser Images Co., Chjna
Wenjun, Luo, Wuhan Packaging and Printing United Co., China
Wemenski, John, Jodon Inc., USA
Wesly, Ed, Wesley, Ed, USA
White, John, Cobum Corporation, USA
White, Steve, Electro Optical Industries, Inc., USA
WillWlt, Fred, Elusive Image, USA
Willard, Susan, Richardson Grating Laboratory, USA
Williams, Sareth, San Jose State University, USA
Wilson, Brett, Lazart Holographics, Australia
Windeln, Wilbert, ETA-Optik Gmbh, Germany
Winopal, Gerhard, Gerhard Winopal Forschungsbedarf, Germany
Wober, Irmfried, Holography Center of Austria, Austria
Wollenweber, Andreas, Magick signs Holografie, Germany
Woodward, Keith, American Bank Note Holographics, USA
Woolford, Jimmy, HDlPanama, Panama
Wootner, Marc O., Transfer Print Foils, Inc., USA

X

Xuzhang, Zeng, Chengdu Xinxing Institute .. of Tech., China

Y

Yamaguchi, Masahiro, Tokyo Institute Of Technology, Japan
Yamaguchi, Mashahiro, Tokyo Institute of Technology, Japan
Yamazaki, Hitoshi, Hyogo Prefectual Museum of Art, Japan
Yao, Wang, Beijing Fantastic Hologram Products, China
Yijun, Qin, Foshan Holosun Packaging Co. Ltd, China
Yokota, Hideshi, Tokai University, Japan
Yoshikawa, Dr. Hiroshi, Nihon University, Japan
Yuan, Wei ben, Tianjin Holdor Optics Inc. China, China
Yuanzhong, Gong, Shanghai Dahua Printing Factory, China

Z

Zec, Peter, Zec, Peter, Dr. phil. , Germany
Zhaoqun, Zhang, Image Technical Development Co., China
Zheng, Zhang, Chongqing Yinhe Laser Products Ltd., China
Zhongming, Lu, Nanjing Sanle Laser Technology R&D, China
Ziping, Fu, Beijing Hologram Printing Tech. Co, China
Ziping, Wu, Zhuhai Xiangzhou Great Wall Laser, China
Zucker, Richard, Bridgestone Technologies, Inc., USA
Zurek, Mr., University of Munich, Germany

A List of Businesses (A - Z), Country and Internet WWW Address

Note: We assume all World Wide Web addresses begin with the prefix ''http://www.''. Standard suffixes include com (commercial), gov (government), edu (educational), org (organization) or country codes. Depending on your Internet browser software program, you may have to enter slight variations of the addresses. Please inform us of any changes, ommisions or mistakes you may find by email;staff@rossbooks.com. Remember to visit our website: www.rossbooks.com!

Business Name	Country	World Wide Web Address
3 Deep Hologram Company	USA	www.3deepco.com
3D Images Ltd.	UK	www.tisco.coml3d-web/3d-images/
3D Technologies & Arts	Slovenia	www.embers.tripod.coml-holography
3DIMAGE	Australia	www.mcm.com.au
3M - Safety and Security Systems	USA	www.mmm.com
A.D. Tech (Advanced Deposition Tech.)	USA	www.adv-dep.com
Accuwave Corp.	USA	www.accuwave.com
AD 2000, Inc.	USA	www.ad2000.comlad20001..
Advanced Optics, Inc.	USA	www.techexpo.com
Advanced Technology Program	USA	www.atp.nist.gov
Ag Electro-Optics Ltd.	UK	www.ageo.co.uk
Ahead Optoelectronics, Inc.	Taiwan	www.ahead.com.tw
AKS Holographie-Galerie GmbH	Germany	www.members.aol.comlakshollhhome_d.htm
Alabama A&M University	USA	www.caos.aamu.edu
Amagic Technologies Inc.	USA	www.thomasregister.coml
American Bank Note Holographics	USA	www.abnh.com
American Laser Corporation	USA	www.amlaser.com
American Societyty for NDT	USA	www.asnt.org/
Andreas Wappelt - Photonics Direct	Germany	www.home.t-online.delhome/wappelt/
Another Dimension Inc. (Spectore/ADI)	USA	www.spectore.com
Applied Holographics, Pic.	UK	www.applied-holographics.com
Art Institute Of Chicago	USA	www.artic.edu
Australian Holographics	Australia	www.mcm.com.au/
Batelle Pacific Northwest National Lab	USA	www.pnl.gov2080
Berkhout, Rudie	USA	www.rudieberkhout.home.mindspring.com
Blue Ridge Holographics, Inc.	USA	www.blueridgeholo.com
Bobst Group Inc.	USA	www.bobstgroup.com
Burleigh Instruments, Inc.	USA	www.burleigh.com
Capilano College	Canada	www2.capcollege.bc.cal-bsimson/
Centre d' Art Holo. et Photonique	Canada	www.cmaisonneuve.qc.calholostar.html
CFC Applied Holographics	USA	www.applied-holographics.com
Checkpoint Security Services Limited	UK	www.checkpoint.co.uk
Coburn Corporation	USA	www.coburn.com
Coherent, Inc. - Laser Group	USA	www.cohr.com
Concordia University	Canada	www.concordia.ca
Control Module Inc.	USA	www.controlmod.com
Control Optics	USA	www.controloptics.com
Corion Corp.	USA	www.Corion.com
Corning Incorporated	USA	www.coming.comlindex.html
Creative Holography Index, The	UK	www.holo.comlpeperlsearch.html
Crown Roll Leaf, Inc.,	USA	www.crownroilleaf.com
CYI Laser Corporation	USA	www.cvilaser.com
Datacard Corporation	USA	www.datacard.coml
Deutsche Gesellschaft fur Holografie	Germany	www.burg-halle.de/dgh
Diamond Images, Inc.	USA	www.Diamondlmages.com
Digital Matrix Co.	USA	www.galvanics.com
Dimension 3	Canada	www.dimension3.net
Dimensional Arts	USA	www.holo.com
Dimensional Foods Co.	USA	www.lightvision.com
Dimuken (GB)	UK	www.dimuken.co.uk
DuPont (E.!. DuPont De Nemours	USA	www.dupont.com
Dutch Holographic Laboratory BY	Netherlands	www.euroweb.comlDHL
Ealing Electro-Optics Inc.	USA	www.ealing.com
Edmund Scientific Company	USA	www.edsci.com
Fisher Scientific	USA	www.fisheredu.com
FLEXcon	USA	www.flexcon.com
Floating Images, Inc.	USA	www.floatingimages.com
Focal Image Ltd	UK	www.focalimage.com
Foil Stamping and Embossing Assn.	USA	www.fsea.com
FoilMark Holographic Images	USA	www.foilmark.com
Foreign Dimension	China	www.dimension.com.hk
Frank Defreitas Holography Studio	USA	www.holoworld.com

Fresnel Technologies Inc.	USA	www.fresneltech.com
Fujitsu Laboratories Ltd.	Japan	www.fujitsu.com
General Design	USA	www.sfo.coml-bk
General Holographics, Inc.	Canada	www2.capcollege.bc.cal-bsimsonl
Geola	Lithuania	www.geola.coml
Glaser - Technical Consulting	Israel	www.weizmann.ac.ill
Glass Mountain Optics	USA	www.glassmountain.com
Harvard Apparatus, Canada	Canada	www.harvardapparatus.com
HDTPanama	Panama	www.sinfo.net/holographic
Heptagon Oy	Finland	www.heptagon.fi
Holo/Source Corporation	USA	www.holo-source.com
HoloCom	USA	www.holo.com
Hologram Fantastic	USA	www.mallofamerica.comldirectlhome.htm
Hologram Land	USA	www.mallofamerica.comldirectlhome.htm
Hologram Research, Inc.	USA	www.hologramres.com
HOLOGRAMAS, S.A. DE c.v.	Mexico	www.holomex.com.mxl
Holograms and Lasers International	USA	www.holoshop.coml
Holograms Unlimited	USA	www.eden.coml-mainlinkltxlsatlartlrailindex.htm
Holographic Design Systems	USA	www.concentric.net/-Museumh/
Holographic Dimensions	USA	www.hgrm.com
Holographic Studios	USA	www.hmt.comlholography!holostudios/index.html
Holographics North Inc.	USA	www.holonorth.com/
Holography and Media Institute	Canada	www.ulaval.cal
Holography Development Co.Ltd.	China	www.holoworld.comlwww/hdc/
Holography Marketplace	USA	www.holoinfo.com
Holonix	USA	www.holonix.coml
Holophile, Inc.	USA	www.connix.com/-barefoot/
Holostar	Canada	wwwcmaisonneuve.qc.calholostar.html
Holoworld.com	USA	www.holoworld.com/
Honeywell Technology Center	USA	www.honeywell.com
Ibsen Micro Structures A/S	Denmark	www.ibsen.dkl
ICI Polyester	USA	www.icipolyester.comlhome.htm
Images Company	USA	www.he.netl-imagesco/
Imagination Plantation	USA	www.iplant.com
ImEdge Teclmology	USA	www.eastview.org/lmEdge/
Industrial Technology Institute	USA	www.iti.org
INETI	Portugal	www.laer.ineti.pt/
Institut fur Angewandte Phys ik	Germany	www.physik.th-darmstadt.de/andreas
James River Products	USA	www.hmt.comlholography/jrp/index.html -
Kaiser Optical Systems, Inc.	USA	www.kosi.com/Products/Ramanl5s1cover.htm
Karas Studios Holografia	Spain	www.webvent.comlkaras
Kimmon Electric Co., Ltd.	Japan	www.kimmon.com
Kinetic Systems, Inc.	USA	www.kineticsystems.com
Krystal Holographics International Inc.	USA	wwwkhiinc.com
Krystal Holographics International Inc.	USA	www.khiinc.com
Krystal Holographics Vertriebs-GmbH	Germany	www.khiinc.com
Laboratory for Optical Data Processing	USA	www.ece.cmu.edul
Lake Forest College	USA	www.lfc.edu
Laser and Motion Development Co.	USA	www.lasermotion.com
Laser Arts Society For Ed, & Research	USA	www.hmt.com/holography/laser/index.html
Laser Focus World	USA	www.lfw.com/WWW/home.htm
Laser Institute Of America	USA	www.laserinstitute.org/
Laser Reflections	USA	www.access.com.coml-hologram
Laser Resa le Inc.	USA	www.laserresale.com
LaserMax, Inc.	USA	wwwlasermax-inc.coml
Laserworks	USA	www.laser-works.com
Lasiris Inc.	Canada	www.lasiris.com/
Lazart Holographics	Australia	www.acay.com.aul-lazartl
Lenox Laser	USA	wwwlenoxlaser.com
LiCONiX	USA	www.liconix.com
Light Impressions International, Ltd.	UK	www.lightimpressions.co.ukl
Light Impressions International, Ltd.	USA	www.lightimpressions.co.uk
Light Wave Gallery	USA	www.hologramsource.com
Lightrix, Inc.	USA	www.lightrix.com
LightVision Confections	USA	www.lightvision.com
Loughborough Univ. Of Tech.	UK	www.lboro.ac.uk
		www.media.mit.edu/groups/spi/
M.LT.	USA	
Man/Environment, Inc.	USA	www.armchair.com
Marks, Gerald	USA	www.vision3d,comlpulltime3d1
Marubun Corporation	Japan	www.newport.com
Mazda Motor Corp.	Japan	www.mazda.coml
Media Interface, Ltd,	USA	www.bway.netl-ronholog
Melissa Crenshaw Holography Studio	Canada	www.capcollege.bc.caldeptlphysics/

INTERNET WWW ADDRESSES

Melles Group Laser Group	USA	www.mellesgriot.com
Menning, Melinda	Australia	www.mpce.mq.edu.au
Meredith Instruments	USA	www.mi-Iasers.com
Metrologic Instruments, Inc.	USA	www.metrologic.coml
Midwest Laser Products	USA	www.midwest-Iaser.coml
Mitsubishi Heavy Industries Ltd.	Japan	www.mitsubishi.coml
Mulhem, Dominique	france	www.alphapix.com
Museu D' Holografia	Spain	www.museuholos.com
Museum Of Holography/Chicago	USA	www.cris.coml-museumh
MWK Industries	USA	www.mwkindustries.coml
Nakamura, Ikuo	USA	www.spacelab.net/-ikuol
National Physical Laboratory	UK	www.npl.co.uk
New Focus, Inc	USA	www.newfocus.com/
New Light Industries	USA	www.iea.coml-nli
New York Hall Of Science	USA	www.nyhallsci.orgl
New York Holographic Laboratories	USA	www.waena.edu/~dan/9999a.htm
Newport Corporation	USA	www.newport.com
Nimbus Manufacturing, Inc.	USA	www.nimbuscd.com/
Norland Products, Inc.	USA	www.norlandprod.com/
NovaVision	USA	www.wcnet.org/-nova/
Ontario College of Art/Holography	Canada	www.ocad.on.ca/
Optical Research Associates	USA	www.optica lres.com
Optical Security Group	USA	www.opticalsecurity.com/
Optical Society of America (OSA)	USA	www.osa.org
Optical Test Equipment	Germany	www.moeller-wedel.com
Optineering	USA	www.primenet.com/-kcreath
Optitek, Inc.	USA	www.optitek.com
Oregon Institute of Technology	USA	www.oit.osshe.edul
Oriel Instruments	USA	www.oriel.com
Oxford Holographics	UK	www.oxfordshire.co.uk
Pasco Scientific	USA	www.pasco.com
Pennsylvania Pulp & Paper Co.	USA	www.holoprism.com
Pepper, Andrew	UK	www.holo.com/peperlsearchJ;:tm I
Photo Research, Inc.	USA	www.photoresearch.com
Photo Sciences	USA	www.photo-sciences.coml
Photonics Spectra	USA	www.laurin.com
Pilkington Optronics	UK	www.thejob.com/pilkington/
Polaris Research Group	USA	www.gwis.com/-polaris/whati s.html
Polaroid Corporation	USA	www.holoroid.com
Potomac Photonics, Inc.	USA	www.potomac-Iaser.com
Process Technologies·	USA	www.execpc.com/-ptilindex.html
PullTime 3-D Laboratories	USA	www.vision3d.com/pulltime3d/
Rainbow Symphony Inc.	USA	www.rainbowsymphony.com/
Ralcon	USA	www.xmission.coml-ralcon
Reconnaissance International Ltd.	UK	www.hmt.comlholography/hnews/index.html
Reynolds Metals Co.	USA	www.rmc.com
Richardson Grating Laboratory	USA	www.spectronic.coml
Richmond Development Group	USA	www.richmondinc.coml
Robert Sherwood Holographic Design	USA	www.h6 Iographicdesign.coml
Rochester Ins!. Of Technology	USA	www.rit.edul
Rochester Photonics Corporation	USA	www.rphotonics.com
Rolls-Royce Pic	UK	www.rolls-royce.co.ukl
Rolyn Optics	USA	www.rolyn.com
Ross Books	USA	www.rossbooks.com
Royal Holographic Art Gallery	Canada	www.islandnet.coml-royal
San Jose State University	USA	www.fire.sjsu.edu
Sandia National Laboratories	USA	www.sandia.gov/
School Of Holography	USA	www.cris.coml-museumh
Sharon McCormack Holography	USA	www.gorge.netlbusiness/holography/
Shuttlecart	USA	www.starlog.com/
Silicon Graphics	USA	www.sgi.com
Sillcocks Plastics International	USA	www.sillcocks.col11/
Sinclair Optics, Inc.	USA	www.sinopt.coml
Sommers Plastic Products	USA	www.sommers.cOl11
Spectra-Physics Lasers Inc.	USA	www.splasers.col11
SPIE	USA	www.spie.orgl
SPIE's Holo.Working Group Newsletter	USA	www.spie.org/
Springer-Verlag New York	USA	www.springer-ny.com
Star Magic	USA	www.starmagic.col11
Starlog	USA	www.starlog.com/
Steinbichler Optotechnik GmbH	Germany	www.steinbichler.com
Steinbichler U.S.A. inc.	USA	www.steinbichler.com
Stensborg Inc.	Denmark	www.catscience.dklcompani/descrip.html#stensborg
Syracuse University	USA	www-che.syr.eduiSponsler.html

TAVEX, Ltd.	Ukraine	www.tav.kiev.ua
Technische Universitaet Berlin	Germany	www.physik.hl-berlin.de
The Hologram Company #1	USA	www.starlog.com!
The Hologram Company #2	USA	www.starlog.com!
The Hologram Company #3	USA	www.starlog.com/
Thorlabs Inc.	USA	www.thorlabs.com
Three Dimensional Imagery	USA	www.3dimagery.com
Toppan Printing Co., Ltd.	Japan	www.toppan.com!
Total Register Inc.	USA	www.totalregister.20m!
Towne Technologies	USA	www.townetech.com
Trace Holographic Art & Design, Inc.	USA	www.traceholo.com!
Transfer Print Foils, Inc.	USA	www.hmt.comiholography/TPF/index.html
Triple-D Laser Imaging	Netherlands	www.home.wirehub.nll-acal
Turing Institute	UK	www.turing.gla.ac.uk
Ultra-Res Corporation	USA	www.acds.com/URIultra-res.html
Unifoil Corporation	USA	www.unifoil.com
Uniphase Lasers	USA	www.uniphase.com
Universite De Neuchatel	Switzerland	www.fsrm.ch
Universite Laval	Canada	www.fsg.ulaval.cal
University Of Dayton	USA	www.udri.udayton.edul
University of Latvia	Latvia	www.cfi.lu.lv/
University Of Rochester	USA	www.optics.rochesteredu
University Of Tokyo	Japan	www.t.u-tokyo.ac.jp/
University Of Wisconsin/Madison	USA	www.engr.wisc.edul
Uvex Safety Inc.	USA	www.uvex.com
Van Leer Metallized Products	USA	www.vanleercom
VinTeq, Ltd.	USA	www.vinteq.com
Virtual Image, a division of Printpack	USA	www.virtimage.com
Voxel	USA	www.voxel.com
Worcester Polytechnic Institute	USA	www.wpi.edu
WYKO-VEECO	USA	www.wyko.com
Zero Gravity/Galaxies Unlimited, Inc.	USA	www.spectore.com

A List of USA Businesses Sorted by Postal Zip Code and State

Zip	State	USA Business	Zip	State	USA Business
01085	MA	Automated Holographic Systems	06611	CT	Carl M. Rodia And Associates
01202	MA	Photonics Spectra	06759	CT	Virtual Image
01420	MA	American Holographic Inc.	06801	CT	Laser Optics, Inc.
01562	MA	FLEXcon	06804	CT	Total Register Inc.
01609	MA	Worcester Polytechnic Institute	07004	NJ	Frank J. Deutsch Inc.
01746	MA	Ealing Electro-Optics Inc.	07015	NJ	Sommers Plastic Products
01752	MA	Shipley Chemical Co.	07022	NJ	Dell Optics Company, Inc.
01776	MA	Laser Resale Inc.	07055	NJ	Unifoil Corporation
01950	MA	FoilMark Holographic Images	07065	NJ	Dri-Print Foils
02038	MA	Corion Corp.	07068	NJ	Bobst Group Tnc.
02038	MA	Van Leer Metallized Products	07111	NJ	Excitek Inc.
02062	MA	Regal Press Inc.	07310	NJ	City Chemical
02109	MA	Dimensional Foods Co.	07435	NJ	Miller, Peter
02109	MA	LightVision Confections ,	07446	NJ	Holographyx inc
02131	MA	Kinetic Systems, Tnc. r	07503	NJ	21 st Century Finishing Inc.
02139	MA	Polaroid Corporation	07503	NJ	Crown Roll Leaf, Inc.
02139	MA	M.I.T. ~-	07647	NJ	Imad, Inc.
02139	MA	M.I.T. Museum ... ; _ '- " .. . . _	07652	NJ	Starlog . ...
02142	MA	Rowland Instltute For SCience '1'>1 ~ • - '	07657	NJ	HolographIC Fmlshmg
02144	MA	Acme Holography	07663	NJ	Lasermetrics, Inc.
02146	MA	Casdin-Silver Holography	07860	NJ	Thorlabs Inc.
02780	MA	A.D. Tech	07866	NJ	Starlog
02917	RI	Uvex Safety Inc.	07922	NJ	Sillcocks Plastics International
03031	NH	Odhner Holographics	07962	NJ	Silhouette Technology
03062	NH	Laser Focus World	08007	NJ	Edmund Scientific Company
03855	NH	Lumenx Technologies, Inc.	08012	NJ	Metrologic Instruments, Inc.
05401	VT	Holographics North Inc.	08057	NJ	Tyler Group
05673	VT	Diffraction Ltd.	08555	NJ	Avant-Garde Studio
06002	CT	Scharr Industries	08701	NJ	Coburn Corporation
06082	CT	Control Module Inc.	08755	NJ	Luminer Printing and Converting
06419	CT	Holophile, Inc.	08816	NJ	Transfer Print Foils, Inc.
06484	CT	Interactive Industries, Inc.	08876	NJ	Towne Technologies
06497	CT	Oriel Instruments	08902	NJ	Norland Products, Inc.
06511	CT	AD 2000, Inc.	10003	NY	Star Magic
06605	CT	Bridgestone Technologies, Inc.	10009	NY	New York Holographic Labs
06605	CT	STI	10009	NY	Pangaea Design

USA BUSINESSES BY ZIP CODE & STATE

10010	NY	Holographic Studios	32789	FL	McMahan Electro-Optic
10010	NY	Marks, Gerald	32809	FL	The Hologram Company #2
10010	NY	PuliTime 3-D Laboratories	32819	FL	Zero Gravity
10010	NY	Springer-Verlag New York	32826	FL	Laser Institute Of America
10011	NY	Berkhout, Rudie	33029	FL	Evolution Design, Inc.
10019	NY	Krystal Holographics International	33065	FL	Holographyx Inc.
10023	NY	Star Magic	33133	FL	Diamond Images, Inc.
10028	NY	Star Magic	33133	FL	Zero Gravity
10169	NY	Hallmark Capital Corp.	33139	FL	Holographic Images Inc .
10301	NY	Art Lab	33139	FL	Larry Lieberman Holography
10301	NY	Bellini, Victor	33166	FL	Holographic Dimensions
10314	NY	Images Company	33442	FL	Another Dimension Inc.
10523	NY	American Bank Note Holographics	33442	FL	Zero Gravity/Galaxies Unlimited
10546	NY	Holographic Optics Inc.	33868	FL	Infinity Laser Laboratories
10601	NY	ImEdge Technology	35762	AL	Alabama A&M University
11101	NY	Holographics Inc.	35899	AL	Univ. Of Alabama at Huntsville
11201	NY	Laser Light Ltd.	38133	TN	American Paper Optics Inc.
11211	NY	Doris Vila Holographics	43228	OH	American Society for NDT
11215	NY	Fornari, Arthur David	43402	OH	NOlo'aVision
11215	NY	Nakamura, Ikuo	44143	OH	Pofaris Research Group
11217	NY	Media Interface, Ltd.	44509	OH	Cnipmagem Inc.
11368	NY	New York Hall Of Science	45469	OH	University Of Dayton
11550	NY	Digital Matrix Co.	45840	OH	Holovision Systems Inc.
11590	NY	Floating Images, Inc.	46556	IN	Saint Mary's College
11721	NY	Linda Law Holographics	47448	IN	Forth Dimension Holographics
11735	NY	Infrared Optical Products	47591	IN	Vincennes University
11768	NY	Holo Art	47715	IN	Southern Indiana Holographics
11788	NY	Continental Optical	48009	MI	Intrepid World Communications
11790	NY	Hologram Research, Inc.	48009	MI	American Propylaea Corp
12534	NY	Louis Paul Jonas Studios, Inc.	48103	MI	Jodon Inc.
12787	NY	David Dann Modelmaking	48103	MI	Applied Optics
13244	NY	Syracuse University	48105	MI	Industrial Technology Institute
14150	NY	Science Kit & Boreal Labs	48106	MI	Kaiser Optical Systems, Inc.
14450	NY	Sinclair Optics, Inc.	48109	MI	University Of Michigan
14453	NY	Burleigh Instruments, Inc.	48150	MI	Holo/Source Corporation
14467	NY	Holotek	48375	MI	Stcillbichler U.S.A. inc.
14623	NY	LaserMax, Inc.	48408	MI	EI Don Engineering
14623	NY	Rochester Photonics Corporation	48710	MI	Saginaw Valley State Univ.
14623	NY	Rochester Inst. Of Tech	48734	MI	Reva's Holographic Illusions
14625	NY	Richardson Grating Laboratory	49038	MI	Witchcraft Tape Products, Inc.
14627	NY	University Of Rochester	49117	MI	Laser Holography Workshop
14650	NY	Eastman Kodak Co.	49456	MI	Textile Graphics, Inc.
14831	NY	Corning Incorporated	49456	MI	Holography Presses On (HPO)
14850	NY	Three-D Light Gallery	50010	IA	Engineering Animation, Inc.
15044	PA	Laser Drive Inc.	50021	IA	Advanced Optics, Inc.
15213	PA	Lab for Optical Data Processing	50068	IA	D. Brooker & Associates
15238	PA	Aerotech Inc.	53154	WI	Process Technologies
17579	PA	Direct Holographics	53186	WI	Letterhead Press, Inc.
18018	PA	Wavefront Research, Inc.	53706	WI	University Of WisconsinlMadison
18102	PA	Frank DeFreitas Holo. Studio	55024	WI	Brandtjen & Kluge, Inc.,
18102	PA	Holoworld.com	55105	MN	Holographic Products
18719	AZ	Optineering	55144	MN	3M - Safety and Security Systems
18972	PA	Laser Technical Services	55418	MN	Honeywell Technology Center
19038	PA	Pennsylvania Pulp & Paper Co.	55425	MN	Hologram Fantastic
19143	PA	Peacock Laboratories, Inc.	55425	MN	Hologram Land
19335	PA	Planet 3-D	55425	MN	Starlog
19372	PA	Keystone Scientific Co.	55439	MN	Holographic Label Converting
19403	PA	Laser Technology, Inc.	55440	MN	Datacard Corporation
19850	DE	ICI Polyester	55441	MN	Hologram World, Inc.
19880	DE	DuPont	60004	IL	MacShane Holography
20009	DC	(ASTI)	60007	IL	Creative Label
20009	DC	The Holography, Laser & ...	60045	IL	Integraf
20036	DC	Optical Society of America	60045	IL	Lake Forest College
20706	MD	Potomac Photonics, Inc.	60048	IL	Holographic Industries, Inc.
20770	MD	Holographic Applications	60068	IL	Capitol Converting Equipment, Inc.
20899	MD	Advanced Technology Program	60115	IL	Northern Illinois University
21057	MD	Lenox Laser	60411	IL	CFC Applied Holographics
21152	MD	OpSec - USA	60423	IL	Midwest Laser Products
21202	MD	Holografica	60504	IL	Mitutoyo Measuring Instruments
22183	VA	Three Dimensional Imagery	60504	IL	Starlog
22901	VA	Robert Sherwood Holographic Design	60521	IL	Fisher Scientific
22901	VA	Trace Holographic Art & Design.	60603	IL	Art Institute Of Chicago
22902	VA	Blue Ridge Holographics, Inc.	60607	IL	Holographic Design Systems
22906	VA	Nimbus Manufacturing, Inc.	60607	IL	Museum Of Holography/Chicago
23230	VA	Reynolds Metals Co.	60607	IL	School Of Holography
23236	VA	James River Products	60611	IL	Light Wave Gallery
27592	NC	VinTeq, Ltd.	60612	IL	Wave Mechanics
28206	NC	Kurz Foils	60614	IL	Light Impressions International, Ltd.
29582	SC	The Hologram Company #1	60616	IL	Illinois Institute Of Technology
32256	FL	Shuttlecart	60618	IL	Holicon Corporation.

60618	IL	Richard Bruck Holography
60618	IL	Wesley, Ed
61761	IL	Merrick, Michael G.
65203	MO	Miller, Neal
66046	KS	Optimation Holographics
66046	KS	Advanced Holographic Laboratories
66206	KS	Laser Images
70112	LA	The Hologram Company #3
75202	TX	Elusive Image
76110	TX	Fresnel Technologies Inc.
77010	TX	Holograms and Lasers International
77010	TX	Holograms and Lasers International
78216	TX	Holograms Unlimited
78216	TX	U.K. Gold Purchasers, Inc.
78727	TX	Tamarack Storage Devices
78746	TX	Lone Star Illusions
78758	TX	Glass Mountain Optics
80112	CO	Navidec Inc.
80202	CO	Optical Security Group
80205	CO	Reconnaissance International Ltd.
80249	CO	Fantastic Holograms
81621	CO	Imagen Holography, Inc.
84047	UT	Optimation
84104	UT	American Laser Corporation
84115	UT	Ion Laser Technology, Inc.
84321	UT	Krystal Holographics International Inc.
84328	UT	Ralcon
84333	UT	Richmond Development Group
85018	AZ	Southwest Holographics
85301	AZ	Meredith Instruments
85308	AZ	Starlog
85704	AZ	Holo Sciences, LLC
85706	AZ	WYKO-VEECO
85721	AZ	University Of Arizona
87185	NM	Sandia National Laboratories
87192	NM	CVI Laser Corporation
87501	NM	Lasart Ltd.
87506	NM	Ana MacArthur
88005	NM	Dimensional Arts
88005	NM	HoloCom
89109	NV	Zero Gravity
89129	NV	Laser Las Vegas
89431	NV	Ultra-Res Corporation
90004	CA	NeoVision Productions
90027	CA	Booth, Roberta
90045	CA	Laser Media, Inc.
90064	CA	ManlEnvironment, Inc.
90066	CA	Spectratek Inc.
90077	CA	World Holographics
90089	CA	Univ. Of Southern California
90402	CA	Chronomotion
90404	CA	Accuwave Corp.
90505	CA	Photo Sciences
90703	CA	MGM Converters Inc.
90723	CA	Wavefront Technology
91012	CA	Photon Cantina Ltd.
91107	CA	Optical Research Associates
91202	CA	Deem, Rebecca ,
91202	CA	Unterseher & Associates 1
91311	CA	Photo Research, Inc.
91335	CA	Rainbow Symphony Inc .. ,r
91355	CA	California Institute of the Arts ,
91406	CA	Holo-Spectra
91601	CA	Zone Holografix
91706	CA	Control Optics
91720	CA	MWK Industries
91722	CA	Rolyn Optics
91769	CA	Panatron Inc.
92008	CA	Melles Group Laser Group
92069	CA	Broadbent Consulting
92108	CA	Hologram, etc.
92124	CA	Jeffery Murray Custom Holography
92606	CA	Amagic Technologies Inc.
92606	CA	Newport Corporation
92614	CA	Rice Systems, Inc.
92629	CA	UAMR
92648	CA	3 Deep Hologram Company
92648	CA	Holograms International
92648	CA	Three Deep Hologram Co.
92653	CA	Voxel

92663	CA	G.M. Vacuum Coating Lab, Inc.
92705	CA	Optics Plus Inc.
92714	CA	Melles Griot
92859	CA	Laserworks
93001	CA	Innovative Technology Associates
93021	CA	Laser Innovations
93108	CA	Stephens, Anait
93111	CA	Electro Optical Industries, Inc.
93566	CA	Starlog
94039	CA	Spectra-Physics Lasers Inc.
94043	CA	Lightwave
94043	CA	Optitek, Inc.
94043	CA	Silicon Graphics
94044	CA	Real Image
94080	CA	A.H. Prismatic, Inc.
94105	CA	Laser Reflections
94107	CA	General Design
94110	CA	Imagination Plantation
94110	CA	Multiplex Moving Holograms
94114	CA	Star Magic
94121	CA	Visual Visionaries
94122	CA	Illuminations
94124	CA	LASER
94124	CA	Holography Institute of SF
94305	CA	Stanford University
94509	CA	Laser Light Designs
94538	CA	Lexel Laser, Inc.
94539	CA	Cambridge Laser Labs
94546	CA	Advanced Precision Technology, Inc.
94577	CA	Lightrix, Inc.
94587	CA	Laser &Motion Devopmt.
94598	CA	Cifelli, Dan
94608	CA	Feroe Holographic Consulting
94608	CA	Wild Style Entertainment
94611	CA	Red Beam, Inc.
94704	CA	Holography Marketplace
94704	CA	Ross Books
94720	CA	Lawrence Berkeley Laboratory
94901	CA	Third Dimension Arts Inc.
94928	CA	Sonoma State University
94956	CA	Kauffman, John
95006	CA	Point Source Productions
95051	CA	New Focus, Inc
95054	CA	Coherent, Inc. - Laser Group
95054	CA	LiCONiX
95060	CA	Simian Co.
95062	CA	Pacific Holographics Inc.
95110	CA	Holographic Impressions
95110	CA	Smith & McKay Printing
95112	CA	Swift Instruments
95120	CA	IBM Almaden Research Center
95134	CA	Uniphase Lasers
95192	CA	San Jose State University
95468	CA	3-D Systems
95472	CA	Cherry Optical Holography
95472	CA	Gorglione, Nancy
95472	CA	Laser Affi liates
95661	CA	Pasco Scientific
96813	HW	Zero Gravity
97212	OR	Foil Stamping and Embossing Assn.
97212	OR	Inside Finishing Magazine
97224	OR	Flight Dynamics
97601	OR	Oregon Laser Consultants
97601	OR	Oregon Institute of Technology
98145	WA	Holonix
98227	WA	SPIE
98672	WA	Sharon McCormack Holography
99224	WA	New Light Industries
99352	WA	Batelle Pacific NW National Laboratory

About ROSS BOOKS

Founded in 1977, Ross Books is a general trade publisher of books in print and electronic format. Our catalog includes books about a variety of topics including health, cooking, music, sports, science, and more! Many of our titles are instructional and educational. We also publish computer software. We are widely known for our books on holography which we have been publishing since 1982.

In addition to our publications, we also provide affordable consulting and research assistance for companies seeking either general or very specific information about the holography industry. We have the world's most comprehensive and extensive listing of companies and individuals involved with commercial applications, research, artistic endeavors, etc. We also have thousands of related and pertinent publications in our database. We can provide reports, database searches and manuscripts custom-tailored to your needs. Call us today for further details about the services we provide.

Franz H. Ross is the President of Ross Books (email: franz@rossbooks.com). He founded Ross Books in April of 1977. The company has published a wide variety of books, as well as computer software. The holography community has especially benefited from Mr. Ross's efforts. In 1982, he published the world's best-selling holography instruction manual The *Holography Handbook - Making Holograms the Easy Way*. In 1989 he published the first edition of the *Holography MarketPlace*. The seventh edition of HMP was released in 1998. Mr. Ross holds a BA degree in physics from the University of California, Berkeley.

Alan E. Rhody is the Managing Editor of Ross Books (email: rhody@rossbooks.com). Mr. Rhody first studied holography as part of his undergraduate curriculum at Clark University in 1978. He has worked in the holography industry consistently since 1985, mainly in the sale and distribution of holographic art and giftware. He also has experience in the design and production of holograms for various commercial applications. In addition, his own holographic limited-edition fine art has been displayed in galleries throughout the United States. Currently, Mr. Rhody writes, edits and publishes holography related articles for Ross Books, including major parts of the *Holography MarketPlace Editions 5, 6, and 7*. His work has also appeared in *Newsweek Magazine, Laser Focus World, International Designer's Network, Signs of the Times, Inside Finishing, Video Toaster User, Holography News* and other publications.

11

Cross-Index Tables

Table 1 - Businesses That Sell or Exhibit Holograms

Country	Name of Business	Whole-sale	Retail	Holo. Gallery Gift Shop	Mail Order Catalog	Artist with Portfolio	Broker /Sales Rep	Exibit - Touring	Gallery/ museum Display
Australia	3D Optical Illusions	X	X		X		X		
Australia	3DIMAGE						X		
Australia	Australian Holographics	X	X	X	X	X	X		X
Australia	Holograms Fantastic and Optical Illus.	X	X		X		X		
Australia	HopSec/dii	X					X		
Australia	Lazart Holographics	X	X	X			X		
Australia	Menning, Melinda					X	X		
Australia	Moonbeamers	X							
Australia	New Dimension Holographics	X	X			X	X		
Australia	Parallax Gallery	X	X			X	X		
Austria	Holography Center of Austria	X			X			X	
Canada	Abrams, Claudette					X	X		
Canada	Capilano College					X			
Canada	Deep Space Holographics	X				X	X		
Canada	Dimension 3	X			X	X	X		
Canada	Fringe Research Holographics		X	X		X			X
Canada	General Holographics, Inc.	X			X	X	X		
Canada	Holocrafts	X	X		X	X	X		
Canada	Hologram Development Corp.					X	X		
Canada	Holography & Media Inst. of Quebec					X			
Canada	Island Graphix					X	X		
Canada	Melissa Crenshaw Holography Studio					X	X		
Canada	Mu's Laser Works						X		
Canada	Ontario Science Centre		X	X		X			
Canada	Page, Michael					X	X		
Canada	Royal Holographic Art Gallery	X	X	X	X				
Canada	Silverbridge Group	X				X	X		
Canada	The Hologram Store, Ltd.		X	X	X				
China	Beijing Fantastic Hologram Products	X	X				X		
China	Beijing Hologram Printing Tech. Co	X							
China	Beijing Sanyou Laser Images Co.	X							
China	Far East Holographics	X							
China	Foreign Dimension	X	X	X	X	X	X		
China	Guangdong Dongguan S. Holoprint	X							
China	Guangzhou Chuntian Indus.Tech.						X		
China	Holography Development Co.Ltd.	X					X		
China	Jiangsu Sida Images, Inc.	X							
China	Minjian Laser Holography	X					X		
China	Morning Light Holograms	X							
China	North Light Holograms Ltd.	X							
China	Qingdao Gaoguang Holography Tech.	X							
China	Qingdao Qimei Images, Inc.	X							
China	Quan Zhou Pacific Laser Images	X					X		

Table 1 - Businesses That Sell or Exhibit Holograms

Country	Name of Business	Whole-sale	Retail	Holo. Gallery Gift Shop	Mail Order Catalog	Artist with Portfolio	Broker /Sales Rep	Exibit - Touring	Gallery/ museum Display
China	Shandong Academy of Sciences								X
China	Shanghai Kanlian S & T Development	X							
China	Taiyuan Shiji Holography Ltd.	X							
China	Tianjin Water Laser Holo. Image	X							
China	Wuxi Light Impressions Inc.	X							
China	Xiamen Grand World Laser Label	X							
China	Yu Feng Laser Images Co.	X							
China	Zhuhai Xiangzhou Great Wall Laser	X							
Columbia	Imagenes Holograficas De Columbia	X					X		
Czech	Lightgate, Ltd						X		
Denmark	Stensborg Inc.						X		
Finland	Starcke, Ky.	X					X		
France	Atelier Holographique De Paris	X					X	X	
france	GEHOL sarl		X				X	X	
France	Hologram Industries	X			X		X	X	
France	Holomedia France	X	X				X		
France	Holomedia France	X	X				X		
France	Magic Laser	X					X	X	
france	Mulhem, Dominique						X		
Germany	3D Vision	X	X		X				
Germany	AB Rueck Holoart	X					X		
Germany	AD HOC Public Relations GmbH						X		
Germany	AKS Holographie-Galerie GmbH	X	X		X	X			X
Germany	AKS Holographie-Gallerie GmbH		X						
Germany	Alfred Dirksen + Sohn,						X		
Germany	Photonics Direct - Andreas Wappelt	X				X	X		
Germany	Arbeitskreis Holografie B.V.						X		
Germany	Armin Klix Holographie	X				X	X		
Germany	Art Agentur Köln						X		
Germany	Artbridge Light Studios						X	X	
Germany	Burgmer, Brigitte						X		
Germany	Curt Abramzik	X			X				
Germany	Decolux GmbH	X					X		
Germany	Deutsche Gesellschaft fur Holografie							X	
Germany	Die Dritte Dimension	X	X		X				
Germany	Dietmar Oehlmann						X		
Germany	Directa GmbH	X	X						
Germany	EPA - Elektro-Physik Aachen GmbH	X					X		
Germany	ETA-Optik Gmbh	X			X				
Germany	Fielmann-Verwaltung KG		X					X	
Germany	Frank Schenker's Aquarius-Vertrieb	X	X						
Germany	Galerie 3D		X						
Germany	Galerie 7								X
Germany	Galerie Westerland/Sylt								X
Germany	H & W, Holographie & Werbung	X	X			X			X
Germany	H. Kallenbach - H.M.V.	X					X		
Germany	Heiß, Peter, Dr., Priv.-Doz.					X			
Germany	Highlite		X						
Germany	HOL 3, Galerie fur Holographie		X						
Germany	Holar Seele KG					X			
Germany	Holarium		X						X
Germany	Holo Service/Service-Druck		X						X
Germany	Holo Time Gericke					X	X		X
Germany	Holo-Idee Reiner Kleinherne						X		
Germany	Holodesign					X			
Germany	Holografie-Hofmann		X			X			X
Germany	Hologramm		X			X			X
Germany	Holographic Laserdesign					X			
Germany	Holographic Studios Rosowski		X			X			X
Germany	Holographie & Design	X				X		X	X
Germany	Holographie Anubis	X			X				
Germany	Holographie Fachstudio Bad Rothen					X	X		
Germany	Holographie Konzept GmbH	X					X		
Germany	Holographie Roth	X				X			
Germany	Holographie-Labor Mielke	X				X			
Germany	Holoptics	X	X				X		

Table 1 - Businesses That Sell or Exhibit Holograms

Country	Name of Business	Whole-sale	Retail	Holo. Gallery Gift Shop	Mail Order Catalog	Artist with Portfolio	Broker /Sales Rep	Exibit - Touring	Gallery/ museum Display
Germany	Holosta Holographie-Galerie		X						X
Germany	HoloVision	X			X				
Germany	K. Thielker & T. Rost	X						X	
Germany	Krystal Holographics Vertriebs	X			X		X		
Germany	Laserfilm Eckard Knuth	X			X				
Germany	Lauk & Partner GmbH	X		X					X
Germany	Leonhard Kurz GmbH						X		
Germany	LichtBlicke Claßen & Voss					X			
Germany	Magick signs Holografie		X						
Germany	Magick Signs Holografie	X	X		X				
Germany	Meulien Odile					X	X		
Germany	Museum 3. Dimension		X			X			X
Germany	Museum für Holographie & neue v.		X			X			X
Germany	Oeserwerk Ernst Oeser & Söhne KG	X					X		
Germany	Ose Holografie-Design					X			
Germany	Phantastica	X	X				X		
Germany	PPM Promotion Products München	X					X		
Germany	Spot Agentur für Holographie	X							X
Germany	Stiletto Studios					X	X		
Germany	topac GmbH, Department Holography	X			X				
Germany	Traumlaboratorium					X			
Germany	Zec, Peter, Dr. phil.						X		
Hungary	Artplay Holographika Studio	X					X		
Hungary	Hologram Varga Miklos	X				X	X		
Hungary	Optopol Panoramic Metrology					X	X		
India	Ojasmit Holographics	X					X		
India	Shriram Holographics						X		
Israel	Glaser - Technical Consulting						X		
Israel	Holography Israel						X		
Japan	Brainet Corporation - Intl. Division	X							
Japan	Hyogo Prefectual Museum of Art								X
Japan	Ishii, Ms. Setsuko					X			
Japan	Japan Communication Arts Co.	X							
Japan	Light Dimension, Inc.	X							
Japan	Ruey-Tung, Miss. Hung					X			
Japan	SAM Museum								X
Lithuania	Geola	X	X				X		
Mexico	HOLOGRAMAS, S.A. DE C.V.	X					X		
Netherlands	3-D Hologrammen	X	X				X		
Netherlands	Dutch Holographic Laboratory BV	X			X				
Netherlands	HOLOTECH -Texel			X					
Netherlands	Triple-D Laser Imaging	X					X		
Norway	Interferens Holografi D.A.								X
Pakistan	Dimensions	X					X		
Poland	Holografia Polska	X					X		
Poland	Hololand S.C.	X					X		
Portugal	INETI - Institute of Info. Tech.						X		
S. Arabia	Wonders of Holography Gallery	X	X	X		X	X		
Singapore	Zero Gravity		X						
Slovenia	Likom								X
S. Africa	Synchron Pty Ltd.	X					X		
Spain	Holosco, Ernest Barnes						X		
Spain	Karas Studios Holografia		X	X		X			
Spain	Museu D' Holografia	X	X	X		X			X
Sweden	Holography Group TEM					X			
Sweden	HoloMedia Ab/Hologram Museum	X	X				X		X
Sweden	Holovision AB					X			
Switzerland	Galerie Illusoria		X			X			X
Switzerland	Holos Art Galerie		X			X			X
Taiwan	Holo Images Tech Co., Ltd.	X					X		
Taiwan	Holo Impressions Inc	X					X		
Taiwan	Holoart Studio	X				X			
Thailand	Electro-Optics Lab, NECTEC						X		
Ukraine	Tair Hologram Company						X		
UAE	Hololaser Gallery		X						
UK	3D Images Ltd.	X	X		X		X		

Table 1 - Businesses That Sell or Display Holograms

Country	Name of Business	Whole-sale	Retail	Holo. Gallery Gift Shop	Mail Order Catalog	Artist with Portfolio	Broker /Sales Rep	Exibit - Touring	Gallery/ museum Display
UK	3D-4D Holographics	X	X			X	X		
UK	A.H. Prismatic, Ltd.	X			X				
UK	Amazing World Of Holograms		X						X
UK	Benyon, Margaret - Holo. Studio					X			
UK	Boyd, Patrick					X	X		
UK	Colour Holographics	X				X			
UK	Holocrafts Europe Limited.	X				X	X		
UK	Holograms 3D						X	X	
UK	Holographic Consulting Agency						X		
UK	Holographics (Uk) Ltd.						X		
UK	John, Pearl					X			
UK	K.C. Brown Holographics					X	X		
UK	LAZA Holograms Ltd	X	X	X	X	X			
UK	Light Impressions International, Ltd.	X			X		X		
UK	London Holographic Image Studio	X			X				
UK	Op-Graphics (Holography) Ltd.	X							
UK	Oxford Holographics	X			X		X		
UK	Pepper, Andrew					X	X		
UK	Spatial Imaging Limited	X				X			
USA	3-D Systems					X			
USA	A.H. Prismatic, Inc.	X			X				
USA	Acme Holography					X	X		
USA	Amagic Technologies Inc.	X			X				
USA	American Paper Optics Inc.	X							
USA	American Propylaea Corporation						X		
USA	Ana MacArthur					X	X		
USA	Another Dimension Inc.	X	X						
USA	Applied Optics						X		
USA	Art Institute Of Chicago								X
USA	Art, Science & Tech. Inst. (ASTI)			X					X
USA	Automated Holographic Systems						X		
USA	Bellini, Victor					X			
USA	Berkhout, Rudie					X			
USA	Booth, Roberta					X			
USA	Broadbent Consulting					X	X		
USA	Carl M. Rodia And Associates						X		
USA	Casdin-Silver Holography					X	X		
USA	Cherry Optical Holography	X				X			
USA	Cifelli, Dan						X		
USA	Coburn Corporation				X				
USA	Control Module Inc.						X		
USA	CVI Laser Corporation				X				
USA	Deem, Rebecca					X			
USA	Diamond Images, Inc.	X				X	X		
USA	Dimensional Foods Co.	X			X				
USA	Direct Holographics	X	X		X				
USA	Doris Vila Holographics					X			
USA	Edmund Scientific Company		X		X				
USA	Elusive Image		X					X	
USA	Evolution Design, Inc.						X		
USA	Fantastic Holograms		X						
USA	Feroe Holographic Consulting						X		
USA	Fornari, Arthur David					X			
USA	Forth Dimension Holographics		X				X		
USA	Frank DeFreitas Holography Studio					X	X		
USA	Gorglione, Nancy					X	X		
USA	Holicon Corporation.					X			
USA	Holo Art				X				
USA	Holo-Spectra						X		
USA	Holografica		X						
USA	Hologram Fantastic		X						
USA	Hologram Land		X						
USA	Hologram Research, Inc.	X				X	X	X	
USA	Hologram World, Inc.	X			X				
USA	Hologram, etc.		X						
USA	Holograms and Lasers International	X	X	X	X	X	X		

Table 1 - Businesses That Sell or Exhibit Holograms

Country	Name of Business	Whole-sale	Retail	Holo. Gallery Gift Shop	Mail Order Catalog	Artist with Portfolio	Broker /Sales Rep	Exibit - Touring	Gallery/ museum Display
USA	Holograms International	X	X		X				
USA	Holographic Applications						X		
USA	Holographic Design Systems	X							
USA	Holographic Dimensions	X				X	X		
USA	Holographic Images Inc.	X				X			
USA	Holographic Impressions	X			X				
USA	Holographic Industries, Inc.	X					X		
USA	Holographic Studios	X	X	X		X	X		X
USA	Holographics Inc.					X	X		
USA	Holographics North Inc.	X				X			
USA	Holography Institute of San Francisco					X			
USA	Holography Presses On (HPO)				X				
USA	Holographyx inc						X		
USA	Holonix						X		
USA	Holophile, Inc.	X				X	X	X	
USA	Holovision Systems Inc.						X		
USA	Illuminations						X		
USA	Infinity Laser Laboratories					X			
USA	Interactive Industries, Inc.	X	X		X				
USA	Jeffery Murray Custom Holography					X	X		
USA	Kauffman. John					X	X		
USA	Krystal Holographics Intl.	X			X	X	X		
USA	Larry Lieberman Holography	X	X			X			
USA	Lasart Ltd.	X			X		X		
USA	Laser Affiliates					X		X	
USA	Laser Light Designs	X					X		
USA	Laser Light Ltd.						X		
USA	Laser Media, Inc.						X		
USA	Laser Reflections			X		X	X		
USA	Laserworks					X	X		
USA	Light Impressions International, Ltd.	X			X				
USA	Light Wave Gallery	X	X		X	X	X		
USA	Lightrix. Inc.	X			X	X			
USA	LightVision Confections	X					X		
USA	Lone Star Illusions		X						
USA	M.I.T. Museum			X					X
USA	Man/Environment, Inc.	X	X		X	X	X		
USA	Marks, Gerald					X	X		
USA	Media Interface, Ltd.						X		
USA	Miller, Peter					X			
USA	Multiplex Moving Holograms	X			X	X			
USA	Museum Of Holography/Chicago		X	X		X			X
USA	Nakamura, Ikuo					X			
USA	NeoVision Productions					X	X	X	
USA	New Light Industries						X		
USA	New York Hall Of Science								X
USA	Oregon Laser Consultants						X		
USA	Pacific Holographics Inc.					X	X		
USA	Photon Cantina Ltd.				X	X	X		
USA	Planet 3-D						X		
USA	Point Source Productions					X	X		
USA	PullTime 3-D Laboratories					X	X		
USA	Rainbow Symphony Inc.	X			X	X	X		
USA	Real Image	X					X		
USA	Red Beam, Inc.						X		
USA	Regal Press Inc.				X				
USA	Reva's Holographic Illusions		X						
USA	Richard Bruck Holography	X				X	X		
USA	Robert Sherwood Holographic Design	X				X	X		
USA	Saint Mary's College					X			
USA	School Of Holography								X
USA	Science Kit & Boreal Labs				X				
USA	Sharon McCormack Holography					X	X		
USA	Shuttlecart		X						
USA	Sommers Plastic Products						X		
USA	Southern Indiana Holographics					X			

Table 1 - Businesses That Sell or Display Holograms

Country	Name of Business	Whole-sale	Retail	Holo. Gallery Gift Shop	Mail Order Catalog	Artist with Portfolio	Broker /Sales Rep	Exibit - Touring	Gallery/ museum Display
USA	Southwest Holographics					X			
USA	Star Magic		X		X				
USA	Starlog		X						
USA	Stephens, Anait					X			
USA	The Hologram Company		X	X					
USA	The Holography, Laser & Photonics			X					X
USA	Third Dimension Arts Inc.	X			X				
USA	Three Dimensional Imagery					X			
USA	Three-D Light Gallery		X	X		X			
USA	Trace Holographic Art & Design, Inc.						X		
USA	Tyler Group						X		
USA	U.K. Gold Purchasers, Inc.	X							
USA	United Assoc. Manufacturer's Reps.						X		
USA	University Of Wisconsin/Madison				X				
USA	Unterseher & Associates					X	X		
USA	Visual Visionaries						X		
USA	Wave Mechanics					X			
USA	Wesley, Ed					X	X		
USA	World Holographics						X		
USA	Zero Gravity		X						
USA	Galaxies Unlimited, Inc.		X						
USA	Zone Holografix					X	X		

Table 2 - Artwork Origination and Hologram Mastering Services

Country	Name of Business	3-D Models: Physical Objects	3-D Models: Digital Artwork	Master for Silver-halide	Master for DCG	Master for Photo-polymer	Master for Embossed replication	Master using Pulsed Laser	Master using Stereogram Printer
Australia	3DIMAGE	X	X	X		X		X	
Australia	Australian Holographics	X		X		X		X	X
Australia	Menning, Melinda	X		X	X	X	X		
Australia	Moonbeamers	X	X	X			X		
Austria	Holography Center of Austria		X	X				X	X
Canada	Dimension 3	X	X	X		X	X		X
Canada	General Holographics, Inc.	X	X	X	X		X		X
Canada	Holocrafts	X	X		X				
Canada	Inspeck, inc	X	X						
Canada	Melissa Crenshaw Holography Studio	X	X	X	X	X	X		
Canada	Royal Holographic Art Gallery			X					
Canada	Silverbridge Group	X	X	X	X				
China	Holography Development Co.Ltd.	X	X	X			X		
Czech	Lightgate, Ltd	X		X			X		
france	GEHOL sarl			X					
France	Hologram Industries	X	X						
Germany	AB Rueck Holoart	X	X	X					
Germany	AKS Holographie-Galerie GmbH	X	X	X					
Germany	Arbeitskreis Holografie B.V.			X					
Germany	Highlite				X				
Germany	Holar Seele KG	X		X				X	
Germany	Holo GmbH	X	X				X		
Germany	Holo Time Gericke			X					
Germany	Holo-Idee Reiner Kleinherne			X					
Germany	Holodesign	X		X					
Germany	Holografie-Hofmann	X		X					
Germany	Hologram Company RAKO GmbH						X		
Germany	Holographic Laserdesign	X		X					
Germany	Holographic Studios Rosowski	X		X					
Germany	Holographic Systems Muenchen GmbH	X	X	X					
Germany	Holographie & Design	X	X	X	X	X	X		
Germany	Holographie Anubis							X	
Germany	Holographie Labor	X	X	X					
Germany	Holographie Roth	X	X	X	X	X	X		
Germany	Holographie-Labor Mielke	X	X	X	X	X	X		

Table 2 - Artwork Origination and Hologram Mastering Services

Country	Name of Business	3-D Models: Physical Objects	3-D Models: Digital Artwork	Master for Silver-halide	Master for DCG	Master for Photo-polymer	Master for Embossed replication	Master using Pulsed Laser	Master using Stereogram Printer
Germany	Holoptics	X	X	X					
Germany	Holotec	X	X	X				X	
Germany	HoloVision	X	X	X	·			X	X
Germany	Holtronic GmbH	X		X	X	X	X		
Germany	Ingenieurbüro Geiger	X		X			X		
Germany	Krystal Holographics Vertriebs-GmbH	X	X			X			
Germany	Laserfilm Eckard Knuth	X	X	X					X
Germany	LichtBlicke Claßen & Voss	X		X					
Germany	Magick Signs Holografie	X	X	X			X		
Germany	mario liedtke pro design	X	X						
Germany	Ose Holografie-Design			X					
Germany	Studio Fuer Holographie	X	X	X					
Germany	Technische Fachhochschule Berlin	X	X	X					
Germany	topac GmbH, Department Holography	X	X	X			X		
Germany	Traumlaboratorium			X					
Hungary	Artplay Holographika Studio		X	X			X		
Hungary	Hologram Varga Miklos	X	X			X			
Hungary	Optopol Panoramic Metrology Consult	X	X						
India	Holostik India Pvt. Ltd.	X					X		
India	Print-M-Boss	X					X		
India	Shriram Holographics	X					X		
India	Spatial Holodynamics (India) Pvt. Ltd.	X	X				X		
Italy	Diavy srl	X	X				X		
Japan	Dai Nippon Printing Co., Ltd.	X	X				X		
Japan	Nippon Polaroid K.K.	X	X			X			
Japan	Toppan Printing Co., Ltd.	X	X				X		
Latvia	University of Latvia						X		
Lithuania	Geola	X		X				X	
Mexico	Corporacion Mexicana De Impresion	X	X				X		
Mexico	HOLOGRAMAS, S.A. DE C.V.	X	X				X		
Netherland	Dutch Holographic Laboratory BV	X	X	X	X	X	X		X
Poland	Holographic Dimensions, Poland S.A.	X					X		
Portugal	INETI - Institute of Information Tech.	X	X				X		
Spain	Holosco, Ernest Barnes	X	X	X			X		
Sweden	Holovision AB	X	X					X	
Taiwan	Fong Teng Technology	X	X				X		
Taiwan	Holoart Studio	X		X					
Taiwan	Infox Corporation						X		
Thailand	Electro-Optics Lab, NECTEC	X		X		X	X		
Ukraine	Feofaniya Ltd.	X	X	X			X	X	
Ukraine	Institute of Applied Optics	X		X					
Ukraine	Tair Hologram Company			X					
UK	3D Holographics	X	X				X		
UK	3D-4D Holographics	X	X	X					
UK	A.H. Prismatic, Ltd.	X	X	X	X	X	X		
UK	Advanced Holographic Laboratories		X				X		X
UK	Applied Holographics, Plc.	X	X				X		
UK	Astor Universal Ltd.	X	X				X		
UK	Colour Holographics	X		X	X	X	X		X
UK	Embossing Technology Ltd	X	X				X		
UK	Holocrafts Europe Limited.	X	X	X	X	X	X		
UK	Holographics (Uk) Ltd.	X	X	X					
UK	Jayco Holographics	X	X	X			X		
UK	K.C. Brown Holographics	X		X				X	
UK	Light Impressions International, Ltd.	X	X				X		
UK	London Holographic Image Studio	X	X	X				X	X
UK	Op-Graphics (Holography) Ltd.			X					
UK	Optical Security Group - England	X	X				X		
UK	Richmond Holographic Studios Ltd							X	
UK	Spatial Imaging Limited	X	X	X		X	X	X	X
USA	3M - Safety and Security Systems		X				X		
USA	Advanced Holographic Laboratories		X	X			X		X
USA	Amagic Technologies Inc.		X	X			X		X
USA	American Bank Note Holographics						X		X
USA	American Paper Optics Inc.		X						

Table 2 - Artwork Origination and Hologram Mastering Services

Country	Name of Business	3-D Models: Physical Objects	3-D Models: Digital Artwork	Master for Silver-halide	Master for DCG	Master for Photo-polymer	Master for Embossed replication	Master using Pulsed Laser	Master using Stereogram Printer
USA	American Propylaea Corporation	X	X	X	X	X	X		X
USA	Ana MacArthur			X					
USA	Automated Holographic Systems			X	X		X		
USA	Avant-Garde Studio	X							
USA	Blue Ridge Holographics, Inc.	X		X		X	X		
USA	Bridgestone Technologies, Inc.	X	X				X		
USA	Broadbent Consulting			X	X	X	X		
USA	California Institute of the Arts	X	X						
USA	Casdin-Silver Holography	X		X					
USA	CFC Applied Holographics	X	X				X		X
USA	Cherry Optical Holography	X		X		X			
USA	Chromagem Inc.					X	X		
USA	Crown Roll Leaf, Inc.,	X	X				X		
USA	David Dann Modelmaking Studios	X	X						
USA	Deem, Rebecca	X	X	X	X	X	X		
USA	Diamond Images, Inc.	X		X	X		X		X
USA	Direct Holographics		X	X		X	X		X
USA	Dri-Print Foils	X					X		
USA	Engineering Animation, Inc.		X						
USA	Feroe Holographic Consulting			X			X		
USA	FLEXcon	X	X				X		
USA	FoilMark Holographic Images	X					X		
USA	Forth Dimension Holographics			X					
USA	Frank DeFreitas Holography Studio			X					
USA	General Design		X						
USA	Holicon Corporation.	X	X	X				X	
USA	Holo Art	X		X					X
USA	Holo Sciences, LLC	X	X	X	X	X	X		X
USA	Holo/Source Corporation	X					X		
USA	HoloCom	X	X				X		X
USA	Hologram Research, Inc.	X	X	X					X
USA	Holograms and Lasers International	X	X	X				X	
USA	Holographic Design Systems	X	X	X	X	X	X		
USA	Holographic Dimensions	X	X				X		X
USA	Holographic Images Inc.	X		X					
USA	Holographic Label Converting (HLC)	X					X		
USA	Holographic Studios	X	X						X
USA	Holographics Inc.	X		X				X	
USA	Holographics North Inc.	X		X					X
USA	Holography Institute of San Francisco			X	X	X	X		
USA	Holophile, Inc.	X	X	X		X			
USA	Imagination Plantation		X						
USA	Infinity Laser Laboratories	X		X		X		X	
USA	Intrepid World Communications	X		X				X	
USA	Jeffery Murray Custom Holography	X	X	X	X	X	X		
USA	Krystal Holographics International Inc.	X	X	X	X	X			
USA	Kurz Foils						X		
USA	Larry Lieberman Holography	X		X					
USA	Lasart Ltd.	X	X	X	X				
USA	Laser Holography Workshop	X							
USA	Laser Images	X	X	X	X	X	X		
USA	Laser Reflections	X		X	X			X	
USA	Light Impressions International, Ltd.	X	X				X		X
USA	Light Wave Gallery	X	X	X					
USA	Lightrix, Inc.	X	X	X		X	X		
USA	Linda Law Holographics		X			X	X		
USA	Louis Paul Jonas Studios, Inc.	X							
USA	Man/Environment, Inc.			X		X	X		
USA	Marks, Gerald	X	X	X		X	X		
USA	Merrick, Michael G.			X					
USA	NovaVision	X	X				X		
USA	OpSec - USA	X	X				X		
USA	Optical Security Group	X	X				X		X
USA	Optimation Holographics	X	X				X		

Table 2 - Artwork Origination and Hologram Mastering Services

Country	Name of Business	3-D Models: Physical Objects	3-D Models: Digital Artwork	Master for Silver-halide	Master for DCG	Master for Photo-polymer	Master for Embossed replication	Master using Pulsed Laser	Master using Stereographic Printer
USA	Pacific Holographics Inc.			X			X		
USA	Pangaea Design	X							
USA	Pennsylvania Pulp & Paper Co.	X	X				X		X
USA	Photon Cantina Ltd.			X					
USA	Polaroid Corporation	X	X			X			
USA	PullTime 3-D Laboratories	X	X	X					
USA	Rainbow Symphony Inc.			X		X	X		
USA	Ralcon	X	X	X	X	X			
USA	Red Beam, Inc.	X	X	X	X	X	X		
USA	Regal Press Inc.	X	X				X		
USA	Reynolds Metals Co.	X					X		
USA	Richard Bruck Holography	X	X	X	X	X	X		
USA	Richmond Development Group	X							
USA	Robert Sherwood Holographic Design	X		X	X	X	X		
USA	Scharr Industries	X	X				X		
USA	Sharon McCormack Holography	X	X	X					X
USA	Silicon Graphics		X						
USA	Simian Co.	X	X				X		X
USA	Sommers Plastic Products	X	X						
USA	Spectratek Inc.						X		
USA	Steinbichler U.S.A. inc.		X						
USA	STI	X	X				X		
USA	Three Dimensional Imagery			X					
USA	Trace Holographic Art & Design, Inc.	X					X		
USA	Transfer Print Foils, Inc.	X	X				X		X
USA	Unterseher & Associates			X	X	X		X	
USA	Van Leer Metallized Products	X	X				X		
USA	Virtual Image (a division of Printpack)	X	X				X		
USA	Visual Visionaries	X	X	X	X	X	X		
USA	Wavefront Technology	X	X				X		
USA	Wesley, Ed	X		X	X	X	X	X	
USA	Wild Style Entertainment		X						
USA	Witchcraft Tape Products, Inc.						X		
USA	Zone Holografix	X	X	X	X	X	X	X	

Table 3 - Mass Production / Replication Services

Country	Name of Business	Silver-halide Glass	Silver-halide Film	DCG & Related	Photo-polymer Film	Security Holograms	Emboss: foils, films	Emboss: adhesive labels	Emboss: paper product
Australia	3DIMAGE	X	X						
Australia	Australian Holographics	X							
Australia	Moonbeamers	X							
Canada	Dimension 3				X		X	X	X
Canada	General Holographics, Inc.	X		X					
Canada	Holocrafts			X					
Canada	Melissa Crenshaw Holo. Studio	X	X	X					
Canada	Royal Holographic Art Gallery	X	X						
Canada	Silverbridge Group			X					
China	Minjian Laser Holography					X			
France	Hologram Industries						X	X	X
Germany	AKS Holographie-Galerie GmbH		X						
Germany	Decolux GmbH					X			
Germany	Highlite			X					
Germany	Holar Seele KG	X							
Germany	Holo GmbH		X						
Germany	Holodesign	X	X						
Germany	Holografie-Hofmann	X	X						
Germany	Hologram Company RAKO							X	
Germany	Holographic Laserdesign	X							
Germany	Holographic Studios Rosowski	X							

Table 3 - Mass Production / Replication Services

Country	Name of Business	Silver-halide Glass	Silver-halide Film	DCG & Related	Photo-polymer Film	Security Holo-grams	Emboss: foils, films	Emboss: adhesive labels	Emboss: paper product
Germany	Holographie Labor		X						
Germany	Holographie-Labor Mielke	X	X	X					
Germany	Holoptics	X							
Germany	Holotec	X	X						
Germany	HoloVision	X	X						
Germany	Holtronic GmbH	X				X	X	X	X
Germany	Krystal Holographics Vertriebs				X				
Germany	Laserfilm Eckard Knuth		X						
Germany	LichtBlicke Claßen & Voss	X	X						
Germany	Ose Holografie-Design	X							
Germany	topac GmbH, Department Holo	X				X	X	X	X
Germany	Traumlaboratorium	X	X						
Hungary	Artplay Holographika Studio					X	X	X	
Hungary	Optopol Panoramic Metrology	X							
India	Holostik India Pvt. Ltd.					X		X	X
India	Print-M-Boss							X	X
India	Shriram Holographics							X	X
India	Spatial Holodynamics (India)							X	X
Italy	Diavy srl						X	X	X
Japan	Dai Nippon Printing Co., Ltd.						X	X	
Japan	Nippon Polaroid K.K.				X				
Japan	Toppan Printing Co., Ltd.						X	X	X
Latvia	University of Latvia							X	
Lithuania	Geola	X	X						
Mexico	Corporacion Mexicana De Imp.					X	X		
Mexico	HOLOGRAMAS, S.A. DE C.V.					X	X	X	X
Netherlands	Dutch Holographic Laboratory	X	X		X		X	X	
Panama	HDIPanama					X			
Poland	Holographic Dimensions, Poland.	X	X				X	X	X
Portugal	INETI - Institute of Info. Tech.					X			
Sweden	Holovision AB	X	X						
Taiwan	Ahead Optoelectronics, Inc.					X			
Taiwan	Fong Teng Technology						X	X	X
Taiwan	Infox Corporation							X	
Thailand	Electro-Optics Lab, NECTEC	X	X			X		X	X
Ukraine	Feofaniya Ltd.							X	X
Ukraine	Institute of Applied Optics	X	X						
Ukraine	Tair Hologram Company	X							
UK	3D Holographics						X	X	X
UK	Action Tapes							X	
UK	Advanced Holographic Lab						X	X	X
UK	Applied Holographics, Plc.						X	X	X
UK	Astor Universal Ltd.						X		X
UK	Embossing Technology Ltd						X	X	X
UK	Holocrafts Europe Limited.			X					
UK	Light Impressions International				X		X	X	X
UK	London Holo Image Studio	X							
UK	Op-Graphics (Holography) Ltd.		X		X				
UK	Optical Security Group - England				X	X	X	X	X
UK	Spatial Imaging Limited					X			
UK	STI - Europe					X		X	
UK	Whiley Foils Limited						X		
USA	3M - Safety and Security Systems					X		X	
USA	A.D. Tech					X			
USA	AD 2, Inc.					X			
USA	Advanced Holographic Lab					X	X	X	X
USA	Amagic Technologies Inc.				X		X	X	
USA	American Bank Note Holo.						X	X	X
USA	American Paper Optics Inc.								X
USA	American Propylaea Corporation				X	X	X	X	
USA	Another Dimension Inc.								
USA	Applied Optics					X			
USA	Bridgestone Technologies, Inc.					X	X	X	X
USA	CFC Applied Holographics				X	X	X	X	X
USA	Cherry Optical Holography				X				

Table 3 - Mass Production / Replication Services

Country	Name of Business	Silver - halide Glass	Silver- halide Film	DCG & Related	Photo- polymer Film	Security Holo- grams	Emboss: foils, films	Emboss: adhesive labels	Emboss: paper product
USA	Chromagem Inc.				X				
USA	Coburn Corporation							X	X
USA	Control Module Inc.					X			
USA	Creative Label						X	X	X
USA	Crown Roll Leaf, Inc.,					X	X	X	X
USA	Datacard Corporation					X			
USA	Dimensional Arts						X		
USA	Dimensional Foods Co.								
USA	Direct Holographics				X				
USA	Dri-Print Foils						X		
USA	FLEXcon					X	X	X	X
USA	FoilMark Holographic Images					X	X	X	X
USA	Forth Dimension Holographics	X							
USA	Holicon Corporation.	X							
USA	Holo/Source Corporation					X	X	X	
USA	HoloCom					X		X	
USA	Holograms and Lasers Intl.	X							
USA	Holographic Design Systems					X			
USA	Holographic Dimensions					X	X	X	
USA	Holographic Finishing						X	X	X
USA	Holographic Label Converting					X	X	X	
USA	Holographic Products					X			
USA	Holographics Inc.	X							
USA	Holographics North Inc.		X						
USA	Intrepid World Communications	X	X						
USA	Krystal Holographics Intl.			X	X	X			
USA	Kurz Foils						X		X
USA	Larry Lieberman Holography	X							
USA	Lasart Ltd.	X		X					
USA	Laser Reflections	X		X					
USA	Light Wave Gallery	X							
USA	Lightrix, Inc.				X			X	
USA	LightVision Confections							X	
USA	NovaVision					X	X	X	
USA	OpSec - USA				X	X	X	X	
USA	Optical Security Group					X	X	X	X
USA	Optimation Holographics							X	X
USA	Pennsylvania Pulp & Paper Co.						X	X	X
USA	Photon Cantina Ltd.	X	X						
USA	Polaroid Corporation				X	X			
USA	Rainbow Symphony Inc.						X	X	X
USA	Ralcon			DCG	X				
USA	Regal Press Inc.						X	X	X
USA	Reynolds Metals Co.								X
USA	Scharr Industries						X	X	
USA	Simian Co.					X	X	X	X
USA	Sommers Plastic Products				X			X	
USA	Spectratek Inc.				X		X	X	
USA	STI					X	X	X	X
USA	Third Dimension Arts Inc.			X					
USA	Three Dimensional Imagery	X	X						
USA	Transfer Print Foils, Inc.					X	X	X	X
USA	Tyler Group					X		X	
USA	Unifoil Corporation							X	X
USA	Van Leer Metallized Products					X		X	X
USA	Virtual Image						X	X	X
USA	Wavefront Technology						X	X	X
USA	Witchcraft Tape Products, Inc.							X	X

Table 4 - Hologram Application / Finishing / Converting Services

Country	Name of Business	Die cutting, Slitting	Over printing Ink	Foil Hot - stamp	Pressure sensitive/ automated	Pressure sensitive / by hand	Laminating	Textiles
Canada	Dimension 3	X	X	X	X	X		
China	Foshan Holosun Packaging Co. Ltd	X	X	X	X	X	X	
China	Shanghai Dahua Printing Factory	X	X	X	X	X		
China	Wuhan Packaging and Printing United Co.	X	X	X	X		X	
France	Hologram Industries	X	X	X	X		X	
Germany	Holographie Fachstudio Bad Rothenfelde		X					
Germany	Kolbe-Druck mit Tochtergesellschaften	X	X	X	X	X	X	
Germany	topac GmbH, Department Holography	X	X	X	X		X	
Germany	W. Cordes GmbH + Co.	X	X	X	X	X	X	
Hungary	Artplay Holographika Studio	X	X					
India	Holostik India Pvt. Ltd.	X	X	X	X		X	
India	I.S. Gill	X	X	X	X			
India	Shriram Holographics	X	X	X	X			
Italy	Diavy srl	X	X	X				
Japan	Dai Nippon Printing Co., Ltd.	X	X	X	X	X		
Japan	Toppan Printing Co., Ltd.	X	X	X	X	X		
Mexico	HOLOGRAMAS, S.A. DE C.V.	X	X	X	X	X	X	
Poland	Holographic Dimensions, Poland S.A.	X	X	X	X	X	X	
S. Africa	Synchron Pty Ltd.			X				
Taiwan	Fong Teng Technology	X	X	X	X		X	
Thailand	Electro-Optics Lab, NECTEC	X	X	X	X		X	
Ukraine	Feofaniya Ltd.	X	X	X	X			
UK	3D Holographics	X	X	X	X	X		
UK	Advanced Holographic Laboratories	X	X	X	X	X	X	X
UK	Applied Holographics, Plc.	X	X	X	X	X	X	
UK	Astor Universal Ltd.	X	X	X	X	X	X	
UK	Embossing Technology Ltd	X	X					
UK	Light Impressions International, Ltd.	X	X	X	X	X		
USA	2Xst Century Finishing Inc.	X	X	X	X	X	X	
USA	Advanced Holographic Laboratories	X	X	X	X			
USA	American Bank Note Holographics	X	X	X	X	X	X	
USA	American Propylaea Corporation	X		X	X	X	X	
USA	Bridgestone Technologies, Inc.	X	X	X	X	X	X	
USA	CFC Applied Holographics	X		X			X	
USA	Creative Label	X	X	X				
USA	Crown Roll Leaf, Inc.,	X	X	X	X	X	X	
USA	D. Brooker & Associates		X					
USA	FLEXcon	X	X	X	X	X	X	
USA	FoilMark Holographic Images	X	X	X	X		X	
USA	Holo Art	X			X	X	X	
USA	Holo/Source Corporation	X	X	X	X	X	X	
USA	Holographic Dimensions	X	X	X	X	X	X	
USA	Holographic Finishing	X	X	X	X	X	X	
USA	Holographic Impressions	X	X	X	X	X	X	
USA	Holographic Label Converting (HLC)	X	X	X	X	X	X	
USA	Holography Presses On (HPO)				X	X		X
USA	Imagen Holography. Inc.							X
USA	Krystal Holographics International Inc.	X	X		X		X	
USA	Letterhead Press, Inc.	X	X	X	X	X	X	X
USA	Luminer Printing and Converting	X	X		X		X	
USA	MGM Converters Inc.	X	X	X	X	X	X	
USA	OpSec - USA	X	X				X	
USA	Optimation Holographics	X	X				X	
USA	Pennsylvania Pulp & Paper Co.	X	X	X	X	X	X	
USA	Polaroid Corporation	X	X		X	X	X	X
USA	Rainbow Symphony Inc.	X	X	X	X	X	X	
USA	Regal Press Inc.	X	X	X	X	X	X	
USA	Reynolds Metals Co.		X					
USA	Smith & McKay Printing Co. Inc.	X	X	X	X	X	X	
USA	Sommers Plastic Products	X	X	X	X	X	X	
USA	Textile Graphics, Inc.							X
USA	Transfer Print Foils, Inc.	X	X	X	X	X	X	
USA	Unifoil Corporation	X			X		X	
USA	Van Leer Metallized Products	X	X	X			X	
USA	Virtual Image (a division of Printpack,)	X	X	X	X	X	X	
USA	Witchcraft Tape Products, Inc.	X	X	X	X	X	X	

Table 5 - Suppliers of Lasers, Optics and Related Studio Equipment

Country	Name of Business	New CW Lasers	New Pulsed Lasers	Used or Surplus Lasers	Laser Repair	Laser Eqipment, accessories	Optics & Related components	Darkroom Supplies, Chemicals
Canada	Harvard Apparatus, Canada						X	X
Canada	Lasiris Inc.						X	
China	Chongqing Yinhe Laser Products Ltd.	X		X	X	X	X	
France	Quantel	X				X		
France	S.O.P.R.A.					X		
Germany	Adlas G.M.B.H. & Co Kg.	X	X			X		
Germany	Andreas Wappelt - Photonics Direct	X				X		
Germany	Bernhard Halle Nachf. GmbH & Co.						X	
Germany	Coherent Luebeck GmbH	X						
Germany	Dr. Steeg & Reuter GmbH						X	
Germany	Gerhard Winopal Forschungsbedarf						X	
Germany	Gigahertz-Optik						X	
Germany	Gresser, E., KG	X				X		
Germany	HMS-Elektronik						X	
Germany	Holostudio Beate Krengel		X				X	X
Germany	Laser-Solution-Management			X		X		
Germany	Mefoma Fototechnik GmbH							X
Germany	Melles Griot GmbH	X				X	X	
Germany	Metrologic Instruments GmbH	X				X		
Germany	NEC Electronics (Europe) GmbH	X				X		
Germany	Newport GmbH	X				X	X	
Germany	Optical Coating Laboratory GmbH						X	
Germany	Optical Test Equipment						X	
Germany	OWIS Gmbh						X	
Germany	Physik Instrumente (PI) GmbH & Co.	X				X	X	
Germany	Polytec GmbH					X		
Germany	Rofin-Sinar Laser GmbH	X						
Germany	Spectra-Physics GmbH	X				X		
Germany	Spindler & Hoyer GmbH & Co.						X	
Germany	Steinbichler Optotechnik GmbH					X	X	
Germany	Technolas Laser Technik Gmbh	X						
Germany	Uniphase Vertriebs-GmbH	X						
Germany	University of Erlanger						X	
Japan	Fuji Electric Co. Ltd	X						
Japan	Kimmon Electric Co., Ltd.	X						
Japan	Marubun Corporation	X					X	X
Korea	Dan Han Optics						X	
Lithuania	Geola		X			X	X	
Mexico	Centro de Investigaciones en Optica, A.C.						X	
Russia	Technoexan Ltd	X				X		
Spain	Lasing S.A.,						X	X
Taiwan	Superbin Co. Ltd	X				X		
Ukraine	TAVEX, Ltd.						X	
UK	INNOLAS (UK) Ltd.		X			X		
UK	Lumonics Ltd.		X			X		
USA	Advanced Optics, Inc.						X	
USA	Aerotech Inc.	X				X	X	
USA	American Laser Corporation	X				X		
USA	Burleigh Instruments, Inc.						X	
USA	Cambridge Laser Labs			X		X		
USA	City Chemical							X
USA	Coherent, Inc. - Laser Group	X				X		
USA	Continental Optical						X	
USA	Control Optics						X	X
USA	Corion Corp.						X	
USA	Corning Incorporated						X	X
USA	CVI Laser Corporation						X	
USA	Dell Optics Company, Inc.						X	
USA	Ealing Electro-Optics Inc.	X				X	X	
USA	Edmund Scientific Company	X				X	X	X
USA	El Don Engineering			X	X	X		
USA	Electro Optical Industries, Inc.						X	
USA	Excitek Inc.			X	X	X		
USA	Fisher Scientific	X				X	X	X
USA	G.M. Vacuum Coating Lab, Inc.						X	
USA	Glass Mountain Optics						X	
USA	Holo-Spectra			X		X	X	

Table 5 - Suppliers of Lasers, Optics, and Related Studio Equipment

Country	Name of Business	New CW Lasers	New Pulsed Lasers	Used or Surplus Lasers	Laser Repair	Laser Eqipment, accessories	Optics & Related components	Darkroom Supplies, Chemicals
USA	Honeywell Technology Center						X	
USA	Images Company	X					X	X
USA	Infrared Optical Products, Inc.						X	
USA	Inrad, Inc.						X	
USA	Ion Laser Technology, Inc.	X				X		
USA	Jodon Inc.	X		X			X	X
USA	Kaiser Optical Systems, Inc.						X	
USA	Ka-lor Cubicle and Supplies							X
USA	Kinetic Systems, Inc.						X	
USA	Laser and Motion Development Company			X		X	X	
USA	Laser Drive Inc.					X		
USA	Laser Innovations			X	X	X		
USA	Laser Las Vegas			X	X	X		
USA	Laser Optics, Inc.			X	X	X	X	
USA	Laser Reflections		X			X		
USA	Laser Resale Inc.			X	X	X		
USA	Laser Technical Services	X	X	X	X	X		
USA	LaserMax, Inc.	X	X			X	X	
USA	Lasermetrics, Inc.				X	X		
USA	Laserworks			X			X	
USA	Lenox Laser	X					X	
USA	Lexel Laser, Inc.	X				X		
USA	LiCONiX	X				X		
USA	Lightwave	X						
USA	Lumenx Technologies, Inc.	X			X	X		
USA	M.O.M. Inc.						X	
USA	Melles Griot	X		X		X	X	
USA	Melles Group Laser Group	X				X		
USA	Meredith Instruments			X		X		
USA	Midwest Laser Products	X		X		X		
USA	Mitutoyo Measuring Instruments (MTI Corp.)						X	
USA	MWK Industries			X		X	X	
USA	Navidec Inc. (formerly ACI Systems, Inc.)	X				X		
USA	New Focus, Inc	X				X	X	
USA	Newport Corporation						X	X
USA	Norland Products, Inc.						X	
USA	Odhner Holographics						X	
USA	Optics Plus Inc.						X	
USA	Optimation						X	
USA	Oriel Instruments						X	
USA	Panatron Inc.			X	X	X		
USA	Pasco Scientific	X					X	X
USA	Photo Research, Inc.		•				X	X
USA	Photo Sciences						X	
USA	Potomac Photonics, Inc.	X						
USA	Rolyn Optics						X	
USA	Science Kit & Boreal Labs						X	X
USA	Spectra-Physics Lasers Inc.	X				X		
USA	Swift Instruments						X	
USA	Thorlabs Inc.						X	
USA	Uniphase Lasers	X				X		
USA	Uvex Safety Inc.							

Table 6 - Holographic Memory & Data Storage

Country	Name of Business	Data processing / storage devices
Canada	Universite Laval	X
Germany	Technische Universitaet Berlin	X
USA	IBM Almaden Research Center	X
USA	Innovative Technology Associates	X
USA	Optitek, Inc.	X
USA	Tamarack Storage Devices	X

Table 7 - Suppliers of Holographic Recording / Replication Materials

Country	Name of Business	Silver-halide Red-sensitive Plates	Silver-halide Red-sensitive Films	Silver-halide Green-sensitive Plates	Silver-halide Green-sensitive Films	DCG & related material	Photo-polymer Films	Photo-resist	Hot-stamp Foil	Plastics & Misc. Carriers
Belgium	Agfa - Gevaert N.V.			X	X					
Canada	Royal Holographic Art Gallery		X							
China	Tianjin Holdor Optics Inc.	X							X	
Denmark	Ibsen Micro Structures A/S							X		
Finland	Starcke, Ky.								X	
Germany	Decolux GmbH								X	
Germany	Holostudio Beate Krengel	X				X				
Germany	HRT Holographic Recording Tec	X	X							
Germany	Oeserwerk Ernst Oeser & Söhne								X	
Germany	Zanders Feinpapiere AG								X	X
Japan	Nippon Polaroid K.K.						X			
Latvia	University of Latvia							X	X	
Russia	Slavich Joint Stock Company	X	X	X	X					
UK	3D Holographics								X	
UK	Advanced Holographic Labs								X	
UK	Astor Universal Ltd.								X	
UK	International Data Ltd.									X
USA	3 Deep Hologram Company	X		X						
USA	Crown Roll Leaf, Inc.,								X	
USA	Dri-Print Foils								X	
USA	DuPont (E.I. DuPont)						X			
USA	Hologram Research, Inc.	X	X	X	X					
USA	ICI Polyester								X	
USA	Images Company	X								
USA	Integraf	X	X	X	X		X			
USA	Jodon Inc.	X	X	X	X					
USA	Kurz Foils								X	
USA	Polaroid Corporation						X			
USA	Process Technologies							X		
USA	Shipley Chemical Co.							X		
USA	Sillcocks Plastics International									X
USA	Sommers Plastic Products									X
USA	Three Deep Hologram Co.	X		X						
USA	Towne Technologies							X		
USA	VinTeq, Ltd.	X	X	X	X					

Table 8 - Suppliers of Hologram-Related Production Equipment

Country	Name of Business	HOPs Standard/ Custom	Embossing Narrow Web	Embossing: Plating,, Shims, etc.	Photo-Polymer Processing	Registra-tion Devices	Hot-stamping Presses	Label Application Equipment
Germany	Baier Praegepressen		X					
Germany	Hologram Company RAKO GmbH		X					
Germany	Leonhard Kurz GmbH		X					
Germany	Steuer KG GmbH & Co.						X	
Netherland	Dutch Holographic Laboratory BV	X	X					
Taiwan	Ahead Optoelectronics, Inc.	X	X					
UK	Global Images		X	X				
UK	Molins PLC						X	X
UK	Spatial Imaging Limited	X						
USA	Bobst Group Inc.					X	X	X
USA	Brandtjen & Kluge, Inc.,					X	X	X
USA	Capitol Converting Equipment, Inc.						X	
USA	Digital Matrix Co.	X						
USA	Dimensional Arts	X	X					
USA	DuPont (E.I. DuPont)				X			
USA	Frank J. Deutsch Inc.					X	X	X
USA	HoloCom	X	X					
USA	James River Products		X			X		
USA	Man/Environment, Inc.	X						
USA	New Light Industries	X	X					
USA	Peacock Laboratories, Inc.			X				
USA	Total Register Inc.					X		
USA	Tyler Group	X						
USA	Ultra-Res Corporation	X						
USA	Unifoil Corporation		X			X		

Table 9 - Industrial Holography

Country	Name of Business	NDT-Stress Analysis	NDT-Fluid Analysis	NDT R&D	HOE Design/Produce	HOE R&D	CGH software	CGH R&D	CGH production	Bio-Medical Displays	Real-time R&D
Australia	Moonbeamers				X						
Belgium	Center for Applied Research in Art					X					
Belgium	Free University Of Brussels.					X					
Belgium	Laboratory Vinckiner			X		X					
Canada	Capilano College							X			
Canada	Lasiris Inc.				X						
Canada	Universite Laval					X		X			
China	Guangzhou Inst.of Electronics					X					
China	Nanjing Sanle Laser Technology				X						
Denmark	Ibsen Micro Structures A/S				X						
Finland	Heptagon Oy				X						
France	Aerospatiale	X		X							
Germany	BIAS	X									
Germany	CHIRON Technolas GmbH		X								
Germany	Daimler Benz Aerospace	X			X						
Germany	Dornier Medizintechnik GmbH	X				X		X		X	
Germany	ETA-Optik Gmbh				X						
Germany	Grau, G., Prof. Dr. techn.							X			
Germany	Holographie-Labor Mielke				X						
Germany	Holtronic GmbH	X			X						
Germany	Institut fur Angewandte Physik	X	X	X							
Germany	Leseberg, Dr. Detlef				X				X		
Germany	Rottenkolber Holo-System GmbH	X	X								
Germany	Steinbichler Optotechnik GmbH	X									
Germany	Technische Fachhochschule Berlin				X						
Germany	Technische Hochschule Darmstadt				X			X			
Germany	Technische Universitaet Berlin			X		X		X			X
Germany	Technolas Laser Technik Gmbh		X								
Germany	University of Erlanger				X	X		X			
Germany	University of Muenster		X							X	
Germany	University of Munich	X	X	X		X		X		X	
Germany	University of Stuttgart	X	X	X		X		X			
Germany	Volkswagen AG	X									
Germany	Zentrum für Kunst und Medientech							X	X		
Hungary	Artplay Holographika Studio				X		X	X			
Israel	Holo-Or Ltd				X						
Italy	Universita Di Roma			X		X					
Japan	Asahi Glass Co.					X					
Japan	Canon Inc. R&D Headquarters					X					
Japan	Central Glass Co., Ltd.					X					
Japan	Chiba University			X		X		X			
Japan	Fujitsu Laboratories Ltd.					X					
Japan	Keio University			X							
Japan	Laboratories of Image Information					X		X			
Japan	Mazda Motor Corp.	X									
Japan	Ministry Of International Trade					X					
Japan	Mitsubishi Heavy Industries Ltd.	X									
Japan	Nihon University			X							
Japan	Nippondenso Co., Ltd.				X						
Japan	Nissan Motor.				X	X					
Japan	Numazu College Of Technology	X									
Japan	Sharp Corp.					X					
Japan	Sophia University					X					
Japan	Tama Art Umversity					X		X			X
Japan	Tokyo Institute of Technology			X		X					
Japan	Topcon Inc.				X						
Japan	Toshihiro Kubota, dept of electronic					X		X			
Japan	Waseda University									X	
Mexico	Centro de Investigaciones en Optica,				X						
Portugal	Universidade Do Porto			X							
Slovakia	Technical University Zvolen	X									
Sweden	Lund Institute Of Tech			X		X		X			
Swiss	Universite De Neuchatel			X		X		X			
Taiwan	Industrial Technology Research Inst.					X					
Taiwan	Institute Of Optical Science			X		X		X			
Ukraine	Feofaniya Ltd.	X									

Table 9 - Industrial Holography

Country	Name of Business	NDT-Stress Analysis	NDT-Fluid Analysis	NDT R&D	HOE Design / Produce	HOE R&D	CGH software	CGH R&D	CGH production	Bio-Medical Displays	Real-time R&D
Ukraine	TAVEX, Ltd.										X
UK	Colour Holographics				X						
UK	Electro Optics Developments Ltd.				X						
UK	Imperial College Of Science			X							
UK	Loughborough Univ. Of Tech.			X		X					
UK	National Physical Laboratory			X							
UK	Pilkington Optronics	X			X		X		X		
UK	Richmond Holographic Studios Ltd					X					
UK	Rolls-Royce Plc	X									
UK	Turing Institute									X	
USA	3-D Systems				X		X				
USA	Accuwave Corp.				X						
USA	Advanced Precision Technology				X						
USA	Advanced Technology Program					X		X			X
USA	Alabama A&M University										X
USA	American Holographic Inc.				X						
USA	American Propylaea Corporation				X						
USA	American Soc. for Nondestruct Test			X							
USA	Applied Optics					X		X			
USA	Automated Holographic Systems				X						
USA	Batelle Pacific Northwest Nat. Labs			X		X				X	
USA	Broadbent Consulting				X						
USA	Chronomotion										X
USA	Control Module Inc.				X						
USA	Datacard Corporation				X						
USA	Diffraction Ltd.				X						
USA	Flight Dynamics				X						
USA	Holographic Optics Inc.				X						
USA	Holographic Products				X						
USA	Holographics Inc.			X		X					
USA	Holonix				X						
USA	Holotek					X					
USA	IBM Almaden Research Center										X
USA	Illinois Institute Of Technology					X		X			X
USA	ImEdge Technology				X						
USA	Industrial Technology Institute	X		X							
USA	Innovative Technology Associates										X
USA	Kaiser Optical Systems, Inc.		X		X					X	
USA	Krystal Holographics International I				X						
USA	Laboratory for Optical Data Process					X		X			
USA	Laser Technology, Inc.	X									
USA	Lawrence Berkeley Laboratory	X	X	X	X	X		X			
USA	M.I.T. (Massachusetts Inst. of Tech)					X		X			X
USA	McMahan Electro-Optic	X									
USA	Metrologic Instruments, Inc.				X						
USA	Nimbus Manufacturing, Inc.				X						
USA	Northern Illinois University	X		X							
USA	Optical Research Associates							X			
USA	Optineering	X	X								
USA	Polaris Research Group	X	X								
USA	Potomac Photonics, Inc.				X						
USA	Ralcon				X						
USA	Rice Systems, Inc.		X								
USA	Richardson Grating Laboratory				X						
USA	Rochester Inst. Of Technology					X		X			
USA	Rochester Photonics Corporation				X						
USA	Rowland Institute For Science			X							
USA	Saginaw Valley State University					X					
USA	San Jose State University					X		X			
USA	Sandia National Laboratories	X	X	X	X	X	X	X	X		X
USA	Silhouette Technology Inc.				X						
USA	Sinclair Optics, Inc.				X				X		
USA	Stanford University			X		X		X			X
USA	Steinbichler U.S.A. inc.	X									
USA	Ultra-Res Corporation				X						X
USA	University Of Alabama at Huntsville			X		X					

Table 9 - Industrial Holography

Country	Name of Business	NDT-Stress Analysis	NDT - Fluid Analysis	NDT R&D	HOE Design / Produce	HOE R&D	CGH soft-ware	CGH R&D	CGH production	Bio-Medical Displays	Real-time R&D
USA	University Of Arizona			X		X		X			
USA	University Of Dayton			X		X		X			
USA	University Of Michigan					X					
USA	University Of Rochester			X		X		X			
USA	University Of Southern California					X		X			
USA	Voxel									X	
USA	Wavefront Research, Inc.				X	X					
USA	WYKO-VEECO	X	X								

Table 10 - Holography-Related Educational Resources, Publications & Organizations

Country	Name of Business	Association, commercial /trade membership	Association, hobbyist /artist membership	News letter	Trade Journal or Magazine	Regularly Published Author/ Book	Book Publisher/ Distributor	Multimedia / Internet Publisher
Germany	AHT 3D-Medien	X	X	X				
Germany	Deutsche Gesellschaft fur Holografie e. V.	X					X	
Germany	Deutscher Drucker Verlagsgesellschaft				X			
Germany	Grau, G., Prof. Dr. techn.						X	
Germany	Heiß, Peter, Dr., Priv.-Doz.						X	
Germany	Holopublic Unbehaun	X					X	
Germany	Zec, Peter, Dr. phil.						X	
Japan	HODIC Holographic Display Artists Club		X					
Japan	Toshihiro Kubota, Dept of eEectronics					X		
Spain	Karas Studios Holografia			X				
Sweden	Hologram Center Holmby		X					
Sweden	Holography Group TEM		X					
Sweden	Royal Institute of Technology					X		
UK	Creative Holography Index, The						X	
UK	International Hologram Manuf. Assn	X						
UK	Reconnaissance International Ltd.			X				
UK	Saxby, Graham					X		
USA	American Society for Nondestructive Test	X			X			
USA	Bjelkhagen, Hans					X		
USA	Foil Stamping and Embossing Association	X						
USA	Frank DeFreitas Holography Studio							X
USA	Holo Art		X					
USA	Holography Marketplace				X	X	X	X
USA	Holoworld.com							X
USA	Images Company							X
USA	Inside Finishing Magazine				X			
USA	Laser Arts Society For Education and Res			X				
USA	Laser Focus World				X			
USA	Laser Institute Of America	X			X			
USA	Optical Society of America (OSA)	X		X	X		X	
USA	Photonics Spectra				X			
USA	Reconnaissance International Ltd.			X				
USA	Ross Books				X	X	X	X
USA	SPIE	X			X	X	X	
USA	SPIE's Holography Working Group News	X		X	X			
USA	Springer-Verlag New York						X	
USA	Unterseher, Fred					X		

Appendix A

Creating the Mona Lisa Hologram that Appears on the Cover of This Book

Artwork Origination & Digital Modeling
Paul Richer (Dimension 3)

The cover image was created by first scanning a copy of the original un-retouched "Mona Lisa" painting in 36bit RGB with an AGFA ArcusII scanner and Photoshop 4.0 on a J28MB 200MHz Pentium. After some brightness, contrast, gamma corrections, detail optimization and scaling, the image was converted to 24bit RGB and saved in TIFF format on our 56GB fileserver.

Still in Photoshop 4.0 but now on an Aspen Durango 256MB/256bit RAM 500MHz DEC A1'tJha AXP, we did layer separations and saved them in 24bit TIFF format. Then starts the artistically intensive task of layers retouching in Photoshop and other special 2-D graphic software.

The resulting layers were saved in TARGA format and imported as texture in a 3-D modeling/rendering software for full detail 3-D modeling. During modeling, we rendered a few stereoscopic tests (in stereogram format) and viewed them with our Holo-Emulator™ system (a 3-D stereoscopic interactive viewer software/hardware we designed to adjust and validate the 3-D picture quality, shape, parallax and depth).

The final rendering was done on four 25dMB 1256bit RAM 500MHz DEC Alpha AXP (Aspen's Dural1go' arilffudk"'22" - hours for the 60 views needed. They were rendered in 24bit RGB TARGA format of 507 x 480 pixel in size.

Since this is a monochrome type (we also do color ones), we had to convert the views to greyscale using special RGB ratios to optimize shading and contrast. Then a final brightness/contrast/gamma optimization specially calibrated for Chromagem's monochrome computer holographic transfer (since we also produce 3-D lenticular printing which uses other correction curves, this is made last so we can use the same files for lenticular printing).

The final step was to compose the 507 x 480 views on 640x480 black canvas to fit Chromagem's imaging device needs, and save the views in 8bit uncompressed TARGA on a 1 GB JAZ disk (could have been on a Syquest since the 60 files only take a bit more than 18MB). Sometimes we send them on CD-R instead, or, if short in time, through direct modem transfer compressed in ZIP.

Hologram Mastering and Reproduction

The HI was produced by Chromagem Inc. from the digital files which were delivered to them. Chromagem, of Youngstown, Ohio has produced some of the highest quality masters for over 15 years. They are extremely experienced at design and mastering for all types of holographic embossing and are now making masters for photopolymer hologram reproduction as well. They specialize in the use of digitized artwork to produce stereograms. Some of the jobs they have worked on include: PepsiCo's Christmas 24-pack carton (which was the largest production run at the time which used holographic packaging materials; Hershey Chocolate's recent "Jurassic Park, Lost World" promotion, Kellog's cereal box inserts (15,000,000 holograms produced) and labels for Miller Brewing Company.

The master hologram was shipped to Krystal Holographics production facility in Logan, Utah, where a production tool was produced and where replication was done. The company specializes in the design and production of photopolymer holograms. They are capable of manufacturing large volumes of holograms using DuPont's photopolymer recording materials and custom built high speed replication equipment. Once the hologram was produced, they were shipped to the bindery where they were hand-applied. (Automatic application is possible, but the quantities produced for this limited edition book, did not require it.)

The artwork which appears on the front cover of the book was designed and produced by Linda Law of Linda Law Holographics. It illustrates how holograms can be integrated with 2-D graphics. A true masterpiece!

Index

DID YOU BORROW THIS COPY?
GET YOUR OWN COPY DELIVERED TO YOUR DOOR
ANYWHERE IN THE WORLD!

Your Cost
(includes all shipping
and handling fees*)

$30 - Continental U.S.

$35 - Alaska, Hawaii,
     Canada, Mexico

$45 - all other countries

All books shipped promptly
upon receipt of payment by
UPS or U.S. Postal Air Mail.

* We are not responsible for any additional
international entry fees that may be imposed.

Ross Books
Toll-free order phone in USA:
1-800-367-0930
Voice phone:
1- 510-841-2474
Fax:
1-510-841-2695
email:
sales@rossbooks.com
Internet Website:
www.holoinfo.com
Mailing Address:
Ross Books P.O. Box 4340
Berkeley, CA 94704 USA

To Order
Fill out and send us the accompanying order form by fax or by mail.
Include a check or money order ($), or your credit card information.
Or phone, email or place your order directly on our Internet Website.

Payment Information:

Check # _____ Money Order # _____ American Express ___ Visa ___ Mastercard __

Credit Card Number _____

Credit Card Expiration Date (month) _____ (year) _____

Name on Card _____ Authorized Signature _____

"Ship to" Information:

Name: _____

Company: _____

Address: _____

City/State/Province: _____

Country: _____

Postal Code: _____

Phone, Fax, or Email: _____

_____Please send me information about advertising rates.

_____Please contact me when additional or related publications become available.

_____Please contact me so my company can be listed in the HMP International Business Directory.

www.ingramcontent.com/pod-product-compliance
Lightning Source LLC
Chambersburg PA
CBHW061526210326
41521CB00027B/2464